Textbook on
Integrated Pest Management
of Horticultural Crops

The Authors

Dr. N. Emmanuel, is currently working as Assistant Professor [Entomology] at Horticultural College and Research Institute, Dr. YSR Horticultural University, Venkatramannagudem, West Godavari Dist., Andhra Pradesh. He obtained Ph.D [Entomology] degree in 2006 from IARI, New Delhi and awarded Dr. Pradhan Gold Medal in Ph.D. programme. Earlier, he received M.S. Venugopal Gold Medal in 2001 for the Best Thesis in M.Sc. Ag. Entomology from TNAU, India. He has attended and delivered research papers presentations in many national and international seminars and conferences including 5th Asia Pacific Congress of Entomology at Jeju, South Korea, 18-21, Oct. 2005. He was awarded best paper presentation award at National Seminar at OUAT, Bhubaneswar in 2010 and at Dr. YSRHU, VRGudem in 2012.

Dr. A. Sujatha is at present working as Associate Dean, at Horticultural College and Research Institute, Dr. YSR Horticultural University, Venkatramannagudem, West Godavari Dist., Andhra Pradesh. She was awarded Meritorious Research Scientist Award (APSA) for the year 2010. She went on for one year deputation to TRINIDAD and TOBAGO Govt. as ITEC Expert by GOI (MFP, CARDI and HCI). Also contributed to Crop protection Compendium in Mango Fruit borer – CAB International, 2007.

Dr. T.S.K.K. Kiran Patro is currently working as Technical Officer to Dean of Horticulture, Dr. YSR Horticultural University, Venkatramannagudem, West Godavari Dist., Andhra Pradesh. She acquired Ph.D [Horticulture] in 2007 from ANGRAU,Hyderabad. She attended and delivered part of her thesis work in the 16th Annual Congress of Tropical Agricultural Research at University of Peradeniya, Srilanka, 18-19th Nov. 2004. She has experience in extension and teaching and contributed many research papers with national and International importance.

Dr. M. Lakshminarayana Reddy, is currently holding the post of Dean PG Studies, Dr.YSR Horticultural University, Venkataramannagudem. He was awarded Ph.D. in Horticulture by UAS, Bangalore in 1993. Received Crop Research Award for the year 1997 by Gourav society of Agricultural Research Information Center, Hissar. He also served in teaching and taught UG and PG courses for more than 10 years and he is also recipient of Andhra Pradesh State Best Teacher Award 2012 and Indira Priyadarshini Award from Health International (NGO) Hyderabad for 2012.

Dr. B. Srinivasulu, Registrar, Dr.Y.S.R. Horticultural University, Venkataramannagudem, West Godavari District, Andhra Pradesh. He received his Ph.D. (Ag.) Plant Pathology in 1991 from TNAU, Coimbatore. He was awarded 'State Best Officer' Ugadi Puraskaram, Govt. of Andhra Pradesh – 2012, Uddaraju Ananda Raju Foundation Award – 2012, Association of Biotechnology and Pharmacy Fellowship Award – 2007, Acharya N.G. Ranga Agricultural University Gold Medal – 1999, International Development Research Centre, Canada Fellowship for the Doctoral Programme. All together he has 267 publications in the form of research papers, popular articles, booklets etc.

Dr. T.S.S.K. Patro is presently working as Senior scientist and Head, ARS Vizianagaram, ANGRAU AP. He obtained his M.Sc (Ag) and Ph.D. degree in Plant Pathology from IARI, New Delhi with 5 Gold medals.He also won two best International Paper Presentation Awards for SAARC countries in Sri Lanka. He was awarded Plant Pathologist award in A.P State in 2009. He visited U.S.A. for advanced leadership skills training at Hawaii in 2008, China for lead paper presentation in2004 and Sri Lanka in 2000 and 2002 for participating in SAARC conference. He was awarded the fellow of Plant protection association of India and Fellow of Applied Society of Biotechnology in 2012. young scientist awards and best. He also recipient of State Young Scientist in 2014.

Textbook on Integrated Pest Management of Horticultural Crops

— Authors —

N. Emmanuel
Assistant Professor [Entomology]

A. Sujatha
Associate Dean

T.S.K.K. Kiran Patro
Tech. Officer to Dean of Horticulture

M. Lakshminarayana Reddy
PG Dean of Horticulture

B. Srinivasulu
Registrar

Horticultural College and Research Institute
Dr. YSR Horticultural University
Venkatramannagudem, West Godavari Dist.
Andhra Pradesh

&

T.S.S.K. Patro
Senior Scientist and Head,
Agriculture Research Station, Vizianagaram, ANGRAU

2015

Daya Publishing House®
A Division of
Astral International Pvt. Ltd.
New Delhi – 110 002

Cataloging in Publication Data--DK
  Courtesy: D.K. Agencies (P) Ltd. <docinfo@dkagencies.com>

Emmanuel, N., author.
Textbook on integrated pest management of horticultural crops / authors, N. Emmanuel [and five others].
   pages   cm
Includes bibliographical references (pages   ) and index.
ISBN 9789351305590 (International edition)

   1. Pests--Integrated control--India.   2. Horticultural crops--Diseases and pests--Control--India.   I. Title.

DDC 632.90954   23

Published by   : **Daya Publishing House®**
                         A Division of
                         Astral International Pvt. Ltd.
                         – ISO 9001:2008 Certified Company –
                         4760-61/23, Ansari Road, Darya Ganj
                         New Delhi-110 002
                         Ph. 011-43549197, 23278134
                         E-mail: info@astralint.com
                         Website: www.astralint.com

Laser Typesetting   : **Classic Computer Services**, Delhi - 110 035

Printed at   : **Thomson Press India Limited**

PRINTED IN INDIA

Dr. K.L Chadha
President,
The Horticulture Society of India,
F-1 Block, NASC Complex, DPS Marg, Todapur,
New Delhi – 110 012, India
E-mail: klchadha@gmail.com

Foreword

Agricultural and Horticultural production in India has increased spectacularly during the last five decades, leading to an age of food self-sufficiency. This noteworthy growth has been achieved through the implementation of several innovative programmes supported with technological advances in the form of high yielding crop varieties, chemical fertilizers and pesticides, as well as the expansion of cropped area.

Insect pests inflict enormous losses to the potential agricultural production in spite of increasing exploitation of chemical pesticides. At the same time, there is an escalating public apprehension about the impending adverse effects of chemical pesticides on the human wellbeing and environment. Pesticide residues in food commodities, resistance to insecticides and resurgence of minor pests while cannot be eliminated altogether; their intensity can be minimized through development, proliferation and encouragement of alternative technologies such as IPM. Keeping this in view, the book entitled **"Text Book on Integrated Pest Management of Horticultural Crops"** has been written. The publications deals with both the basic and applied aspects of IPM based on the technologies developed so far. The chapters present a precise and comprehensive review of concepts and components of Integrated Pest Management, thus making the text an excellent coverage of tactics and strategies employed in integrated pest management (IPM). The approach is targeted for students with the objective of leading these to a sound foundation and understanding of the why's and how's of pest management in horticultural crops. This book is expected to meet the growing needs of undergraduate and post graduate students, KVK, development workers and progressive farmers regarding the IPM strategies related to horticulture crops.

(K L Chadha)

Preface

Insects continue to exist in all types of environment and inhabit more than two thirds of the identified species of animals in the planet earth. They are the tiny sophisticated living creatures which can produce silk, honey, wax and lac that man has not been able to make as yet. They fly on their own wings, produce 'cold light' and carry loads many many times heavier that its own weight, live in extreme temperatures and environments. Insects, particularly the bees, butterflies, moths and thrips are the pollinators of our grain crops, fruits, vegetables, cotton and flowers. They serve as food for animals from Amphibia to Mammalia and even man. They are the destroyers of weeds, improve soil fertility and acts as scavengers and aid in scientific research programmes. On the contrary, insects impinge on human beings in numerous ways. Many of them feed on all kinds of plants including agricultural and horticultural plants, medicinal plants, forest trees, and weeds. They also infest the food and other stored products in storage structures, packages, godowns, and bins causing enormous amount of loss to the stored food and also deterioration of food quality. Insects cause injury to plants and stored products either directly or indirectly in their attempts to secure food. Insects that cause less than 5 per cent damage are not considered as pests. The insects which cause damage between 5 - 10 per cent are called minor pests and those that cause damage above 10 per cent are considered as major pests. Insects that cause injury to plants and stored products are grouped into two major groups namely chewing insects and sucking insects. The former group chews off plant parts and swallow them in this manner causing damage to the crops. Sucking insects pierce through the epidermis and suck the sap. Several sucking pests serve as vectors of plant diseases and inject their salivary secretions containing toxins that cause severe damage to the crop.

Introduction of high yielding varieties, increase in irrigation facilities and haphazard use of increased rates of agrochemicals such as fertilizers and pesticides in recent years with a view to increase productivity has resulted in heavy crop losses due to insect pests in certain crops. This situation has risen mainly due to elimination of natural enemies, resurgence of pests, and development of insecticide resistance and out-break of secondary pests. Distribution, nature of damage, life history of important key pests of horticultural crops and their management strategies are outlined here under.

We have struggled very hard to bring out this comprehensive book called ***Textbook on Integrated Pest Management of Horticultural Crops*** with a view to meet the long - felt need of the students of Horticultural and Agricultural Universities.

N. Emmanuel

Contents

Part II:
Pest Management in Horticultural Crops

Part I

Integrated Pest Management: Concepts and Components

Chapter 1

Essentials of Integrated Pest Management

1.1 Introduction

India is an affluent country with varied agro-climates, which is highly encouraging for growing a great number of horticultural crops such as fruits, vegetables, root tuber, aromatic and medicinal plants and spices and plantation crops. At present India is next to China in area and production of fruits and vegetable crops and has been contributing 10 per cent of fruits and 14 per cent of vegetable of the total world production. India leads the world in the production of mango, banana, sapota, acid lime and cauliflower while the highest productivity of grape is also recorded here. India occupies second position in production of onion and third in cabbage production globally. Fruits such as mango, banana, citrus, guava and apple account for 75 per cent of the total fruit production in the country. India produces about 70 different varieties of various vegetables. The horticulture sector constituted nearly 20 per cent of agricultural GDP and contributes 4 per cent in the national economy. In spite of having all the favourable factors, the desired level of growth in horticulture has not been achieved because of a number of constraints like low productivity of many fruits and vegetables than international averages, non availability of good planting material, lack of post harvest management and less value addition, pests, diseases and weed problems.

Insect pests are one of the chief limiting factors for agricultural/horticultural productivity growth. It is estimated that herbivorous insects eat about 26 percent of the potential food production. India loses about 30 per cent of its crops every year due to pests and diseases.The insect pests inflict crop losses to the tune of 40 percent in vegetable production [Sharma and Rao,2012]. The production losses have revealed an increasing tendency over the years. The losses due to insect pests in 1983, were estimated to the tune of Rs 6,000 crores and further increased to Rs 20,000 crores in 1993 and alarmingly to 29,000 crores in 1996 [Dhaliwal and Arora,1996]. Inception of green revolution technologies in mid- 1960's gave an impetus to pesticide use, and by 1975-76, it amplified to 266 g/ha, and attained a pinnacle of 404 g/ha in 1990-91. Nearly, 96,000 tonnes of technical grade pesticides are currently produced in the country, of which two-thirds are used in agriculture [Khan,1996].

Continuous use of chemical inputs such as pesticides has resulted in damage to the environment, caused human ill-health, negatively impacted on agricultural production and reduced agricultural sustainability. Fauna and flora have been adversely affected. Numerous short- and long-term human health effects have been recorded. Human deaths are not uncommon. The decimation of beneficial agricultural predators of pests has led to the proliferation of several pests. There is an urgent need to find viable alternatives to pesticides so as to minimize the pesticide residues. According to the noted agricultural scientist, M.S. Swaminathan, agriculture production systems in the 21st century need to be based on the appropriate use of biotechnology, information technology, and eco-technology Integrated Pest Management (IPM) is one such technology. There should be promotion of IPM practice which is an eco-friendly approach which employs available alternate pest control methods such as mechanical, biological control with greater emphasis on use of crop rotation, biopesticides and plant origin pesticides like neem formulations to keep pest population low. Pesticides should be used only when pest population crosses economic threshold level. The need for pesticide free agriculture is multidimensional. But alternative pest control technologies available at present are not very successful and popular among the farmers in different production environments and across crops. There is significant gap between awareness and adoption of IPM practices for different crops.

Insects have been afflicting humans since prehistoric times and in order to fight back the situation, man has come up with various techniques in pest management. The history of insecticides may go back as far as 2500 B.C., when ancient Sumerians used sulphur to control insects and mites. In China, about 1200 B.C., chalk and wood ash were used to control insects in enclosed spaces, and various plant extracts were used for treatment of stored grain. In addition, arsenic sulphide was used to control human lice. The ancient Greeks and Romans utilized sulphur, fumigants, oil sprays, oil and bitumen sticky bands, oil and ash and other preperations for insect control. In Japan, Whale oil was used to control insects in rice paddies.

An account of insecticides used in the 19th century would include sulphur, arsenicals, fluorides, soaps, kereosene and various botanicals, of which nicotine, rotenone, pyrethrum, sabadilla and quassia appear to have been most widely used. DDT was first synthesized in 1874 by Othmar Zeidler. In 1939 a Swiss entomologist, Paul Muller of J.R.Geigy Company, found its insecticidal property for the first time. This discovery brought the 'Nobel Prize' for medicine to Paul Muller in 1948 for the life saving discovery for controlling malaria and typhus by Western Allies during World War II. This marked the era of the chlorinated hydrocarbon insecticides, with the subsequent synthesis of HCH and the cyclodiene compounds. These molecules were generally persistant and had long residual properties. They were welcomed in the beginning, but their stability and hydrophobicity resulted in the contamination of the environment and bioconcentration in the body of many animals; thus they were restricted in use or banned later on. The Germans, on the other hand, experimented with the synthesis of organophosphorous compounds to replace nicotine. Three of these compound, HETP, Parathion and Schradan, attributed to Gerhard Schrader, were subsequently used on a worldwide scale despite their high toxicity to mammals. They were not persistant, and in attempts to lower the mammalian toxicity and increase the efficacy, hundreds of other OP insecticides have since been synthesized. The existence of another class of insecticides, the carbamates, was foreshadowed by swiss workers in 1940, but the first major success as the introduction of American insecticide Carbaryl in 1950. It was followed by many other carbamates. The period between 1949 and early 1970's saw the development of a number of synthetic pyrethrin analogs like allethrin, resmethrin etc.

By 1962, when "Silent Spring" by Rachel Carson was published, serious concerns about the disadvantages of pesticide use were widely raised. Rachel Carson challenged the practices of agricultural scientists and the government. Carson was attacked by the chemical industry and some in government as an alarmist, but courageously spoke out to remind us that we are a vulnerable part of the natural world subject to the same damage as the rest of the ecosystem. Testifying before Congress in 1963, Carson called for new policies to protect human health and the environment. Rachel Carson died in 1964 after a long battle against breast cancer. Her witness for the beauty and integrity of life continues to inspire new generations to protect the living world and all its creatures. Carson and others suggested that pest control methods other than chemical pesticides should be used in order to protect wildlife, human health, and the environment. Public pressure led to government legislation restricting pesticide use in many countries, causing agriculturists to reconsider the heavy use of persistent pesticides such as DDT.

DDT was soon followed by number of other molecules *viz.*, HCH, Aldrin, Dieldrin [Organochlorine group], Parathion, Schradan, Toxaphene [Organophosphorous group], Carbary, propoxur [Carbamate group], allethrin [synthetic pyrethroid group] all through the 1950's and a great number of other widely used organophosphates and carbamates in the subsequent decades. Recent advances in the understanding of insect ecology, biology, physiology and biochemistry are providing new impetus and opportunities for insect pest control. The introduction of resistant genes into commercial plant varieties, use of antisense technology, virus coat proteins or satellite RNA to prevent spread of viral diseases in plants and many other technologies through modern biotechnology is envisioned by some to be the answer to crop protection problems. Alternative tactics such as male sterilization technique, male annihilation technique, use of sex pheromones of important crop pests, use of chitin synthesis inhibitors, IGR's, juvenile hormone analogues, anti juvenile hormones, non steroidal ecdysteroids and non terpenoid juvenile hormone compounds, Avermectins, *Bacillus thuringiensis,* NPV in pest management can significantly reduce pesticide use, particularly when used in combination with other control tactics.

Owing to their effectiveness and economy, these insecticides played a key role in increasing crop production.The triumph of high yielding varieties of wheat and rice that ushered in the "green revolution" was partially due to the protection by umbrella of pesticides (Pradhan,1983).The intensive and extensive use, misuse and abuse of pesticides over the years caused widespread damage to the human health and environment.

1.2 Toxic Consequences of Indiscriminate Use of Pesticides, Pesticide Poisoning and Pesticide Residues in India

"The application of insecticides is not an ecological offence if used sensibly. It is the misuse of these chemicals that create the Hazards"

In 1961 India was on the brink of mass famine, Norman Borlaug was invited to India by the adviser to the Indian minister of agriculture M. S. Swaminathan. Despite bureaucratic hurdles imposed by India's grain monopolies, the Ford Foundation and Indian government collaborated import wheat seed from CIMMYT. Punjab was selected by the Indian government to be the first site to try the new crops because of its reliable water supply and a history of agricultural success. India began its own Green Revolution program of plant breeding, irrigation development, and financing of agrochemicals.

A. Pesticide Residues in the Blood of Punjab Farmers

The green revolution has been lauded for what it did for our country. We went from a country begging for food to being an exporter. However, along with it came several drawbacks too, the foremost being the increased cancer prevalence in states like Punjab which contributed majorly to the revolution. Natural pesticides and fertilizers used earlier gave way to the more 'effective' but deadly artificial ones. Little did anyone know the perils it would expose them and the generations to come. According to the latest statistics released by the Punjab Government, Punjab has over 90 cancer patients per one lakh population. This is much higher than the national average of 80 per lakh. *The Malwa region, also known as the 'cancer belt', has the highest average of 136 cancer patients per one lakh of population.* Data over the last five years has shown that 18 people die of cancer every day, on an average. The connection is hard to ignore – Punjab with just 2.5 per cent of the agriculture land of the country consumes around 18 per cent of pesticides used in India, a very high number by any standard. There are high subsidies provided by the Government on pesticides in the state and this has led to their indiscriminate use.

B. Endosulfan: A Monster Killer in Kasargod

In 1980's, PCK started its death spray thrice a year, in the cashew plantation hills of Kasargod district of Kerala. For the state-run company it was cost-effective annihilation of tea mosquito bug which causes yield losses in cashew plantations-a major forex churner for the government. And since then Kasargod villagers who initially watched the jaw-dropping spraying of chemical by the helicopters never realized the slow-death that was enveloping their skin, water ways, food and even fuel wood. A random epidemiological survey carried out in Kerala in 2010 jointly by the state government's health and agriculture department, revealed that 2,210 victims of Endosulfan poisoning in Kasargod, with nearly 200 deaths due to cancer in last eight years.For over a decade, Kasargod has been witnessing a silent genocide through the spraying of endosulfan on its fields. Though a popular pesticide amongst farmers growing cashew crops, the toxicity of the chemical leads not just in a slow painful fatality and deformation amongst adults, but the exposure attacks neo-natal infants by mutating the growth pattern. Countless researches, surveys and plethora of reports have indexed the adverse effect of the usage of Endosulfan - children born with stag-horn limbs, scale-like skin, protruding tongues, eye deformities, extra fingers and toes, cleft palates, club feet and harelips; of those suffering from hydrocephalus (progressive enlargement of the head, convulsion and mental disability), dermatitis, renal diseases, respiratory disorders, cognitive and emotional deterioration, memory loss, impairment of visual-motor coordination, blindness, cerebral palsy, epilepsy and infertility; of young girls and boys who have undergone multiple surgery.

C. Bhopal Gas Tragedy

The Bhopal disaster was a gas leak incident in India, considered the world's worst industrial disaster. It occurred on the night of 2–3 December 1984 at the Union Carbide India Limited (UCIL) pesticide plant in Bhopal, Madhya Pradesh. Over 500,000 people were exposed to methyl isocyanate gas and other chemicals. The toxic substance made its way in and around the shantytowns located near the plant. The government of Madhya Pradesh confirmed a total of 3,787 deaths related to the gas release. Others estimate 8,000 died within two weeks and another 8,000 or more have since died from gas-related diseases. A government affidavit in 2006 stated the leak caused 558,125 injuries including 38,478 temporary partial injuries and approximately 3,900 severely and permanently disabling injuries. UCC's Sevin [Carbaryl] production plant was built in Madhya Pradesh not to avoid environmental regulations in the U.S. but to exploit the large and growing Indian pesticide market. However the manner in which the project was executed suggests the existence of a double standard for multinational corporations operating in developing countries. Enforceable uniform international operating regulations for hazardous industries would have provided a mechanism for significantly improved in safety in Bhopal. Even without enforcement, international standards could provide norms for measuring performance of individual companies engaged in hazardous activities such as the manufacture of pesticides and other toxic chemicals in India. National governments and international agencies should focus on widely applicable techniques for corporate responsibility and accident prevention as much in the developing world context as in advanced industrial nations. Specifically, prevention should include risk reduction in plant location and design and safety legislation.

D. Pesticide Residue in Agri/Horticultural Commodities

In 2010, the European Union rejected three okra consignments from India due to high levels of Monocrotophos, Acephate and Triazaphos. All three of these pesticides can cause headaches, vomiting, nausea, abdominal cramps and cardiac problems. EU Maximum Residue Limit is 0.03mg/kg, but tests revealed levels of 0.13mg/kg. India's MRL for Monocrotophos is considerably higher at 0.2mg/kg, but it is recommended only for use on cotton crops, as it is toxic to birds and humans. Neverthless, levels detected in food for sale on the domestic market are far higher than for exports. The Indian Ministry of Commerce and Industry's drive to increase grape exports from 37,000 to 44,000 tonnes is being hampered by differing MRLs in exporting countries.In 2012, exports to the EU were threatened by a deadlock caused by Chlormequat, just one of 98 pesticides for which grape consignments to the EU are tested. The UK and Sweden allowed import of Indian grapes by introducing their own MRL.A child weighing 16.15kg needed to eat just 211.5g of grapes to be at risk, No warning was issued in the UK. Aldrin, was detected in brinjal, cauliflower, tomato, okra, banana, apple, wheat and milk. Chlordane, which is banned in 47 countries was found in apples, bananas and cabbage. Chlorfenvinfos was detected in bitter gourd, cabbage, cauliflower, tomatoes, rice and wheat. Heptachlor was detected in brinjal, okra, tomatoes, rice, milk and butter. These four substances are among the persistent organic pollutants (POPs) identified by the Stockholm Convention as the 'dirty dozen.' DDT is not supposed to be used on vegetable crops, was found in tomatoes in Uttar Pradesh at over 100 times the MRL. Fenpropathrin is not recommended for use on tea plants, was detected in Assam tea at more than twice the CODEX MRL of 2ppm.

E. Pesticide Residue in Soft Drinks in India

Centre for Science and Environment (CSE) in 2003 found high levels of toxic pesticides and insecticides, high enough to cause cancer, damage to the nervous and reproductive systems, birth defects and severe disruption of the immune system. Market leaders Coca-Cola and Pepsi had almost similar concentrations of pesticide residues.

F. Mid-Day Meal Poisoning by *Monocrotophos* in Bihar

On 16 July 2013, at least 23 students died and dozens more fell ill at a primary school in the village of Dharmashati Gandaman in the Saran district of the Indian state of Bihar after eating a Mid-day Meal contaminated with monocrotophos. Angered by the deaths and illnesses, villagers took to the streets in many parts of the district in violent protest. India, continues to use monocrotophos and other highly toxic pesticides that rich and poor nations alike, including China, are banning on health grounds.

1.3 Origin of IPM

The over reliance on pesticides with indiscriminate use over last four decades has resulted in many negative consequences, mainly the infamous 3 R's *viz.*, Resurgence, Resistance and Residual aspects. Pest resurgence occurs after using a broad spectrum pesticide to control a target pest. Broad spectrum pesticides usually kill all arthropods, so not only is the target pest killed but so are beneficial arthropods.

Shortly after World War II, when synthetic insecticides became widely available, entomologists in California developed the concept of "supervised insect control" Around the same time, some entomologists in the Cotton Belt region of the United States were advocating a similar approach. Under this scheme, insect control was "supervised" by qualified entomologists, and insecticide applications were based on conclusions reached from periodic monitoring of pest and natural-enemy populations. This was viewed as an alternative to calendar-based insecticide programs. Supervised control was based on a sound knowledge of the ecology and analysis of projected trends in pest and natural-enemy populations.

Supervised control formed much of the conceptual basis for the "integrated control" that University of California entomologists articulated in the 1950s. Integrated control sought to identify the best mix of chemical and biological controls for a given insect pest. Chemical insecticides were to be used in manner least disruptive to biological control (Stern *et al.,* 1959). The term "integrated" was thus synonymous with "compatible." Chemical controls were to be applied only after regular monitoring indicated that a pest population had reached a level (the economic threshold) that required treatment to prevent the population from reaching a level (the economic injury level) at which economic losses would exceed the cost of the artificial control measures.

IPM extended the concept of integrated control to all classes of pests and was expanded to include tactics other than just chemical and biological controls. Artificial controls such as pesticides were to be applied as in integrated control, but these now had to be compatible with control tactics for all classes of pests. Other tactics, such as host-plant resistance and cultural manipulations, became part of the IPM arsenal. IPM added the multidisciplinary element, involving entomologists, plant pathologists, nematologists, and weed scientists In the United States, IPM was formulated into national policy in February 1972 when President Richard Nixon directed federal agencies to take steps to advance the concept and application of IPM in all relevant sectors. In 1979, President Jimmy Carter established an interagency IPM Coordinating Committee to ensure development and implementation of IPM practices.

The IPM concept was explicitly defined in 1967 at a symposium sponsored by the Food and Agriculture Organization, of the United Nations, held in Rome, Italy (FAO, 1967.). The concept of "Integrated Control" (Smith and Bosch, 1967), originally limited to the combination of chemical and biological control methods (Michelbacher and Bacon 1952), was greatly expanded in that symposium, and redefined to become synonymous with what we presently consider IPM. Thus the concept of "integration" stemmed from foundations established in the U.S.A. Concurrently, however, the concept of "Pest Management" that had been proposed by Australian ecologists in 1961 (Geier and Clark 1961), started receiving greater recognition in the U.S.A and a report by the US National Academy of Sciences and the proceedings of a conference held in North Carolina which included participation by the original proponents of pest management from Australia provided the impetus for that recognition. The convergence of the concepts of integrated control and pest management, and the ultimate synthesis into integrated pest management, opened new era in the protection of agricultural crops.

1.4 Definitions of IPM

1. Stern, *et al.,* 1959—"Integrated control is defined as: 'Applied pest control which combines and integrates biological and chemical control. Chemical control is used as necessary and in a manner which is least disruptive to biological control. Integrated control may make use of naturally occurring biological control as well as biological control effected by manipulated or induced biotic agents'."

2. Smith and Reynolds (1966)—"Integrated pest control is a pest population management system that utilizes all suitable techniques in a compatible manner to reduce pest populations and maintain them at levels below those causing economic injury."

3. FAO (1967)—"Integrated control is a pest management system that in the context of the associated environment and the population dynamics of the pest species, utilizes all suitable techniques and methods in as compatible a manner as possible and maintains the pest populations at levels below those causing economic injury."

4. Smith and Bosch (1967)—"Integrated control is a pest population management system that utilizes all suitable techniques either to reduce pest populations and maintain them at levels below those causing economic injury or to so manipulate the populations that they are prevented from causing such injury."

5. National Academy of Science (1969)—"Utilization of all suitable techniques to reduce and maintain pest populations at levels below those causing injury of economic importance to agriculture and forestry, or bringing two or more methods of control into a harmonized system designed to maintain pest levels below those at which they cause harm—a system that must rest on firm ecological principles and approaches."

6. FAO. (1980)—"Integrated pest management (IPM) is an interdisciplinary approach incorporating the judicious application of the most efficient methods of maintaining pest populations at tolerable. Recognition of the problems associated with widespread pesticide application has encouraged the development and utilization of alternative pest control techniques. Rather than employing a single control tactic, attention is being directed to the coordinated use of multiple tactics, an approach known as integrated pest management."

7. Annual Report–University of California (1997)—"Integrated Pest Management (IPM) is an ecosystem-based strategy that focuses on long-term prevention of pests or their damage through a combination of techniques such as biological control, habitat manipulation, modification of cultural practices, and use of resistance varieties. Pesticides are used only after monitoring indicates they are needed according to established guidelines, and treatments are made with the goal of removing only target organism. Pest control materials are selected and applied in a manner that minimizes risks to human health, beneficial and nontarget organisms, and the environment."

8. Kogan (1998)—"IPM is a decision support system for the selection and use of pest control tactics, singly or harmoniously coordinated into a management strategy, based on cost/benefit analyses that take into account the interests of and impacts on producers, society, and the environment."

9. Food Quality Protection Act of U.S. (1998)—"Integrated Pest Management is a sustainable approach to managing pests by combining biological, cultural, physical, and chemical tools in a way that minimizes economic, health, and environmental risks."

10. The Children's Health Act of US. (2000)—"An approach to the management of pests in public facilities that combines biological, cultural, physical, and chemical tools in a way that minimizes economic, health, and environmental risks."

1.5 Phases of PM

Smith. R.F (1969) classified World wide patterns of crop protection in cotton agro ecosystem into the following phases which are also applicable to other crop ecosystems.

1. Subsistence Phase

The crop is frequently grown under non irrigated conditions. Crop does not enter the global market and is consumed in the rural community or exchanged in the local bazaars. Crop yields are ever low. Crop protection is through natural control, hand picking, host plant resistance, other cultural practices and seldom insecticides are used.

2. Exploitation Phase

The agricultural production was greater than before from survival level to privileged position so as to reach the market. Pest management exclusively depended on chemical pesticides and these are used intensively at fixed intervals. Chemical control measures were exploited to the maximum extent wherein new synthetic insecticides, new methods of application, intensive use of pesticides resulted in higher yields.

3. Crisis Phase

Subsequent to few years in exploitation phase, more frequent applications of pesticides and higher doses are needed to obtain successful control. Insect populations often resurge rapidly after treatments and the pest population gradually becomes tolerant to the pesticide. Another pesticide is substituted and pest population becomes tolerant to it too. Occasional feeders become serious pests. Excessive use of insecticides over a number of years show the way to severe problems like: Pest resistance to insecticides, Pest resurgence and Pesticide residues in food commodities

4. Disaster Phase

As a result of all deleterious effects, the cost of cultivation got increased and the crops were not grown profitably. There were frequent encounters of crop failures and produce not acceptable at market (rejection of the produce due to residues), and finally collapse of the existing pest control system.

5. Integrated Control Phase

In this phase it is aimed to give the control measures to the optimum and not to the maximum. Pest management concept is followed to avoid crisis and disaster phases by (a) Combination of the resources; (b) analysis of eco- factors; (c) optimization of techniques) recognizing or restoring the pest at manageable level.

1.6 Concepts of IPM

Integrated Pest Management (IPM) is the judicious use and integration of various pest control tactics in the context of the associated environment of the pest in ways that compliment and facilitate the biological and other natural controls of pests to meet economic, public health, and environmental goals. Wherever applicable, IPM uses scouting, pest trapping, pest resistant plant varieties, sanitation, various cultural control methods, physical and mechanical controls, biological controls, and precise timing and application of any needed pesticides. With IPM, the decision to use pesticides is made when an action threshold for a pest is reached and no other alternative management methods are available that will provide effective control. When pesticides are needed, the safest and most effective materials should be selected for use. The goals of IPM are to achieve the effective management of pests in the safest manner.

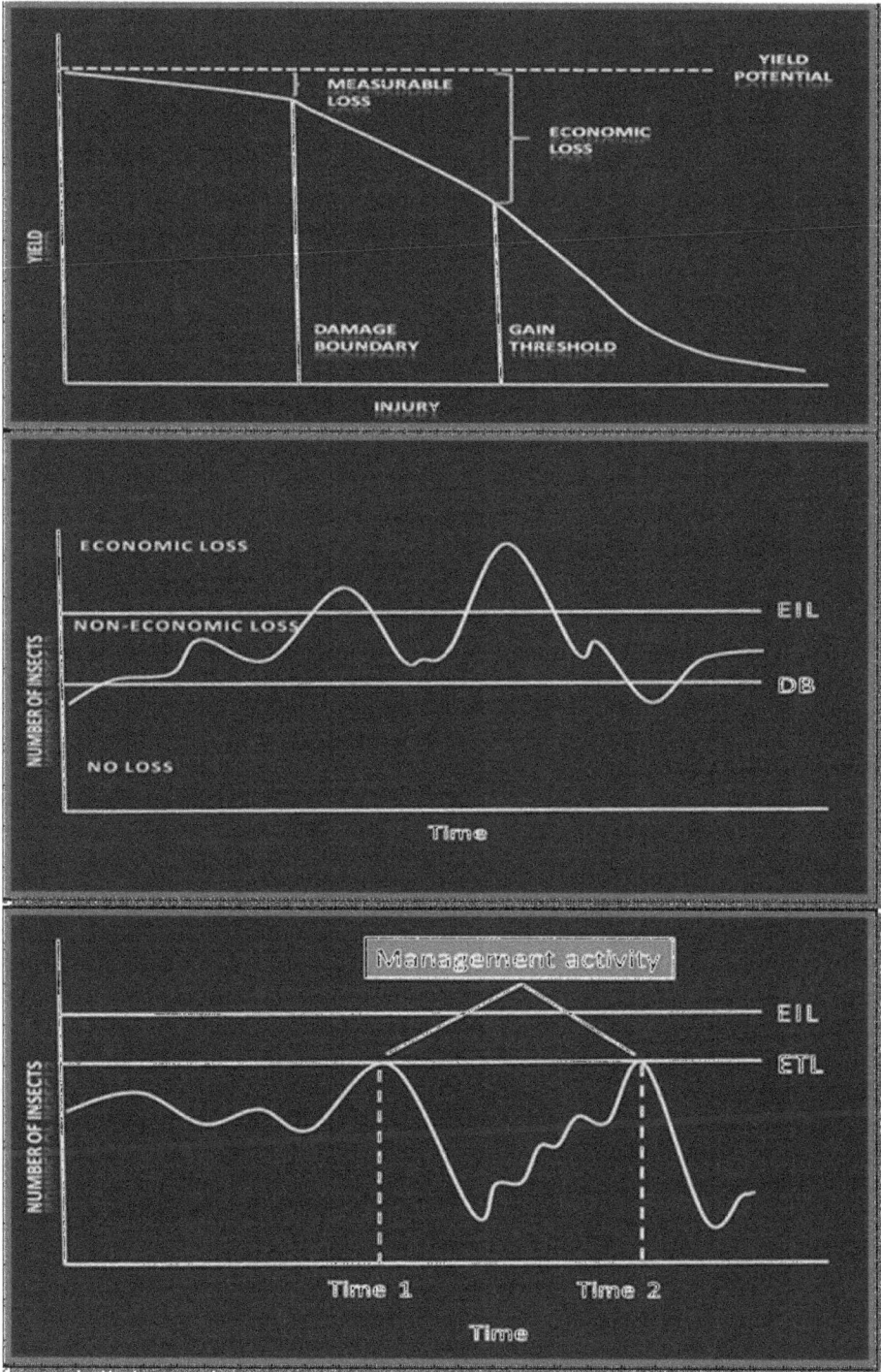

IPM Concepts.

Good pest management decisions can be made only after answering questions such as:

1. What pests are present, and in what numbers and stages of development?
2. What conditions exist that may increase or decrease pest problems?
3. What natural enemies of the pests, such as parasites, predators, and diseases, are present that my play an important role in control?
4. What amount and type of damage is being caused or may soon be caused by pests?
5. What is the stage of development, condition, and value of the crop?
6. What is the potential for economical injury? How much damage is tolerable? Has the action threshold been reached?
7. What is the history and severity of previous infestations at the site? How were those infestations managed? What were the results?
8. What pest management options are available, and how do the advantages and disadvantages of each apply to the situation?
9. If alternatives are not available, is a pesticide treatment justified for the situation? If so, what is the material of choice?
10. If a pesticide is not justified, what approaches, if any, should be taken?

Field scouting, insect trapping, and action thresholds can be used to provide much of the information needed to help answer most of these questions.The action threshold is the level of pest infestation at which treatment is justified to keep an increasing pest population from causing economical losses. Fields should not be treated when pest populations are below the action threshold. Applying a pesticide treatment for such infestations would not be an economic or qualitative benefit.

The action threshold is a key IPM decision-making tool. Thresholds are based on considerable amounts of research and field experience. If an action threshold is approached, but not reached, do not apply a pesticide at that time. Instead, rescout the field within a few days to determine the status of the infestation. Pest populations can decline naturally due to mortality from natural enemies and unfavorable weather conditions. Also, many pests, such as caterpillars, change from an active feeding (larva) to a non-feeding stage (pupa) during their development. Such changes will often produce a natural decline in infestations as pupation occurs. Precise timing of needed pesticide applications is extremely important to achieve good pest control. Pest monitoring, action thresholds, and a a good knowledge of the life cycles of pests are used to determine the best timing of needed treatments.

Economic decision levels usually are expressed as the number of insects per area, plant or animal unit, or sampling procedure. Less commonly, such levels are given as degree of plant damage or combinations of numbers and damage. The levels are unique in that they have both biological and economic attributes, and they are most often used for management decisions for private concerns. In 1959, V. M. Stern and colleagues formally proposed the concepts and terminology of bioeconomics that we use today. Specifically, they developed the ideas of economic damage, economic-injury level, and economic threshold, collectively called the economic-injury level concept. Although this concept was originally pro posed in 1959, some of the ideas expressed had been discussed years earlier (1934) by W D. Pierce in a particularly farsighted article. Pierce raised questions that became one incentive for developing economic-injury levels (EIL "Is all insect attack to be computed as assessable damage? If not. at what point does it become assessable? is control work warranted when damage is below that point?" Although fundamental to the concept, such questions may not have been the ultimate impetus for EIL development. The concept actually emerged as an encouragement for more rational use of insecticides. In discussing the EIL concept, Stern and colleagues emphasized the concerns of many persons regarding excessive and other inappropriate uses of insecticides. They highlighted problems of insecticide resistance, residues, and effects on non-target organisms. These ideas were a critical part of the concept of integrated control, a new approach at the time, recommended as a replacement for the overly simplistic strategy of "identify and spray."

Ecosystems are self-sufficient habitats where living organisms and the non-living environment interact to exchange energy and matter in a continuing cycle. Ecosystems are entities, such as forests, ponds, and fields, and in general they are self-regulating. Agricultural ecosystems (agroecosystems) contain a lesser diversity of animal and plant species than do natural ecosystems such as forests and prairies. Usually there are a few major species and numerous minor species and, in a pest outbreak, usually only one pest species at a time (often a major species) is present in large numbers. A typical agricultural unit may contain only 1 to 4 major crop species and 6 to 10 major pest species. The agroecosystem is intensively manipulated by man and subject to sudden alterations such as plowing, mowing, and treatments with pesticides. Agronomic practices are critical in pest management, since the need for pest control or the intensity of a pest problem is often directly related to agronomic practices. Agroecosystems can be more susceptible to pest damage and catastrophic outbreaks owing to the lack of diversity in species of plants and species of insects and the sudden alterations imposed by weather and man. It is important in pest management to recognize the existence of complex biological systems in the agroecosystem.

Any IPM programme should focus to minimize the disadvantages associated with use of pesticides and maximizing socio, economic and ecological advantages in the following ways:

1. **Understanding the agricultural ecosystem :** An agro ecosystem contains a lesser diversity of animal and plant species than natural ecosystem like forests. A typical an agro ecosystem contain only 1-4 major crop species and 6-10 major pest species. An agro ecosystem is intensively manipulated by man and subjected to sudden alterations

such as ploughing, inter cultivation and treatment with pesticides. These practices are critical in pest management as pest populations are greatly influenced by these practices. Agro ecosystem can be more susceptible to pest damage and catastrophic outbreaks owing to lack of diversity in species of plants and insects and sudden alternations imposed by weather and man. However, agro ecosystem is a complex of food chains and food webs that interact together to produce astable unit.

2. **Planning of agricultural ecosystem :** In IPM programme the agricultural system can be planned in terms of anticipating pest problem and also the ways to reduce them that is to integrate crop protection with crop production system. Growing of susceptible varieties should be avoided and related crops shouldn't be grown. Bhendi followed by cotton increases incidence of the spotted borer. Groud nut followed by soybean increases incidence of the leaf miner.

3. **Cost benefit ratio :** Based on the possibility of pest damage by predicting the pest problem and by defining economic threshold level, emphasis should be given to cost benefit ratio. The crop life table to provide solid information analysis of pest damage as well as cost benefit ratio in pest management. Benefit risk analysis comes when a chemical pesticide is applied in an agro ecosystem for considering its impact on society as well as environment relevant to its benefits.

4. **Tolerance of pest damage :** The pest free crop is neither necessary in most cases for high yields nor appropriate for insect pest management. Castor crop can tolerate upto 25 per cent defoliation. Exceptions occur in case of plant disease transmission by vectors. The relationship between density of pest population and profitability of control measures is expressed through threshold values.

5. **Leaving a pest residue :** Natural enemy population is gradually eliminated not only in the absence of their respective insect hosts because of the indiscriminate use of broad spectrum insecticides, which in turn also eliminate natural enemies. Therefore, it is an important concept of pest management, to leave a permanent pest residue below economic threshold level, so that natural enemies will survive.

6. **Timing of treatments :** Treatment in terms of pesticide spray should be need based, with minimum number of sprays, timely scheduled, combined with improved techniques of pest monitoring and crop development *E.g.*: Use of pheromone traps for monitoring of pest population

7. **Public understanding and acceptance:** In order to deal with various pest problems special effort should be made for effective communication to the people for better understanding and acceptance of pest management practices. The IPM practices followed should be economical and sustainable

Economic Decision Levels of Pest Population

Economic decision levels are the foundation of insect pest management programs. Such levels are vital because they designate the course of action to be taken in any given pest situation. Economic decision levels usually are expressed as the number of insects per area, plant or animal unit, or sampling procedure. Less commonly, such levels are given as degree of plant damage or combinations of numbers and damage. The levels are unique in that they have both biological and economic attributes, and they are most often used for management decisions

A. Economic Damage and the Damage Boundary

Economic damage was originally defined as "the amount of injury which will justify the cost of artificial control measures." To understand this term, we distinguish between injury and damage. Injury is the effect of pest activities on host physiology that is usually deleterious. Damage is a measurable loss of host utility, most often including yield quantity, quality, or aesthetics. Therefore, injury is centered on the pest and its activities, and damage is centered on the crop and its response to injury. As the concept applies to pest management, economic damage begins to occur when money required for suppressing insect injury is equal to the potential monetary loss from a pest population. The term gain threshold has been used to express this beginning point of economic damage.

The gain threshold can be expressed as follows:

$$\text{Grain threshold} = \frac{\text{Management costs (Rs./ha)}}{\text{Market value (Rs/kg)}} = \text{kg/ha}$$

B. Economic Injury Level

The economic-injury level (EIL) is defined as the lowest number of insects that will cause economic damage, or the minimum number of insects that would reduce yield equal to the gain threshold (Pedigo, 1991). Although expressed as numbers of insects per unit area, the EIL, as its name implies, is really a level of injury. Because injury is usually difficult to measure in a field situation, numbers of insects are used as an index of that injury. For example, it is usually easier to count insect numbers than it is to estimate the area of foliage removed by a pest population or the amount of juices sucked from plants. In some instances, particularly when several pest species causing similar injury are present, insect equivalents may he considered instead of insect numbers. If management action can be taken quickly and loss can be averted entirely, then the EIL can be expressed as follows:

$$V \times I \times P \times D = C$$

where,

V =
   Market value per unit of produce (for example, Rs./kg)

I = Injury units per insect per production unit (for example, percent

P = Density or intensity of insect population (for example, insects/ha)

D = Damage per unit injury (for example, kg lost/ha/percent defoliation)

C = Cost of management per area (for example, Rs./ha)

then

$$P = \frac{C}{V \times I \times D}$$

EIL = P., In instances where some loss from the insect is unavoidable, for example, if damage or injury can be reduced only 80 percent, then the relationship becomes

$$EIL = \frac{C}{V \times I \times D \times K}$$

where,

K = proportionate reduction in potential injury or damage (for example, 0.8 for 80 percent) With some insect pests, particularly piercing-sucking insects, the separation of the land D variables presents a problem. This is because the I variable for plants would represent photosynthate (sap) removed per insect and the D variable would represent yield loss per unit of photosynthate removed.

C. Economic Threshold

The economic threshold (ET) is perhaps the best-known term and most extensively used index in making pest management decisions. The ET indicates the number of insects (density or intensity) that should trigger management action. For this reason, it is sometimes called the **action threshold**. Although expressed in insect numbers, the ET is really a time parameter, with pest numbers used as an index for when to implement management. Just as with EILs, ETs can also be expressed in insect equivalents. If a pest population is growing as the season progresses, growth rates are predicted, and the ET is set below the EIL. By setting the ET at a lower value, we are predicting that once the population reaches the ET, chances are good that it will grow to exceed the EIL. Therefore, it is appropriate for us to take action on an earlier date, before we accumulate losses in reaching the EIL. The relationships between the EIL and the ET demonstrate that action taken when a population level exceeds the ET forces down the population before it can reach the EIL. No action is taken at levels below the ET.

D. General Equilibrium Position (GEP)

It is the average population density of insect over a long period of time unaffected by temporary interventions of pest control.Nonetheless, the economic injury level may be at any level well above or below the general equilibrium. The EIL may be at any level from well below to well above the GEP. Based on this pests can be grouped into four classes.

1. **Negligible pest:** Population density by no means raise high to cause economic injury.

2. **Occasional pest:** Occasionally their density reaches EIL when their population is affected by abnormal weather conditions or the indiscreet use of insecticides. At their zenith of population density, some sort of intervention usually an insecticide is vital to lessen their numbers to tolerable level.

3. **Perennial pest:** EIL's are slightly higher than the GEP and intervention is compulsory at nearly every upward population instability. The common exercise is to intervene with insecticides whenever necessary to generate a modified average population density.

4. **Severe pest:** They have EIL below the GEP. Regular and continuous interventions with insecticides are required to produce marketable crops.EIL decreases as the value of crop increases. It also depends on the stage of the crop, stage of the pest etc.

1.7 Components of IPM

Integrated Pest Management (IPM) is an ecosystem-based approach that focuses on long-term prevention of pests or their damage through a combination of techniques such as biological control, habitat manipulation, modification of cultural practices, and use of resistance varieties. Pesticides are used only after monitoring indicates they are needed according to established guidelines, and treatments are made with the goal of removing only target organism. Pest control materials are selected and applied in a manner that minimizes risks to human health, beneficial and non-target organisms, and the environment."

Indiscriminate and injudicious use of chemical pesticides in agriculture has resulted in several associated adverse effects such as environmental pollution, ecological imbalances, pesticides residues in food, fruits and vegetables, fodder, soil and water, pest resurgence, human and animal health hazards, destruction of biocontrol agents, development of resistance in pests etc. Therefore, Govt. of India has adopted Integrated Pest Management (IPM) as cardinal principle and main plank of plant protection in the overall Crop Production Programme since 1985. IPM is an eco-friendly approach which encompasses cultural, mechanical, biological and need based chemical control measures. The IPM approach is being disseminated through various schemes/ projects at national and state level with the following objectives: Maximise crop production with minimum input costs, Minimise environmental pollution in soil, water and air due to pesticides. Minimise occupational health hazards due to chemical pesticides, Preserve ecosystem and maintain ecological equilibrium, No or less use of chemical pesticides for minimum pesticide residues, To improve farming systems. The various components of IPM are discussed in detail underneath.

1. Cultural methods
2. Mechanical methods
3. Physical methods
4. Biological methods
5. Legislative methods and
6. Chemical methods

1.7.1 Cultural Methods of IPM

Definition

"Cultural practices" refers to that broad set of management techniques or options which may be manipulated by agricultural producers to achieve their crop production goals or "the manipulation of the environment to improve crop production." "Cultural control" on the other hand, is the deliberate alteration of the production system, either the cropping system itself or specific crop production practices, to reduce pest populations or avoid pest injury to crops. The manoeuvring of cultural practices at an apt time for reducing or evade pest damage to crops is known as cultural control.

The cultural practices create an environment less encouraging for the pests and or more favourable for its natural enemies. It is the economical of all pest control methods.

1. **Deep ploughing**: Deep ploughing has been recommended as a strategy to kill insect pests that live in the top 20 cm of soil. By exposing the larvae or pupa to sunlight, and by physical crushing in the ploughing process, pest numbers can be reduced. *e.g.*roots grubs and Pupae of moths.

2. **Weed control**: Many of the insect pests are polyphagous, capable of feeding and reproducing on crops and weeds in numerous plant families. Weeds found on field margins and ditch banks can provide insect pests with suitable resources needed for rapid population growth which subsequently can lead to insect infestations occurring in adjacent vegetable crops. Many weed species are important for economic insect pests by providing host plants that serve as a bridge between cropping seasons when vegetables crops are not in production. Weeds also serve as alternate host plants for a three important groups of viruses that affect vegetable and melons; tospoviruses, potyviruses and crini viruses. These viruses utilize a number of important weeds as hosts and are vectored by key insect pests. For example, one of the primary insect vectors of the tospovirus, tomato spotted wilt virus, is the western flower thrips. two common weed species (lambsquarters/ Pigweed *Chenopodium album* serve as hosts to both the vector and the virus. Hence Exclusion of such weeds which act as alternate hosts will limit the population build up of the pests. Removal of weeds of

Menispermaceae family controls Fruit sucking moth larvae *Eudocima ancilla.*

3. **Pruning/Removal of infested parts**: Successive infestations are maintained at lower levels. *e.g.*Cutting and removal of infested parts of brinjal attacked by *Leucinodes orbonalis,* Pruning of dried branches of citrus eliminates scales and stem borer. Clipping of leaf lets in coconut reduces the black headed caterpillar. Pests like coccids get carried over to the next season through stubbles, which should be promptly removed.

4. **Alteration in system of cultivation**: Alteration of banana crop from perennial to annual crop reduces the infestation of banana rhizome weevil *Cosmopolitus sordidus* in addition to giving increased yields.

5. **Crop rotation**: Crop rotation is the practice of growing a series of dissimilar/different types of crops in the same area in sequential seasons. Crop rotation also mitigates the build-up of pathogens and pests that often occurs when one species is continuously cropped, and can also improve soil structure and fertility by alternating deep-rooted and shallow-rooted plants.

Crop rotation is also used to control pests and diseases that can become established in the soil over time. The changing of crops in a sequence tends to decrease the population level of pests. Plants within the same taxonomic family tend to have similar pests and pathogens. By regularly changing the planting location, the pest cycles can be broken or limited. For example, root-knot nematode is a serious problem for some plants in warm climates and sandy soils, where it slowly builds up to high levels in the soil, and can severely damage plant productivity by cutting off circulation from the plant roots. Growing a crop that is not a host for root-knot nematode for one season greatly reduces the level of the nematode in the soil, thus making it possible to grow a susceptible crop the following season without needing soil fumigation.

It is also difficult to control weeds similar to the crop which may contaminate the final produce. For instance, ergot in weed grasses is difficult to separate from harvested grain. A different crop allows the weeds to be eliminated, breaking the ergot cycle.This principle is of particular use in organic farming, where pest control may be achieved without synthetic pesticides.

Crop rotation is largely successful practice which is not in favour of pests that have a narrow host range and dispersal capacity. If a non-host crop is grown after a host crop, it decreases the pest population. *e.g.*Cereals followed by pulses. Cotton should be rotated with non hosts like ragi, maize, rice to minimize the incidence of insect pests. Groundnut with non leguminous crops is recommended for minimizing the leaf miner incidence.

6. **Intercropping**: Intercropping of two or more crops has been followed by our farmers from a very long time. Sorghum/red gram is one of the popular systems which is followed even today. While intercropping was practiced mainly from economic angle, it has also an effective bearing on pest preference and incidence. Many intercropping systems have been evolved over years which has become an important arm of pest management. Certain intercropping like peanut/coriander, maize/soyabean etc. are helpful in enhancing activity of beneficial predators like lady bird beetles, spiders etc. and reducing weed population. Anticipated for attaining some income when the main crop is attacked by pest, the other escapes. *e.g.*Garden peas and sunhemp

7. **Use of resistant varieties**: Cultivation of resistant varieties helps in suppressing pest population specially in endemic areas. This is one of the cost effective method of pest management without disturbing ecosystem. *e.g.* Pusa Purple Long, Pusa Purple Cluster, Pusa Purple Round, Banaras Long Purple, Arka Kesav, Arka Kusmakar, have been reported to be tolerant or resistant to brinjal fruit and shoot borer.

8. **Adjusting planting or sowing or harvesting times**: The manoeuvring of planting time facilitate to minimize pest damage by generating asynchrony between host plants and the pest or synchronizing insect pests with their natural enemies. *e.g.* Early sown sorghum in kharif reduces the infestation of shoot fly Timely and synchronous planting has been found to reduce bollworm damage in cotton and stem borer damage in sugarcane.

9. **Trap cropping**: Some plants are more preferred by some pests. These plants can be grown as "trap plants" to reduce pest incidence on the main crop. Marigold attracts fruit borer (Heliothis) pest. Growing of these plants in main crops of chillies will reduce fruit borer damage in these crops. It is also reported that soil nematodes are attracted to the root zone of marigold plants. Growing castor plants on borders in chillies, cotton etc. will reduce fruit borer (Spodoptera) in these crops. Similar trap crop can be identified and cultivated to reduce pest infestation.Tomato in Citrus crop against Fruit sucking moths.

10. **Flooding the field**: Summer flooding of the fields will reduce the infestation of Cutworms, army worms, termites, root grubs

11. **Alley ways**: Leaving alley ways for every 2 metres at the time of sowing, would facilitate good air movement and reduce the humidity and reduces infestation of rice BPH *Nilaparvata lugens.*

12. **Raking up and Hoeing**: *Fruit fly* pupates in the soil around the trees of the host plants and *Raking up* and *hoeing* of soil around these trees helps in killing the pupa.

13. **High seed rate**: Application of high seed rate of sorghum seed @10 kgs/ha,limits sorghum shoot fly (*Atherigona soccata*)

14. **Mulching**: Silver, aluminized or metalized mulch reduce the population of thrips, aphids and silverleaf whitefly by deterring adult insects from landing on plants. Silver, aluminized or metalized mulch can delay tomato yellow leaf curl (TYLC) by at least 2 weeks and reduce tomato spotted wilt (TSW) significantly.Trash mulching @ 3 t/ha 3 days after planting or earthing up at a month or two after planting minimize early shoot borer (*Chilo infuscatellus*) attack in sugarcane.

15. **Destruction of crop residue**: Crop residues serving as overwintering sites for pests should be destroyed. Ploughing and burning Stubbles of sugarcane and paddy controls the borers that harbour

16. **Sun drying**: Harnessing the power of the sun to reduce pest damage in the stored products is simple and effective and appropriate exposure of stored products to the sun leads to a temperature rise that kills most if not all of the pests' eggs, larvae and adults - on and inside the grains.

1.7.2 Mechanical Methods of Pest Management

Mechanical control can be effectively practiced either by individual farmers or on community basis. Suppression or reduction or insect pest population by contribution manual strategies. Mechanical controls directly remove or kill pests. Mechanical control methods can be rapid and effective, but many are mostly suited for small acute pest problems. Importantly, mechanical controls have relatively little impact on natural enemies and other non-target organisms, and are therefore well suited for use with biological control in an integrated pest management approach. Egg masses, larvae or nymphs and sluggish adults can be handpicked and destroyed. Some of the examples are listed below:

1. Egg masses of paddy stem borer (*Scirpophaga incertulas*) and groundnut hairy caterpillar can be collected and destroyed.

2. The moringa caterpillars, which collect at tree trunks in the mornings can be burnt.

3. Collection and destruction of fallen fruits is effective against fruit flies and fruit borers.

4. Uprooting virus infected plants.

5. Bagging/wrapping of pomegranate and mango fruits in paper bags avoids the infestation of pomegranate butterfly *Virachola isocrates* and mango fruit fly *Bactrocera dorsalis*.

6. Tin bands are fixed over coconut palms and Construction of rat proof godowns to prevent damage by rats.

7. Digging of 30 -60 cm wide and 60 cm deep trenches or erecting 30 cm height tin sheets barriers around the fields is useful against pests like hairy caterpillars.

8. Use of arrow headed Iron rod/hook for extraction of adult Rhinoceros beetle (*Oryctes rhinoceros*) from the crown of coconut trees.

9. Make use of of an alkathene band around the tree trunks of mango to check the migration of first instar nymphs of mealybugs and red ants.

10. Shaking of trees during evening hours to dislodge and destroy the adults of root grubs and chaffer beetles.

1.7.3 Physical Methods

Physical Pest Control is a method of getting rid of insects and rodents by removing, attacking, or setting up barriers that will prevent further destruction of one's plants. These methods are used primarily for crop growing, Physical methods that are used to decrease the pests are listed below:

1. **Vapour Heat Treatment (VHT):** Vapour heat treatment technologies can be used for post-harvest insect control for perishable commodities such as fresh fruits (*e.g.*, mangos, papaya, persimmon, citrus, bananas, carambola), fresh vegetables (*e.g.*, peppers, eggplant, tomatoes, cucumber, and zucchini squash), bulbs, and cut flowers. Heat treatments for disinfestation of fruit have been used since 1929 when Baker and co-workers developed a vapor heat treatment against the Mediterranean fruit fly (Couey 1989). Against the mango against fruit flies, heated air is saturated with water (>RH 90 per cent) for 6 to 8 hours for raising pulp temperature to 43-44.5°C.

2. **Light traps:** Light traps, with or without ultraviolet light, attract certain insects. Light sources may include fluorescent lamps, mercury-vapor lamps, or black. Designs differ according to the behavior of the insects being studied. Light traps are widely used to survey nocturnal moths. Total species richness and abundance of trapped moths may be influenced by several factors such as night temperature, humidity and lamp type. Grasshoppers and some beetles are attracted to lights at a long range but are repelled by it at short range. Farrow's light trap has a large base so that it captures insects that may otherwise fly away from regular light traps. Light traps can attract flying and terrestrial insects, and lights may be combined with other methods. Light traps are useful for monitoring the population of important insect pests in an area. *e.g*: Most of the moths and beetles.

3. **Diatomaceous Earth (DE)**: The diatomaceous earth wholesale, normally called diatomite is a desiccant dust made from silica-based skeletons of microorganisms called diatoms. The chemical composition of DE is approximately 90 per cent silica, 2 per cent alumina and 2 per cent iron oxide. DE is used in an all-natural non-chemical pest-control therapy. The tiny diatoms within the powder adhere to the outside of the insect's exoskeleton and scratch it off until it dehydrates and dies. This really is a straightforward successful and long term residual cure for bed bugs. It's safe to spread the dust across

the ridges of the bedding, beneath the baseboard and within the box-spring and joints of furniture.This physical treatment kills the pest whenever it crawls on the dust. But, it's not just a fast killing treatment like others. It will the task gradually. The therapy is best suited in low humid environment and is safe to combine it with other insecticide dusts.

4. **Silica gel :** Silica-gel is a desiccant. The definition of serum can be a misnomer since they are truly beads that absorb moisture in the air and are hard and brittle. Standard silica gel is clear and it contains silicon dioxide. Some types contain metal oxide or alumina (a product much like sandpaper). Bed bugs are killed by silica gel dust when they interact with it. It is safe to make use of crushed silica gel or sachets of silica gel purchased in almost any product. It is possible to mix the dust using an pesticide. Nevertheless, the dust is effective when used alone. Silica gel is available in the industrial market. It is a really cheap and effective therapy to kill bed bugs. Silica-gel is definitely an irritant so use a disguise when using it. Silica-gel is non-hazardous and non-toxic when handled. It's broadly speaking recognized as safe, but shouldn't be swallowed. It is non-reactive, non-flammable, non-toxic and stable with standard use. Shop fits in away from young ones and animals.

5. **Kaolin Clay :** Kaolin is a naturally occurring clay resulting from weathering of aluminous minerals such as feldspar with kaolinite as its principal constituent. It is also used as an "inert" carrier in some pesticides, and enhances the performance of some microbial products. It is also used as an "inert" carrier in some pesticides, and enhances the performance of some microbial products. The clay minerals absorb the lipoid layer of the insect cuticle by which the insects lose their body moisture and die due to desiccation. Kaolin clay is available as a wettable powder to be mixed with water. *e.g.* Surround W. P. is made from 95 per cent kaolin clay, a naturally occuring mineral. When applied to fruit trees, crops, and other plants, it forms a white film. Surround suppresses a wide range of pests, especially those which damage fruit crops including pears, apples, grapes, berries, and some vegetables. It can be applied up to day of harvest and is easily rubbed off when the fruit or produce is ready to eat.

6. **Steam sterilization of soil**: Soil steam sterilization (soil steaming) is a farming technique that sterilizes soil with steam in open fields or greenhouses. Pests of plant cultures such as weeds, bacteria, fungi and viruses are killed through induced hot steam which causes their cell structure to physically degenerate.

7. **Controlled atmosphere:** A controlled atmosphere is an agricultural storage method in which oxygen, carbondioxide and nitrogen concentrations as well as temperature and humidity are regulated. In air tight containers small volume of air is enclosed, the available

oxygen is quickly utilized by insects and raise concentration of carbon dioxide. High concentration of carbon dioxide leads to death of stored products insects. Two major classes of commodity can be stored in controlled atmosphere :

A) Dry commodities such as grains, legumes and oilseed. In these commodities the primary aim of the atmosphere is usually to control insect pests. Most insects cannot exist indefinitely without oxygen or in conditions of raised (greater than approximately 30 per cent) carbon dioxide. Controlled atmosphere treatments of grains can be a fairly slow process taking up to several weeks at lower temperatures (less than 15°C). A typical schedule for complete disinfestation of dry grain (<13 per cent moisture content) at about 25°C, with carbon dioxide, is a concentration above 35 per cent (v/v) carbon dioxide (in air) for at least 15 days. These atmospheres can be created either by: adding pure gases carbon dioxide or nitrogen or the low oxygen exhaust of hydrocarbon combustion, or using the natural effects of respiration (grain, moulds or insects) to reduce oxygen and increase carbon dioxide Hermetic storage.

B) Fresh fruits, most commonly apples and pears, where the combination of altered atmospheric conditions and reduced temperature allow prolonged storage with only a slow loss of quality.

8. **Sterile Insect Technology:** Male insects can be made sterile by exposing them to gamma radiation or by using chemicals. When sterile males are released in normal population they compete with normal males in copulation and to that extent reductive capacity of the population are reduced. By sterilizing the pupae of screwworm, livestock pest (*Cochliomyia hominivorax*) with radiations, sterile males were obtained. They were released @ 400/sq mile for 7 weeks. By this method total eradication was achieved in South East parts of America and in the Curacao islands in case of screwworm.

1.7.4 Legislative Methods of Pest Control

Plant Quarantine

Legal enforcement of the measures aimed to prevent pests from spreading or to prevent them to multiply further in case they have already gained entry and have established in new restricted areas is Plant Quarantine. The importance of imposing restrictions on the movement of pest-infested plants or plant materials from one country to another was realized when the grapevine phylloxera got introduced into France from America by about 1860 and the San jose scale spread into the USA in the later part of the 18th century and caused severe damage. The first Quarantine Act in USA came into operation in 1905. While Govt. of India passed an Act in 1914 entitled "Destructive Insect and Pests Act of 1914" to prevent the introduction of any insect, fungus or other pests into our country. This was later supplemented by a more comprehensive act in 1917.

The Directorate of Plant Protection, Quarantine and Storage (DPPQS) is established under Department of Agriculture and Cooperation of Ministry of Agriculture, Government of India, Faridabad, Haryana is an apex plant protection organisation in the country enforcing Plant Quarantine Regulations issued under The Destructive Insects and Pests Act, 1914 and amendments issued there under to prevent introduction and spread of exotic pests. Plant quarantine aims at preventing the introduction of exotic pests through imported seeds, plants and plant material and to contain the spread of exotic pest that are accidentally got introduced to the country by implementing the provisions of The Destructive Insects and Pests Act, 1914 and the regulations issued there under. The Joint Director (PP) heads the Plant Quarantine scheme. There are five regional plant quarantine stations at Amritsar, Kolkata, Chennai, New Delhi and Mumbai and 21 minor PQ stations functioning at various sea ports/airports and land borders.

National Bureau of Plant Genetic Resources (NBPGR) undertakes the quarantine processing of all germplasm including transgenic planting material under exchange for research purposes. NBPGR also deals with testing for absence of terminator technology which is mandatory as per national legislation. Presently, there is a provision to restrict the inter-state movement of nine pests *viz.*, fluted scale, San José scale, coffee berry borer, codling moth, Banana bunchy top virus, Banana mosaic virus, potato cyst nematode, potato wart and apple scab.

In several countries numerous dangerous pests have frequently been found to be foreign pests and they cause greater damage than the native ones. Potato tuber moth *Pthorimea operculella*, cotton cushiony scale *Icerya purchasi,* wooly aphis on apple *Eriosoma lanigerum,* san jose scale *Quadraspidiotus perniciosus*, golden cyst nematode *Globodera rostochinesis* and the giant African snail, *Achatina fulica* (Predatory snail, *Eugladina rosea),* serpentine leaf miner *Liriomyza trifolii,* Spiralling whitefly, *Alerodicus dispersus,* Coconut mite *Aceria guerreoronis* etc,are some exotic pests introduced into our country.

The legislative measures in force now in different countries can be grouped into five classes:

1. Legislation to prevent the introduction of new pests and weeds etc from foreign countries (International quarantine).

2. Legislation to prevent the spread of already established pests, diseases and weeds from one part of the country to another (Domestic quarantine).

3. Legislation to enforce upon the farmers regarding the application of effective control measures to prevent damage by already established pests.

4. Legislation to prevent the adulteration and misbranding of insecticides and determine their permissible residue tolerance levels in food stuffs.

5. Legislation to regulate the activities of men engaged in pest control operations and application of hazardous insecticides.

Phytosanitary Certification

Phytosanitary certificates are issued to indicate that consignments of plants, plant products or other regulated articles meet specified phytosanitary import requirements and are in conformity with the certifying statement of the appropriate model certificate. Phytosanitary certificates should only be issued for this purpose. Phytosanitary certification of plants and plant products is carried out in accordance with article IV of International Plant Protection Convention (IPPC) to meet the legal obligations of the member countries. The Phytosanitary certificates are issued in the model formats set out under article V of the IPPC in consistence with current PQ regulations of the importing country. Such certificates are issued after careful inspection and treatment of plants and plant products by technically qualified and duly authorized officers at the country of export and shall include additional declarations as may be required by the importing country and also particulars of treatment, if any given by the duly authorized officer. Accordingly inspecting and certifying authorities are notified by Ministry of Agriculture for undertaking export inspection and phytosanitary certification. The Phytosanitary issuing authority may refer Export-Import Policy as well as provisions of the CITES (Convention on International Trade in Endangered Species of wild flora and fauna) for the detailed list of prohibited/restricted plant species. While entering into trade negotiations (Letter of Credit (LC)/Agreement etc.,), the exporter should ensure to reflect in the contract which they entered with the importer about the current Plant Quarantine Regulations in respect of the Commodity being exported from the importing country or a copy of the permit issued by the importing country wherever applicable. The phytosanitary certificate is issued as per the requirements of the importing country duly reflected in the contract or the permit issued by the importing country. Importing countries should only require phytosanitary certificates for regulated articles. These include commodities such as plants, bulbs and tubers, or seeds for propagation, fruits and vegetables, cut flowers and branches, grain, and growing medium etc.

Chapter 2

Biological Control

The successful management of a pest by means of another living organism like parasitoids, predators and pathogens that is encouraged and disseminated by man is called biological control. In such programme the natural enemies are introduced, encouraged, multiplied by artificial means and disseminated by man with his own efforts instead of leaving it to nature. Biological control is a component of an integrated pest management strategy. It is defined as the reduction of pest populations by natural enemies and typically involves an active human role. Keep in mind that all insect species are also suppressed by naturally occurring organisms and environmental factors, with no human input. This is frequently referred to as natural control. Natural enemies of insect pests, also known as biological control agents, include predators, parasitoids, and pathogens. The conservation of natural enemies is probably the most important and readily available biological control practice available to growers. The biological control involves the supplemental release of natural enemies also.

2.1 Techniques in Biological Control

Biological control practices involve three techniques *viz.,* Introduction, Augmentation and Conservation.

1. Introduction or Classical Biological Control

It is the deliberate introduction and establishment of natural enemies to a new locality where they did not occur or originate naturally. When natural enemies are successfully established, it usually continues to control the pest population. *e.g.* The predator *Rodolia cardinalis* was imported and introduced in Nilgiris from California in 1929 and from Egypt in 1930 and multiplied in the laboratory and released towards Control of cottony cushion scale, *Icerya purchasi* on fruit trees and within one year the pest was effectively checked.

Neo-classical Biological Control

A form of biological control in which natural enemies are imported from elsewhere and released in small numbers in attempt to establish a permanent population to control a native pest with which they have not co-evolved. Example:

1. Control of glassy winged sharpshooter [GWSS: *Homalodisca vitripennis*] egg masses in California by *Gonatocercus tuberculifemur* from Argentina.

2. Control of Levuana iridescens [coconut moth of Fiji] tachinid fly *Bessa remota*, introduced from Malaya.

2. Augmentation

It is the rearing and releasing of natural enemies to supplement the numbers of naturally occurring natural enemies. There are two approaches to augmentation.

Inoculative Releases

Individuals bio-agents are released only once during the season and natural enemies are expected to reproduce and increase its population for that growing season. Hence biological control is expected from the progeny and subsequent generations and not from the release itself. *e.g. Release of Chrysoperla* spp. @ 6 second instar larvae/plant

Inundative Releases

This technique involves mass multiplication and periodic release of natural enemies when pest populations approach damaging levels. Natural enemies are not expected to reproduce and increase in numbers. Control is achieved through the released individuals and additional releases are only made when pest populations approach damaging levels. *e.g.* Realease of *Trichogramma japonicum* @50,000 – 1,00,000 parasitized eggs/ha for control of rice yellow stem borer.

3. Conservation

Conservation is defined as the actions to preserve and release of natural enemies by environmental manipulations or alter production practices to protect natural enemies that are already present in an area or non use of those pest control measures that destroy natural enemies.

Important conservation measures are :

a. Use selective insecticide which is safe to natural enemies.

b. Avoidance of cultural practices which are harmful to NE

c. Cultivation of varieties that favour colonization of natural enemies

d. Providing alternate hosts for natural enemies.

e. Preservation of inactive stages of natural enemies.

f. Provide pollen and nectar for adult natural enemies

Parasite

A parasite is an organism which is usually much smaller than its host and a single individual usually doesn't kill the host. Parasite may complete their entire life cycle (eg. Lice) or may involve several host species. Or Parasite is one, which attaches itself to the body of the other living organism either externally or internally and gets nourishment and shelter at least for a shorter period if not for the entire life cycle. The organism, which is attacked by the parasites, is called hosts.

Parasitoid

Parasitoid is an insect parasite of an arthopod, parasitic only in immature stages, destroys its host in the process of development and free living as an adult. *e.g.,* Braconid wasps

2.2 Qualities of a Successful Parasitoid in Biological Control Programme

A parasitoid should have the following qualities for its successful performance.

1. Should be adaptable to environmental conditions in the new locally.
2. Should be able to survive in all habitats of the host.
3. Should be specific to a particulars sp. of host or at least a narrowly limited range of hosts.
4. Should be able to multiply faster than the host.
5. Should be having more fecundity and have high sex ratio.
6. Life cycle must be shorter than that of the host.
7. Should have good searching capacity for host.
8. Should be amendable for mass multiplication in the labs.
9. Should bring down host population within 3 years.
10. There should be quick dispersal of the parasitoid in the locality.
11. It Should be free from hyperparasitoids.

2.3 Some Successful Examples of Biological Control

1. ***Rodolia cardinalis***: Control of cottony cushion scale, *Icerya purchasi* on fruit trees by its predatory vedalia beetle, *Rodolia cardinalis* in Nilgiris. The predator was imported from California in 1929 and from Egypt in 1930 and multiplied in the laboratory and released. Within one year the pest was effectively checked.

2. ***Cyrtobagous salviniae***: For the biological suppression of Water Fern, *Salvinia molesta,* the weevil, *Cyrtobagous salviniae*, was

imported from Australia in 1982. Exotic weevil, *C. salviniae* was released for the control of water fern, S. molesta in a lily pond in Bangalore in 1983-84. Within 11 months of the release of the weevil in the lily pond the salvinia plants collapsed and the lily growth, which was suppressed by competition from salvinia resurrected.

3. ***Neochetina bruchi***: Biological Control of Water Hyacinth, *Eichhornia crassipes,* three exotic natural enemies were introduced in India *viz.,* hydrophilic weevils – *Neochetina bruchi* and *N. eichhorniae* (Argentina) and galumnid mite *Orthogalumna terebrantis* (South America) in 1982 for the biological suppression of water hyacinth.

4. ***Aphelinus mali***: Apple woolly aphis, *Eriosoma lanigerum* in Coonor area by *Aphelinus mali* (parasitoid)

5. ***Trichogramma australicum***: Control of shoot borers of sugarcane, cotton bollworms, stem borers of paddy and sorghum with the egg parasitoid, *Trichogramma australicum* @ 50,000/ha/week for 4-5 weeks from one month after planting.

6. ***Cryptolaemus montrouzieri***: *Centrococcus isolitus* on brinjal; Pulvinaria *psidi* on guava and sapota; *Meconellicoccus hirsutus* on grape and *Pseudococcus carymbatus* on citrus suppressed by *Cryptolaemus montrouzieri.*

7. ***Acerophagous papayae***: Papaya mealy bug *Paracoccus marginatus,* an introduced pest into India recently was successfully managed by the hymenopteran parasitoid *Acerophagous papaya* that was Imported and introduced from Mexico to India in 2010-11 by NBAII Bangalore which stands as a latest successful example of Classical Biological Control in India.

8. ***Goniozus nephantidis***: Biological suppression of Coconut black headed caterpillar by the larval parasitoid *Goniozus nephantidis.*

Based on the stage of the host attacked, parasites are grouped as [*e.g.* Parasites of Coconut black headed caterpillar]

1. Egg parasite: *Trichogramma australicum*
2. Early larval parasite: *Apanteles taragama*
3. Mid larval parasite: *Bracon hebtor*
4. Prepupal parasite: *Gonizus nephantidis*
5. Prepupal parasite: *Elasmus nephantidis*
6. Pupal parasite: *Brachymeria nosotoi, Stomatoceros sulcatiscutellum, Trichospilus pupivora, Testrastichus israeli.*

2.4 Kinds of Parasitism/Parasite

1. **Simple parasitism**: Irrespective of number of eggs laid the parasitoid attacks the host only once. *e.g. Apanteles taragamae* on the larvae of *Opisina arenosella, Goniozus nephantids.*

2. **Super parasitism**: phenomenon of parasitization of an individual host by more larvae of single species that can mature in the host. *e.g. Apanteles glomeratus* on *Pieris brassica, Trichospilus pupivora* on *Opisina arenosella*.

3. **Multiple parasitism**: Phenomenon of simultaneous parasitization of host individual by two or more different species of primary parasites at the same time. *e.g., Trichogramma, Telenomous and Tetrastichus* attack eggs of paddy stem borer *Scirpophaga incertulas*. Super parasitism and multiple parasitisms are generally regarded as undesirable situations since much reproductive capacity is wasted

4. **Hyper parasitism**: When a parasite itself is parasitized by another parasite. *e.g. Goniozus nephantidis* is parasitized by *Tetrastichus israeli,* Most of the Bethylids and Braconids are hyper parasites.

5. **Primary parasite:** A parasite attacking an insect which itself is not a parasite (Beneficial to man.) But in few cases, a primary parasitoid becomes harmful in case of productive insects like silkworms, *Bombyx mori* and lac insect *Kerria lacca*.

6. **Secondary parasite**: A hyperparasite attacking a primary parasite (Harmful to man).

7. **Tertiary parasite**: A hyperparasite attacking a secondary parasite (Beneficial to man).

8. **Quaternary parasite**: A hyperparasite attacking tertiary parasite (Harmful to man).

2.5 Brief Description of Important Parasitoids

1. *Trichogramma* spp.

Trichogramma are extremely tiny wasps. Adults are approximately 1/25 inch (1 mm) or less,. They often have wing hairs (setae) arranged in rows. Their body is relatively compact and the antennae are short. *Trichogramma* species are difficult to identify due to their minute size and generally uniform morphological features. The female Trichogramma lays an egg within a recently laid host egg, and as the wasp larva develops, the host egg turns black. Each female parasitizes about 100 eggs and may also destroy additional eggs by host feeding. The short life cycle of 8-10 days allows the wasp population to increase rapidly. *Trichogramma* turns the eggs of some caterpillar species black. This is the best way to detect parasitization by *Trichogramma*. Gram pod borer is effectively controlled by releasing *Trichogramma chilonis* 1 lakh/ha. Egg Parasitoids *(T. evanescens)* of tobacco caterpillar, bihar caterpillar. Parasiting on egg of capitulum borer. Eggs of Spotted bollworm, American and pink bollworm are parasitised.

2. *Apanteles* sp.

Adults have a predominantly black body with some yellow colouring on the abdomen and legs, and are 2.0-2.5mm long. Females have a short, pointed ovipositor through which eggs are injected into host caterpillars. Eggs are elongate and translucent, with a stalk at the posterior end, and are about 0.3mm long. Larvae hatch from eggs within 3 days of oviposition. Host caterpillars die within 24 hours of parasite emergence. Larvae of tobacco caterpillars controlled by releasing Apanteles africanus Larvae stage of Spotted bollworm, American and pink bollworm are attacked by releasing Apanteles angaleti. Larval parasitoid of tobacco caterpillar by releasing Apanteles prodeniae. Larva of gram caterpillar and stem fly are parasitized.

3. *Bracon* spp.

Braconids are small wasps, usually dark-colored with 4 transparent wings. The external cocoons resemble insect eggs, but are made of silk. Braconids are diverse and parasitize many insects. Some attack the host internally, others feed from the outside of a host insect. Larva of gram caterpillar and tobacco caterpillar are parasitized. Larvae stage of Spotted bollworm, American and pink bollworm are attacked by releasing *Bracon brevicornis, B. greeni, B. gelechidophagus.* Parasites on leaf roller by releasing. Parasitoid on capitulum borer in sunflower.

4. *Campoletis chloridae*

The adult wasp is black, slender and about 1/4- inch long. The silken cocoon is white, oblong, and about 1/4 inch long. Antennae very long, with 16 or more segments. These are internal parasitoids of immature holometabolous insects. Larva of gram caterpillar is attacked.

5. *Cotesia* spp.

Cotesia adults are small (about 7 mm), dark wasps and resemble flying ants or tiny flies. They have two pairs of wings, the hind wings being smaller than the forewings, and chewing-lapping mouth parts. The antennae are about 1.5 mm long, and curved (not elbowed) upward. The abdomen of the female narrows to a downward curving extension called the ovipositor with which she lays eggs. The pupae are in an irregular mass of yellow silken cocoons attached to the host larva or to plant leaves. Larval parasitoid of tobacco caterpillar. Red hairy caterpillar is parasitized

6. *Diadegma insulare*

Diadegma insulare is a small (6 mm long) ichneumonid wasp with reddish-brown legs and abdomen. It pupates inside the cocoon made by the mature diamondback moth larva replacing the host pupal covering with its own cocoon which may have a distinctive white band. Diamondback moth cocoons are white inside (green when the larvae first form the cocoon), *D. insulare* wasps are visible as dark bodies inside the cocoon, before the adult *D. insulare*emerges. Adults can be seen searching in the crop foliage. It is the most important parasitoid of the diamondback moth. Limiting insecticide use and using Bacillus thuringiensis (Bt) where possible, allowing wildflowers (especially wild brasiccas) to grow around crop fields, and allowing diamondback moth to colonize wild brassicas and crops will increase the abundance and effectiveness of *D. insulare* for management of diamondback moth. *D. insulare* females require nectar sources. A nectar source can increase *D. insulare* female longevity from 2-5 days to more than 20 days.

7. *Eretmocerus mundus*

Both males and females are lemon-coloured. The males are only dark yellow on the upperside of the thorax, a part of their underside is brown. Eretmocerus can develop in any larval stage of the whitefly, but it prefers the second and early third stage. It lays its eggs under the whitefly larva. After 3 days the translucent eggs turn brown. Larvae will not develop before the whitefly larva has reached the second larval stage. The complete life cycle takes 17 to 20 days, depending on temperature and the larval stage of whitefly. Two weeks after parasitation, the pupa will turn yellow. In order to leave its host, Eretmocerus makes a small round hole in the parasitised whitefly, just asEncarsia. Nymph and pupa stages of white fly are parasitized.

8. *Goniozus nephantidis*

These are primary ectoparasitoids of the larvae, and occasionally pupae, of Coleoptera or Lepidoptera, both as larvae and adults. Females search for hosts in concealed niches and sting them to permanent or temporary paralysis. Larvae are often gregarious. Males almost always fully winged; females alate, with reduced wings or without wings. Head elongate and depressed. Antennae 12- or 13-segmented, inserted close to clypeus. Seven or eight abdominal tergites visible. Larvae of coconut back headed caterpillar, pink bollworm, spotted bollworm and american bollworm are attacked.

9. *Opius exigua*

Adults are small black braconid wasps. The eggs are laid in leafminer larva. Larva develops in leafminer larva but does not complete development until pupation of host. Pupate in leafminer host puparium. The most abundant braconid parasite attacking leafminers completes it development in the leafminer puparia.

10. *Telenomus* spp.

The Telenomus wasp is a shiny, smooth, black wasp slightly larger than an adult Trichogramma. They are generally small black species with keeled abdomens, short clubbed antennae, and no wing venation. They are parasitic on insect and spider eggs. Eggs of American bollworms are attacked by Telenomus heliothidae. Eggs of tobacco caterpillar is parasitised by egg parasitoid of *Telenomus remus*. Larva of gram caterpillar are parasitized.

11. *Trichospilus pupivora*

They are parasites of a wide variety of hosts including lepidoptera larvae, some diptera larva, some are egg parasites, some are hyperparasites. Body with or without metallic luster. These small species (1-3 mm) have segmented tarsi and are diverse. Pupal parasitoid of tobacco caterpillar.

12. *Encarsia formosa*

Encarsia formosa, an endoparasitic wasp, is the most important parasite of the greenhouse whitefly. Adult female *Encarsia formosa* are tiny wasps (<1 mm in length) with a dark brown to black head and thorax and a bright yellow abdomen. Males are dark in color, but are rare. Adult females host feed on all immature stages by puncturing the body with their ovipositors and consuming the exuding blood. Eggs are laid into third and fourth-instar whiteflies and hatch into larvae that feed within the whitefly nymph and grow through three larval instars before killing the host. Greenhouse whitefly nymphs turn dark brown or black approximately one week after being parasitized and their skin forms a black pupal case for the parasite. Silverleaf whiteflies parasitized by *E. formosa* stay lighter in color and do not turn black. Like many whitefly parasites, *E. formosa* leaves a circular hole and black feces in the host remains. In contrast, emerging whiteflies leave a ragged or T-shaped emergence hole in their mostly clear or whitish pupal skin. *E. formosa* is used for whitefly control in greenhouses on tomatoes, strawberries and in floricultural and nursery plants. Biological control of the greenhouse whitefly can often be provided in enclosed areas by introducing sufficient numbers of commercially available *E. formosa.* Release programs of *Encarsia formosa* are most effective when the initial population of whiteflies is quite low (only a few whiteflies per plant) and long-residual insecticides have not been applied in advance of the parasite release. For biological control to be successful use more selective and less persistent insecticides, and control ants since they disrupt the oviposition of *E. formosa.*

13. *Brachmeria* sp.

Adults robust, coloration mainly black or brownish, with yellow, reddish or white markings. Head and thorax heavily sclerotized and usually coarsely and densely punctate. Antennae 13-segmented, with 1 or 2 ring segments and with club segments not markedly different from the funicle. Parapsidal sutures usually well developed, sometimes incomplete. Ovipositor horizontal. Coconut black headed caterpillar, American and Spotted bollworm are attacked. Larvae and pupal stages is attacked.

14. Tachinid Flies

The family Tachinidae is the most important family of parasitic flies providing biological control. Tachinid larvae are internal parasites of immature beetles, butterflies, moths, sawflies, earwigs, grasshoppers, or true bugs. Adults measure between 3 and 14 mm (<1/2 inch), are often dark, robust, hairy and resemble houseflies, but with very stout bristles at the tips of their abdomens. Egg laying varies considerably. In some species, eggs are deposited on foliage near the host insect, and the maggots are ingested during feeding by the host after they hatch. In other species, the adult fly glues eggs to the body of the host, and the maggots penetrate into the host's body after the eggs hatch. Some female tachinids possess a piercing ovipositor and insert their eggs into the host body. In all cases, tachinid maggots feed internally in their hosts and exit the host body to pupate. Pupae are commonly oblong and dark reddish. Tachinid flies complete one to several generations per year. Colorful *Trichopoda pennipes,* a parasite of squash bugs, lays its oval pale eggs singly or in groups on the sides of large nymphs or adults of several species of true bugs including the southern green stinkbug and the squash bug. The larvae burrow into the bug's body where one larva will survive. When ready, the large maggot will exit the host's body and drop to the ground to pupate. The host dies soon after the maggot leaves the body.

15. *Diglyphus isaea*
Diglyphus isaea is a black parasitic wasp of 2-3 mm long. The parasitic wasp lays its eggs in or next to leaf miner larvae of the second and third instar. The young parasite larvae hatch from these eggs, and will then feed on the body fluids of the larvae. Diglyphus isaea larva has 3 stages. The first instar larva is transparent, the second one is yellowish, and the third one is bluish green. In the last stage the larva crawls a little bit back in the mine to pupate. It parasites about 65 to 86 per cent on pea leaf miner.

16. *Cryptochaetum iceryae*
The parasitic fly, *Cryptochaetum iceryae,* and the vedalia beetle were imported from Australia to California in the late 1880s to control cottony cushion scale. They provide complete biological control of cottony cushion scale on most hosts in California and elsewhere in the world where they have been introduced, unless disrupted by adverse conditions such as pesticide applications. *Cryptochaetum* apparently predominates in coastal areas; vedalia is most abundant in the Central Valley and desert areas of California. Both species can occur in interior areas of California if cottony cushion scale is present. The *Cryptochaetum* adult is a dark blue or green to black fly, about 1/12 inch (2 mm) long, with short rounded, grayish wings. One fly generation requires about 1 month in summer, with up to five or six generations developing per year. The female *Cryptochaetum* parasite lays one egg in small scales and a dozen or more in larger hosts. The larva feeds and usually pupates inside its host. Pupae are black with two tiny protruding breathing tubes (spiracles). Metamorphosis is complete.

2.6 Predators
A predator is one which catches and devours smaller or more helpless creatures by killing them in getting a single meal. It is a free living organism through out its life, normally larger than prey and requires more than one prey to develop.

Insect Predator Qualities
1. A predator feeds on many different species of prey, thus being a generalist or polyphagous nature
2. A predator is relatively large compared to its prey, which it seizes and devours quickly

3. Typically individual predator consumes large number of prey in its life time. *e.g.*: A single coccinellid predator larva may consume hundreds of aphids

4. Predators kill and consume their prey quickly, usually via extra oral digestion

5. Predators are very efficient in search of their prey and capacity for swift movements

6. Predators develop separately from their prey and may live in the same habitat or adjacent habitats

7. Structural adaptation with well developed sense organs to locate the prey

8. Predator is carnivorous in both its immature and adult stages, feeds on the prey in both the stages

9. May have cryptic colourations and deceptive markings, *e.g.* Preying mantids and Robber flies

Predatism

Based on the degree of use fullness to man, the predators are classified as on

1. Entirely predatory, *e.g.* lace wings, tiger beetles lady bird beetles except *Henosepilachna* genus

2. Mainly predator but occasionally harmful. *e.g.* Odonata and mantids occasionally attack honey bees

3. Mainly harmful but partly predatory. *e.g.* Cockroach feeds on termites. Adult blister beetles feed on flowers while the grubs predate on grass hopper eggs.

4. Mainly scavenging and partly predatory. *e.g.* Earwigs feed on dead decaying organic matter and also fly maggots. Both ways, it is helpful

6. Stinging predators. In this case, nests are constructed and stocked with prey, which have been stung and paralyzed by the mother insect on which the eggs are laid and then scaled up. Larvae emerging from the egg feed on paralyzed but not yet died prey. *e.g.* Spider wasps and wasps.

2.7 Description of Important Insect Predators

1. Australian Lady Bird Beetle, *Cryptolaemus montrouzieri*

Lady beetles are easily recognized by their shiny, convex, half-dome shape and short, clubbed antennae. Most lady beetles, including this species, are predaceous as both larvae and adults. Young lady beetle larvae usually pierce and suck the contents from their prey. Older larvae and adults chew and consume their entire prey. Larvae are active, elongate, have long legs, and resemble tiny alligators. Many lady beetles look alike and accurate identification requires a specialist. The adult mealybug destroyer is small, measuring 3-4 mm (1/6 inch) long and is mostly dark brown or blackish with an orangish head and tail. Larvae grow up to 1.3 cm (1/2 inch) long and are covered with waxy white curls making it difficult to see their legs. Larvae resemble mealybugs except that they are larger and more active. The wax can be scraped off larvae to reveal the pale, alligator-shaped beetle larvae. *C. montrouzieri* eggs are yellow and are laid among the cottony egg sacks of mealybugs. Pupation occurs in sheltered places on stems or other substrate. The mealybug destroyer undergoes complete metamorphosis and has about 4 generations per year. Both adults and larvae feed on exposed mealybug species and other hompterans such as the green shield scale. *C. montrouzieri* are most effective at controlling mealybugs when the mealybug population is high. Eggs and larvae are the preferred food for both adults and larvae. *C. montrouzieri* does not survive very well in cold weather and in some situations (citrus orchards and greenhouses) adults are bought and released in the spring in order to establish populations. When purchasing beetles, be sure you have an adequate ratio of females to males. Females have dark brown forelegs; males' forelegs are light brown.

2. Vedalia Beetle, *Rodolia cardinalis*

Adult vedalia beetles are small, measuring 2-4 mm (<3/16 inch) long, and are red and black with a covering of fine hairs which often gives them a grayish appearance. The larvae are reddish in color. Red and black pupae develop within the grayish skin of the last larval instar and occur among or near scale colonies. Oblong, red eggs are laid singly or in groups on or near cottony cushion scales. *R. cardinalis* undergoes complete metamorphosis and has 8 or more generations per year. Both adults and larvae feed exclusively the cottony cushion scale on a variety of plants including rose, acacia, magnolia, olive, and citrus. Adults and mature larvae feed on all stages of the scale while young larvae feed only on eggs. The vedalia beetle is extremely sensitive to some pesticides and care should be used when applying pesticides in areas where the beetle is relied upon for control of the cottony cushion scale.

3. Assassin Bugs, *Zelus renardii*

Assassin bug adults and nymphs are slender, colorful insects, often blackish, reddish, or brown. They have long legs; a long narrow head, round beady eyes, and an extended, 3-segmented, needle-like beak. Nymphs are quite small, 5 mm (1/4 in) in length when they hatch and grow to an adult size measuring approximately 2 cm (3/4 inch). Insects in this order undergo incomplete metamorphosis. Eggs of Zelus spp. are barrel-shaped, dark brown with a white cap, and are laid openly in groups on plant surfaces. Adults are poor fliers, and both adults and nymphs move rapidly when disturbed. All assassin bugs are predators, some species feed on insects while others feed on the blood of mammals. Insect-feeding species eat a wide variety of small to medium-sized insect prey including caterpillars, leafhoppers, other bugs, and aphids. They also feed on beneficial species such as lacewings. Nymphs and adults are often seen stalking or laying in wait for their prey, which they inject with venom once they have caught. Assassin bugs are common natural enemies on many plants, including row and tree crops.

4. Green Lacewings, *Chrysopa* spp., *Chrysoperla* spp.

Green lacewings are generalist predators and are commonly found in agricultural, landscape, and garden habitats. Adult green lacewings are soft-bodied insects with four membranous wings, golden eyes, and green bodies. Adults often fly at night and are seen when drawn to lights. Some species of green lacewing adults are predaceous, others feed strictly on honeydew, nectar, and pollen. Females lay their tiny, oblong eggs on silken stalks attached to plant tissues. Depending on the species, eggs are laid singly or in clusters, each on an individual stalk. Eggs are green when laid, then darken before hatching. Lacewings undergo complete metamorphosis with eggs hatching about 4 days after being laid, and larvae develop through three instars before pupating. Larvae, which are pale with dark markings, look like tiny alligators. Larvae are flattened, tapered at the tail, measure 3-20 mm (1/8 to 4/5 of an inch) long, have distinct legs, and possess prominent mandibles with which they attack their prey. Larvae prey upon a wide variety of small insects including mealybugs, psyllids, thrips, mites, whiteflies, aphids, small caterpillars, leafhoppers, and insect eggs. Pupation occurs in loosely woven, spherical, silken cocoons attached to plants or under loose bark. All stages of lacewings can survive mild winters and can be found throughout the year in many agricultural areas of California. Green lacewings are commercially available and are among the most commonly released predators.

5. Seven Spotted Lady Bird Beetle, *Coccinella septempunctata*

Lady beetles are easily recognized by their shiny, convex, half-dome shape and short, clubbed antennae. Most lady beetles, including this species, are predaceous as both larvae and adults. Young lady beetle larvae usually pierce and suck the contents from their prey. Older larvae and adults chew and consume their entire prey. Larvae are active, elongate, have long legs, and resemble tiny alligators. The adult *Coccinella septempunctata* is relatively large, 0.28 to 0.31 inch (7-8 mm), and has a white to pale spot on either side of the head. Its thorax is black with white along the front margin. There are seven large black spots on its red or orangish wing covers, which may have some white near the front. Larvae are alligator shaped and range from 0.28 to 0.31 inch (7-8 mm) in length. Metamorphosis is complete. The pupal stage duration is temperature dependent, lasting between 3 and 12 days. Eggs are spindle shaped and small, about 0.04 inch (1 mm long). *C. septempunctata* undergoes complete metamorphosis. In spring, overwintering adults emerge from protected sites near fields where they fed and reproduced in the previous season. After feeding on aphids, a female will start depositing eggs, generally laying them near prey, in small clusters on protected sites found on leaves and stems. In a one to three month period the female can lay from 200 to over 1,000 small (about 0.04 inch or 1 mm) eggs.

6. Six Spotted Thrips, *Scolothrips sexmaculatus*

Thrips are tiny, 2-3 mm (less than 1/8 inch) in length, slender insects with long fringes on the margins of their wings. Thrips undergo incomplete metamorphosis, have multiple generations per year, and can be phytophagous or predaceous. Six spotted thrips adults can be distinguished from other species by the three dark spots on each wing cover of the mostly pale-yellow adult. Nymphs are translucent white to yellow and difficult to discern from other thrips species. Adults and larvae are entirely predaceous, feeding most commonly on mites such as the European red mite, cyclamen mite, and *Tetranychus* spp. spider mites. Sixspotted thrips can rapidly reduce high mite populations, but often don't become numerous until after mites have become abundant and damaging.

7. Syrphid, Flower, or Hover Flies

Syrphid flies are regularly found where aphids are present in agricultural, landscape, and garden habitats. Adults of this stingless fly hover around flowers, have black and yellow bands on their abdomen and are often confused with honeybees. Syrphid flies undergo complete metamorphosis with 3 larval instars. Females lay their whitish to gray oblong eggs, each measuring 1 mm (1.32 inch), singly on their sides usually near aphids or within aphid colonies. Larvae are legless and maggot shaped and vary in color and patterning but most have a yellow longitudinal stripe on the back. They can be distinguished from caterpillar larvae by their tapered head, lack of legs and their opaque skin, through which internal organs can be seen. Larvae vary in length from 1 to 13 mm (1/32 to 1/2 inch) depending upon their developmental stage and species. Pupa are oblong, pear-shaped, and green to dark brown in color. Pupation occurs on plants or on the soil surface. Adult syrphid flies feed on pollen and nectar, while it is the larval stage that feeds on insects. Larvae of predaceous species feed on aphids and other soft-bodied insects and play an important role in suppressing populations of phytophagous insects. Larvae move along plant surfaces, lifting their heads to grope for prey, seizing them and sucking them dry and discarding the skins. A single syrphid larva can consume hundreds of aphids in a month. Not all syrphid fly larvae are predaceous, some species feed on fungi.

8. Twicestabbed Lady Beetle, *Chilocorus orbus*

The twicestabbed lady beetle is one of the most common scale-feeding lady beetles, and both larvae and adults feed on a variety of species. Adults measure 3-5 mm (1/10-1/5 inch) long, are shiny black with two red spots on their elytra and have reddish undersides. Larvae are grey to black and are distinct from other coccinellids as they have prominent spines with multiple branches that look like spines on spines. Larvae of this and other scale-feeding coccinellids are often overlooked because they frequently feed hidden underneath the scale body or cover. Eggs measure about 1 mm (1/32 inch) in length and are laid either singly or in groups on their sides. This beetle undergoes complete metamorphosis and in warmer climates has several generations per year. This species can be confused with several other coccinellids that are also shiny black with two red spots on their elytra: *Axion plagiatum, Chilocorus kuwanae,* and *Olla v-nigrum.* When compared with these other species, the spots of *C. orbus* are located closer to the head.

9. Predatory Mite, *Phytoseiulus persimilis, Metaseiulus occidentalis*

other mites, predatory mites do not have antennae, segmented bodies, or wings. They pass through an egg stage, a six-legged larval stage and 2 eight-legged immature nymphal stages before becoming adults. Western predatory mites are about the size of twospotted spider mites, but lack spots, range in color from cream to amber red (depending on what they just recently consumed), and are shinier and more pear-shaped than their prey. The shiny, oval eggs of the western predatory mite are larger than spider mite eggs. In addition, predatory mites are more active than pest mites, only stopping to feed. Under magnification the mouthparts of predatory mites can be seen extending in front of their body while pest mite mouthparts extend downward to feed on plants. The preferred foods of western predatory mites are mites of all stages, including eggs, but they also feed on pollen and other food. The western predatory mite is commercially available and is commonly released against *Tetranychus* spp. spider mites such as the Pacific spider mite and the twospotted spider mite. Effective control of spider mite pests has been documented in various many crops and ornamentals. The western predatory mite tolerates hot climates as long as the relative humidity is above about 50 percent. Most mites, such as this *Phytoseiulus persimilis,* hatch from eggs and develop through three immature stages before becoming an adult. The adult female lays a light-orangish egg near prey that darkens before hatching. A six-legged mite larva emerges and in many predaceous species is relatively inactive, apparently not feeding before molting to the nymph stage. The eight-legged protonymph, deutonymph, and adult are active searchers and begin feeding almost immediately after molting.

10. Predaceous Soldier Beetles, Leather-Winged Beetles

Cantharidae Adults are long and narrow. Common species are often about 1/2 inch (13 mm) long with a red, orange or yellow head and abdomen and black, gray or brown soft wing covers. Adults are often observed feeding on aphids or on pollen or nectar on flowering shrubs and trees. Metamorphosis is complete. Larvae are dark, elongate, and flattened. They feed under bark or in soil or litter, primarily on eggs and larvae of beetles, butterflies, moths, and other insects. There are over 100 species of soldier beetles in California.

11. Predaceous Midge (Aphid Midge), *Aphidoletes aphidimyza*

Adults are delicate flies with long, slender legs. They often stand with their antennae curled back over their head. Larvae have two projecting anal spiracles (small tubes) relatively close together at their rear ends. Metamorphosis is complete. Eggs are orangish, oval, and only about 0.12 inch (3 mm) long. Larvae develop through three instars, are pale yellow to red or brown, and at maturity are about 2.5 mm (1/10 inch) long. Pupae are orange to brown, about 2 mm (1/12 inch) long, and occur beneath plants in litter where they may form cocoons made from soil particles, excrement, and aphid cast skins.

12. Predaceous Ground Beetle

Predaceous ground beetles are medium to large soil-dwelling beetles, often about 1/3 to 2/3 inches (8–16 mm) long. Over 2,500 species are known in

North America. Their shape and color varies greatly. Adults are often black or dark reddish, although some species are brilliantly colored or iridescent. Most species have a prominent thorax that is narrower than their abdomen. Their long antennae have 11 segments and are not clubbed at the end. They have long legs, are fast runners, and rarely fly. Carabids resemble plant-feeding darkling beetles. Unlike darkling beetles, carabids have enlarged basal segments (trochanters) on their hind legs. Darkling beetles' antennae are attached beneath a distinct ridge on each side of their head; carabids lack this ridge. Carabid adults and larvae feed on soil dwelling insect larvae and pupae, other invertebrates such as snails and slugs, and sometimes on seeds and organic litter. Eggs are laid in moist soil. Larvae dwell in litter or in soil. Larvae are elongate and their heads are relatively large with distinct mandibles. Most species complete their life cycle from egg to adult in one year. Metamorphosis is complete.

13. Bigeyed Bugs, *Geocoris* spp.

Lygaeid bug adults and nymphs are oval, somewhat flattened, about 4 mm (1/6 of an inch) long, usually brownish or yellowish, and have a wide head with prominent bulging eyes. Bigeyed bugs can be confused with other hemipterans in the same family (Lygaeidae) as well as insects in the Miridae family. Other Lygaeids are more slender and have smaller eyes when compared to *Geocoris* spp. Bugs in the Miridae family do not have their eyes spaced widely apart, generally have longer antennae and only have one or two closed cells in the tip of their forewings. Bigeyed bugs undergo incomplete metamorphosis with 5 nymphal instars. Females lay oblong, pale-colored eggs singly on leaf surfaces which develop reddish eyespots shortly after being laid. Bigeyed bugs are common on low-growing plants including many field and row crops in which they stalk their prey. Their widely separated eyes give them an extensive field of vision for spotting their prey which includes insect eggs, other bugs, small caterpillars, flea beetles, and mites

14. Brown Lacewings, *Hemerobius* spp.

Adult brown lacewings are soft-bodied insects with four membranous wings and light brown bodies. Adults fly predominately at night and are often seen when drawn to lights. Brown lacewings are less common than green lacewings, and adults are about half the size, measuring approximately 1 cm (3/8 inch) long. Females lay their tiny, oblong eggs singly on their side onto plant tissues. Brown lacewing eggs look similar to syrphid fly eggs but are smoother and have a small protrusion on one end. Lacewings undergo complete metamorphosis with eggs hatching about 4 days after being laid and larvae developing through three instars before pupating. The larvae are creamy-brown with dark reddish-brown stripes and spots and move their heads from side to side when walking. Larvae look like tiny alligators, they are flattened, tapered at the tail, have distinct legs and prominent mandibles with which they attack their prey. Pupation occurs in loosely woven, spherical, silken cocoons attached to plants or under loose bark. Both adults and larvae prey upon a wide variety of small insects including mealybugs, psyllids, thrips, mites, whiteflies, aphids, small caterpillars, leafhoppers, and insect eggs.

15. Damsel bugs, *Nabis* spp.

Adult damsel bugs are slender insects that are mostly yellowish, gray, or dull brown, measure about 10 mm (2/5 inch) in length, and have elongated heads and long antennae. Damsel bugs undergo incomplete metamorphosis with nymphs that look similar to adults. Both adults and nymphs move rapidly when disturbed. Females insert their eggs into plant tissues where they are difficult to detect. Damsel bugs are generalist predators appearing later in the season than some other predators. Damsel bugs are common on many plants including row and tree crops where they prey upon thrips, mites, aphids, other bugs, small caterpillars, and leafhoppers.

16. Mantids

Adults are 2 to 4 inches (5-10 cm) long and are usually yellowish, green, or brown. Mantids have incomplete metamorphosis and one generation per year. Overwintering eggs are laid in groups in hard, grayish egg cases which are glued to wood, bark, or other plant material. Adults and immatures have an elongated thorax and grasping forelegs, which they have the habit of holding up while waiting for prey. Mantids are wholly predaceous, feeding on many kinds of insects including beneficial insects and other mantids. They often wait for prey at flowers where they capture nectar- and pollen-feeding insects. Mantids grasp their prey with spined front legs and hold them while they eat. As mantids consume both pests and beneficials, they are difficult to use reliably for biological control.

17. Pirate Bugs, *Orius* spp. and *Anthocoris* spp.

Adult minute pirate bugs are small, 2-5 mm (1/12 to 1/5 inch) long, oval, black to purplish with white markings, and have a triangular head. Adults can be confused with plant bugs in the family Miridae, which are generally larger, have longer antennae, and only have one or two closed cells in the tip of their forewings. Minute pirate bugs undergo incomplete metamorphosis, and nymphs are usually pear-shaped and yellowish or reddish brown with red eyes. Eggs are inserted into plant tissues where they are difficult to detect. Developmental time for minute pirate bugs is very short, only 3 weeks from egg to adult. They are generalist predators and are often the first and most common predaceous insects to appear in the spring. Minute pirate bugs are common insect predators in many crops including alfalfa, corn, small grains, cotton, soybeans, and tomatoes as well as on ornamentals and landscapes. Adults and nymphs feed on insect eggs and small insects such as psyllids, thrips, mites, aphids, whiteflies, and small caterpillars. Commercially available Orius spp. are sometimes released in greenhouses to control thrips.

2.8　Mass Multiplication of Parasitoids and Predators in the Laboratory

1. *Bracon hebetor* Say

This parasitoid can be mass multiplied in the laboratory on the alternate host *Corcyra cephalonica*. Grown up caterpillars are to be spread on stretched cloth over the mouth of jar and a glass plate is placed over the larvae to immobilize. Adult parasitoids are to be released in the jar. The female parasitoid paralyses the host before depositing its eggs on the body of the caterpillar. The eggs hatch out into grubs and feed on the body of the caterpillar. The grubs before spinning cocoons are to be collected on to a clean piece of paper (2"x1"). This piece of paper is sufficient for grubs to spin cocoons and attach themselves to the paper. When parasitoids are to be dispatched by post, it is convenient to send the paper strips containing cocoons. Emergence of parasitoids start from 7th day of parasitisation.

2. *Goniozus nephantidis* Mues

This parasitoid can also be reared on the alternate host, *Corcyra cephalonica*. Parasitoids on emergence are to be placed together in tube (6"x1") for mating and fed on honey solution. After 24 hours, the mated female is separated into a specimen tube (3"x1") and a single well developed caterpillar introduced. The parasitoid at first paralyses the caterpillar by stinging it once or twice after which it starts oviposition. In this case also the grubs are to be removed on to strips of paper (2"x1"). After oviposition, the female parasitoid has to be removed, mated and fed on honey solution and used once again after 24 hours, for setting in a new cage. This process has to be repeated until such time the parasitoids die.

3. *Brachymeria nosatoi* Habu

Adults of *B. nosatoi* comprising both sexes are released into a cylindrical glass jar 17.5 cm x 6.75 cm, the mouth of which is covered with muslin cloth. Honey dipped cotton swab is used as food for the parasitoids. The jar containing parasitoids is to be kept in dim sunlight for 10 – 15 minutes daily for three days after which only the host pupae are to be offered for Parasitisation. Pupae of *O. arenosella* reared in the laboratory are to be carefully removed. Leaf bits containing pupae within cocoons and silken galleries are to be exposed. The host pupae are to be exposed for a period of 4 – 6 hours for parasitisation. The parasitized pupae have to be transferred to a similar glass jar and kept for emergence of parasites, which commences, from 12 days after oviposition and continue upto 20 days in the laboratory.

4. Rearing *of Corcyra cephalonica*

Rice moth larvae is a potential host/prey insect for rearing number of parasitoids and predators. The larvae of *C. cephalonica* can be reared on cumbu grain. Heat sterilized broken cumbu grain @ 2.5 kg along with 100g of groundnut powder and 5 g of powdered yeast tablet are taken in a wooden or plastic tray (45 cm x 30 cm x10 cm). Streptomycin sulphate 0.05 per cent spray is given @ 10 to 20 ml per tray to prevent bacterial infection. Sulphar WP is added @ 5 g per tray to prevent storage mite infection. *Corcyra* eggs @ 0.5 cc (8000-9000eggs)/tray are uniformly mixed in cumbu medium and the trays are covered with kada cloth secured by rubber band. The hatching larvae feed on the grain by webbing and larval period lasts for 30-35 days. The pupation takes place inside the web itself pupal period last for 5-7 days and adult moths emerge after 30-45 days from the date of egg inoculation. The emerging *Corcyra* adults are collected every morning and transferred to a specially designed mating drum made of G.I. with wire mesh bottom where they are provided with honey solution as food. The eggs are collected at the bottom on a blotting paper kept in a tray. The eggs are cleaned with sieves or egg separator. One cc of eggs will contain approximately 16,000 to 18,000 eggs. A bout 100 pairs of *Corcyra* moth will produce 1.5 cc of eggs during the four days of egg laying period. From each culture tray a maximum of 2500 moths can be obtained. A tray can be kept for about 90 days for collection of adult moths due to staggered development.

5. *Trichogramma* spp.

The eggs of *Corcyra* are sterilized by exposing to UV light (15 W for half an hour) to kill embryo and are sprinkled uniformly on large egg cards (30 cm x 20 cm) divided into 30 rectangles (7 cm x 2 cm) by drawing lines containing a thin layer of gum @ 6 cc/card at a ratio of 1: 6 to fresh eggs and exposed for 2 days. They are kept for another 2 days at room temperature and on fourth day parasitized eggs turn black in colour. At this stage, the egg and can be used for field release or stored at 10 C for a fortnight.

6. *Chelonus blackburni*

C. blackburni parasitizes the egg stage but life cycle is completed in larval stage. *Corcyra* eggs are sparsely sprinkled on white cards on a thin layer of diluted gum. After drying the parasitoid adults are allowed at one per 100 eggs into a plastic container and covered with muslin cloth. After exposing for 24 hr, the cards are transferred to another plastic container containing 250 g of broken cumbu grain. The parasitoids develop inside *Corcyra* larvae and spin small white cocoons the adult emerge in 15-20 days.

7. *Eriborus trochanteratus*

E. trochanteratus a larval parasitoid can be reared on *Corcyra* larvae under laboratory condition. The adult parasitoids (1: 1 male: female) are released into mating cage (30 x30 x 30 cm) with adult food kept in a sponge. Next morning the females are separated and transfer to glass or plastic containers. *Corcyra* larvae @ 10/female parasitoid are allowed for parasitisation. The container is kept upside down on a sheet of paper for 3 hr. The parasitoids inject their eggs into host larval body. After 3 hr, the parasitized larvae are transferred into a container with broken cumbu grains for further development. The parasitized cocoons are collected after 10 days from the rearing containers and kept separately for adult emergence.

8. *Cryptolaemus montrouzieri*

It is a promising predator on mealybugs, scale insects and aphids. This exotic predator is used in large scale to control grapevine mealybugs, *Maconellicoccus hirsutus*. Red pumpkin is used for the multiplication of grape vine mealy bug in the laboratory. Select the well ripened pumpkin having a small stalk and sterilize the outer surface with 0.1 per cent fungicide (Mancozeb

Table 1: Important Examples of Successful Biological Control of Insect Pests by Use of Introduced Natural Enemies in Asian Countries.

Year	Country	Crop	Insect Pest	Natural Enemy	Imported From
1925-26	Kgushu Island (Japan)	Citrus	Spiny black fly, Aleurocanthus spiniferus	Encarsia smithi (Silvestri)	Japan mainland
1925-29	Fiji	Coconut	Coconut moth, Levuana iridiscens	Bessa remota Aldrich	Malaysia
1928-29	Fiji	Coconut	Coconut scale, Aspidious destructor	Cryptognatha nodiceps Marshall	Trinidad
1920s	Assam	Apple	Wooly apple aphid, Eriosoma lanigerum (Hausmann)	Aphelinus mali (Haldeman)	England
1929-31	India	Wattle of Commerce, Acacia decurrens	Cottony cushion scale, Icery purchase Maskell	Rodolia cardinalis Mulsant	USA
1933-34	Fiji	Coconut	Coconut leaf-mining beetle, Promecotheca coeruleipennis Blanchard	Pediobius parvulus (Ferriere)	Java
1934-35	Korea	Apple	Wooly apple aphid, E. lanigerum	A. mali	Japan
1958-60	India	Apple	San Jose scale, Quadras pidiotus Perniciosus (Comstock)	Prospaltella perniciosi (Tower)	China
1960	India	Apple	Q. pernicious	Aphytis diaspidis	USA
1964	India	Castor	Achae janata	Telenomus sp.	New Guinea
1965	India	Coconut	Oryctes rhinoceros (Linnaeus)	Platymeris laevicollis (Distant)	Zanzibar
1980-86	Japan	Citrus	Arrowhead scale, Cinaspis vanonensis (Kuwana)	Physcus fulvus Compere and Annecke Aphytis yanonensis De Bach and Rosen	Taiwan

Source: Sankaran (1974); DeBach and Rosen (1991).

Table 2: List of Potential Bio-control Agents and their Dosage.

Crop	Insect Pest	Biotic Agent	Dosage
Rice	Scirpophaga incertulas	Trichogramma japonicum, T.chilonis	50,000 – 1,00,000 parasitized eggs/ha
	Nilaparvatalugens	Cyrtorhinus lividipennis Lycosa pseudoannulata (Spider)	100 adults or 50-75 nymphs/-/-m² at 10 days interval
Cotton	Helicoverpa armigera, Pectinophora gossypiella, Spotted bollworm, Earias insulana	Trichogramma chilonis	1,50,000 parasitized eggs/ha
	Helicoverpa armigera (Hubner)	Ha – NPV	$1.5\text{-}3.0 \times 10^{12}$ POB/ha (250)-500 LE)
	Tobacco caterpillar, Spodoptera litura (Fabricius)	Sl-NPV	$1.5\text{-}3.0 \times 10^{12}$ POB/ha (250)-500 LE)
Sugar-cane	Shoot tissue borer, Chilo spp.	Trichogramma chilonis	50,000 parasitized eggs/ha
	Chilo infuscatellus	Granulovirus	250 LE or 750 virosed larvae ($10^6 – 10^7$)
	Scale insect, Melanaspis glomerata	Pharoscymnus horni, Chilocorus nigrita, Sticholotis madgassa	1500 adult beetles or grubs
	Pyrilla perpusilla	Epiricania melanoleuca Fletcher	2-3 egg masses or 5-7 cocoons in 40 selected spots or 4000 to 5000 viable cocoons/ha
	Leucopholis lepidophora	Paenibacillus popilliae	0.5 kg/ha once at the time of planting
		Metarhizium anisopliae	2.5×10^{10} spores/m³ once at the time of planting
Pulses	Helicoverpa armigera	Ha – NPV	$1.5\text{-}3.0 \times 10^{12}$ POB/ha (250-500 LE)
	Lab lab Pod borer, Adisura atkinsoni	A – NPV	$1.5\text{-}3.0 \times 10^{12}$ POB/ha (250-500 LE)
Oil seeds	Mustard aphid, Lipaphis erysimi	Verticillium lecanii	10×10^6 spores/ml
	Groundnut white grub Holotrichia consanguinea	Paenibacillus popilliae	0.5 – 1 kg/ha once at the time of planting
		Metarhizium anisopliae	42.5×10^{10} spores/m³ once at the time of planting
Tobacco	Tobacco caterpillar, Spodoptera litura (Fabricius)	Sl-NPV	$1.5 – 30 \times 10^{12}$ POB/ha (250-500 LE)
		Telenomus remus	1,20,000/ha five times at weekly interval
		Chrysoperla spp.	60,000 second instar larvae/ha
		Chrysoperla spp.	6 second instar larvae/plant

Table 3: List of Insect Parasitoids and Predators along with Pests Targeted.

Sl.No.	Parasitoid	Insects Pests Targeted	Family
I.	**Parasitoids**		
A.	**Hymenoptera**		
1.	**Egg parasitoids**		
A.	*Trichogramma chilonis*	Bollworms of cotton, stem and shoot borers of rice, maize, sorghum, sugarcane pulses and vegetables	Trichogrammatidae
B.	*T. brasiliensis*	Rice stem borer, cotton bollworms	Trichogrammatidae
C.	*T.achaeae*	Castor semi-looper, rice stem borer	Trichogrammatidae
D.	*T.japonicum*	Rice stem borer and leaf folder	Trichogrammatidae
E.	*T.dendrolini*	Borers in cotton, rice and maize	Trichogrammatidae
F.	*T.embryophagum*	Coddling moth of apple	Trichogrammatidae
G.	*T.partisum*	Coddling moth on fruits	Trichogrammatidae
H.	*Telenomus remus*	Tobacco caterpillar	Scleoniidae
I.	*Evania appendigaster*	Ootheca of cockroaches	Evaniidae
J.	*Aphidius colemani*	Aphids	Braconidae
2.	**Egg-larval parasotoids**		
A.	*Chelonus blackburnii*	Spotted bollworm of cotton	Braconidae
B.	*Capidosoma koehleri*	Potato tuber moth	Encrytidae
3.	**Larval parasitoids**		
A.	Goniozus nephantidis	Coconut black headed caterpillar	Bethylidae
B.	*Bracon brevicornis* and *Bracon hebetor*	Shoot borers of rice, cotton, pulses, oilseeds and sugarcane, Coconut BHC	Braconidae
C.	*B.kirkpatrici*	Pink bollworm, sugarcane top shoot borer	Braconidae
D.	*Chelonus blackburni*	Cotton bollworms, potato tuber moth	Braconidae
E.	*Cotesia plutella*	Diamond back moth	Braconidae
F.	*Cotesia flavipes*	Cotton bollworms	Braconidae
G.	*Elasmus nephantidis*	Coconut black headed caterpillar	Elasmidae
H.	*Tetrastichus Israeli*	Coconut black headed caterpillar	Eulopidae
I.	*Campolestis chloridae*	American bollworm	Ichneumonidae
J.	*Platygaster oryzae*	Rice gall midge	Platygasteridae
4.	**Larval pupal parasitoids**		
A.	*Brachymeria nephantidis*	Coconut black headed caterpillar	Chalcididae
B.	*Isotima javansis*	Top shoot borer of sugarcane	Ichneumonidae
5.	**Pupal parasitoids**		
A.	*Trichospilus pupivora* and *Tetrastichus Israeli*	Coconut black headed caterpillar	Eulopidae
B.	*Xanthopimpla punctata*	Coconut black headed caterpillar	Ichneumonidae

Contd...

Table 3–Contd...

Sl.No.	Parasitoid	Insects Pests Targeted	Family
B.	**Lepidoptera**		
	Ecto parasitoid		
A.	*Epiricania melanoleuca*	Sugarcane leafhopper	Epipyropidae
II.	**Predators**		
1.	**Hymenoptera**		
	Tetraponera rufonigra	Termites	Formicidae
2.	**Lepidoptera**		
	Dipha aphidivora	Sugarcane wooly aphid	Pyralidae
3.	**Neuroptera**		
A.	*Chrysoperla carnea*	Sucking pests on rice pulses, vegetables and cotton	Chrysopidae
B.	*Chrysoperla scelestis*	Sucking pests on sugarcane and cotton	Chrysopidae
C.	*Mallada boninensis*	Mealy bugs on pineapple, papaya	Chrysopidae
4.	**Coleoptera**		
A.	*Encarsia perniciosi*	Sucking pests on apple, cotton, sugarcane	Aphelinidae
B.	*Encarsia haitiensis* and *Encarsia guadeloupae*	Whitefly	Carabidae
C.	*Omphra pilosa*	All castes of termites	Coccinellidae
D.	*Cryptolaemus montrouzieri*	Mealy bugs on grapes, guava, mango, apple	Coccinellidae
E.	*Coccinella septempunctata*	Aphids and mealy bugs on fruit crops, vegetables, sugarcane, cotton	Coccinellidae
F.	*Menochilus sexmaculatus*	Aphids and mealy bugs on fruit crops, vegetables, sugarcane, cotton	Coccinellidae
G.	*Rodolia cardinalis*	Cottony cushion scale	Coccinellidae
H.	*Chilocorus nigritus*	Sucking pests on oilseeds, cereals, sugarcane and cotton	Coccinellidae
I.	*Brumoides suturalis*	Sucking pests on oilseeds, cereals, sugarcane and cotton	Coccinellidae
J.	*Scymnus coccivora*	Aphids and mealy bugs on fruit crops	Coccinellidae
5.	**Hemiptera**		
A.	*Orius insidiosus*	Thrips, mites, aphides	Anthocoridae
B.	*Geocoris* spp.	Epilachna beetle	Lygaeidae
C.	*Cyrtorhinus lividipennis*	Rice brown plant hopper	Miridae
D.	*Eucanthecona furcelleta*	Red hairy caterpillar	Pantatomidae
E.	*Platymeris laevicollis*	Coconut Rhinoceros beetle	Reduviidae
F.	*Sycanus collaris; Cydnocoris gilvus*	Termites	Reduviidae
G.	*Rhinocoris marginatus*		
H.	*Harpector costalis*	Red cotton bug, *Helicoverpa armigera*	Reduviidae

Contd...

Table 3–*Contd...*

Sl.No.	Parasitoid	Insects Pests Targeted	Family
6.	**Diptera**		
A.	*Robber flies*	Small insects	Asilidae
B.	*Crossopalpus* sp.	Leafhoppers	Empididae
C.	*Ishiodon scutellaris*	Aphids of cotton, pulses and vegetables	Syrphidae

75WP) and air dried. Release the crawlers of mealy bug on the pumpkin and allow them to multiply on the pumpkin in a dark place. A fully infested pumpkin with mealy bug is placed in a cage (30 x 30 x 30 cm) covered with cloth on all sides having a glass door in front. Expose the fruits to adult beetle for oviposition and remove the fruit after 48 hours. The hatching larvae feed and develop on the mealy bug. The fully grown larvae pupate on the folded paper placed on the floor of cage and collect the pupae in a separate cage for emergence of adults.

9. *Chrysoperla carnea*

Larval rearing can be done in round plastic basins at 250 larvae/basin covered with kada cloth. The eggs of *Corcyra* are given as food to the larvae of aphid lion. For rearing 500 larvae 25 cc of *Corcyra* eggs are required @ 5 cc/ feeding on alternate days. The *Chrysoperla* larvae pupate (white silken cocoons) in 10 days and green coloured adults with transparent lace like wings emerge in 9-10 days. The cocoons are transferred with fine brush to 1 litre plastic containers with wire mesh window for adult emergence. The adults are transferred to pneumatic glass trough or G.I. round trough (30 x 12 cm) wrapped with brown sheet acting as oviposition substratum. About 250 adults (60 per cent females) are allowed into each trough and covered with white nylon cloth. Three bits of moist foam sponge and protein rich diet in semi solid form are placed over the nylon cover as food for the adults. The stalked eggs are laid on the brown sheet. The egg sheets can be stored at 10 C for 21 days. When the eggs are required for field release, the egg sheet is kept at the room temperature and the eggs turn brown and hatch on third day. The eggs are destalked by brushing with a sponge piece on second day. The first instar larvae are either taken for culture (or) field release.

Chapter 3

Microbial Control

Microbial control deals with exploitation of disease causing organism to reduce the population of insect pest lower than the damaging levels. Steinhaus (1949) Coined the term 'Microbial Control' when microbial organisms or other products (toxins) are employed by man for the control of pests on plants, animals or man.

3.1 Bacterial Pathogens

Bacillus thuringiensis (or *Bt*) is a Gram-positive, soil-dwelling bacterium that produces crystal proteins (proteinaceous inclusions), called δ-endotoxins, that have Insecticide action. This has led to their use as insecticides, and more recently to genetically modified crops using *Bt* genes.

3.1.1 History of *Bt*

Japanese biologist, Shigetane Ishiwatari was investigating the cause of the sotto disease (sudden-collapse disease) that was killing large populations of silkworms when he first isolated the bacterium *Bacillus thuringiensis* (*Bt*) as the cause of the disease in 1901. Ernst Berliner isolated a bacteria that had killed a Mediterranean flour moth in 1911, and rediscovered *Bt*. He named it *Bacillus thuringiensis*, after the German town Thuringia where the moth was found. Ishiwatari had named the bacterium *Bacillus sotto* in 1901 but the name was later ruled invalid. In 1915, Berliner reported the existance of a crystal within *Bt*, but the activity of this crystal was not discovered until much later. Farmers started to use *Bt* as a pesticide in 1920. France soon started to make commericialized spore based formulations called Sporine in 1938. Sporine, at the time was used primarily to kill flour moths.

More products containing *Bt* were marketed, but many of these products had limitations. *Bt* products such as sprays are rapidly washed away by rain, and degrade under the sun's UV rays. Also, there were many insects that are not susceptible to any of the limited number of *Bt* strains known at the time. All the *Bt* strains known at the time were toxic to lepidopteran (moth) larvae only. There were also some insects that live within the plant or underground where the *Bt* sprays could not reach. Since synthetic insecticides were readily avaliable and often very efficient in killing insects, *Bt* was not used widly. In 1956, researchers, Hannay, Fitz-James and Angus found that the main insecticidal activity against lepidoteran (moth) insects was due to the parasporal crystal. With this discovery came increased interest in the crystal structure, biochemistry, and general mode of action of *Bt*. Research on *Bt* began in ernest.

In the US, *Bt* was used commercially starting in 1958. By 1961, *Bt* was registerd as a pesticide to the EPA. Up until 1977, only thirteen *Bt* strains had been described. All thirteen subspecies were toxic only to certain species of lepidopteran larvae. In 1977 the first subspecies toxic to dipteran (flies) species was found, and the first discovery of strains toxic to species of coleopteran (beetles) followed in 1983.In the 1980's use of *Bt* increased when insects became increasingly resistant to the synthetic insecticides and scientists and environmentalists became aware that the chemicals were harming the environment. *Bt* is organic and it affects specific insects and does not persist in the environment. Because of this, governments and private industries started to fund research on *Bt*.

Today, there are thousands of strains of *Bt*. Many of them have genes that encode unique toxic crystals in their DNA. With the advancement in molecular biology, it soon became feasible to move the gene that encodes the toxic crystals into a plant. The first genetically engineered plant, corn, was registered with the EPA in 1995. Today, GM (genetically modified) crops including, potato and cotton are planted throughout the world.

Genetic Engineering Approval Committee (GEAC), Ministry of Environment and Forests, Govt of India, at its 32nd meeting held in New Delhi on 26th March 2002 approved three *Bt*-cotton hybrids for commercial cultivation. This is a historic decision as *Bt*-cotton became the first transgenic crop to receive such an approval in India. These transgenic hybrids were developed by MAHYCO (Maharashtra Hybrid Seed Company Limited) in collaboration with Monsanto.

3.1.2 Characteristics of *Bacillus thuringiensis*

1. Highly pathogenic to lepidopterous larvae
2. Non-toxic to man
3. Non-phytotoxic
4. Safer to beneficial insects
5. Compatible with number of insecticides
6. So far no resistance is developed in insects

7. Synergistic in combination with certain insecticides like carbaryl
8. Available in different formulations (Thuriocide, Delfin, Bakthane, Biobit, Halt, Dipel etc).
9. Formulation is so standardized that 1 gm of concentration spore dust contains 100 million spores *Bacillus popillae* (available as Doom) causes milky disease on Japanese beetle, *Popillia japonica*

3.1.3 Mode of Action of *Bacillus thuringiensis*

B. thuringiensis is a common gram positive, spore-forming, soil bacterium. When resources are limited, vegetative *Bt* cells undergo sporulation, synthesizing a protein crystal, during spore formation. Proteins in these crystals are called Cry (from Crystal) endotoxins and have been known for decades to display insecticidal activity against specific insect groups. Even though insecticidal formulations based on *Bt* toxins have been used for many years, it was the development and commercialization of insect-resistant transgenic *Bt* crops expressing Cry toxins that revolutionized the history of agriculture. Benefits of this technology include high specificity and potency, reduction in chemical pesticide applications, and increased crop yield. This multi-step toxicity process includes ingestion of the Cry protein by a susceptible insect, solubilization, and procesing from a protoxin [130 kDa] to an activated toxin core [65 -67 kDa] in the insect digestive fluid. The toxin core travels across the peritrophic matrix and binds to specific receptors called cadherins on the brush border membrane of the gut cells. Toxin binding to cadherin proteins results in activation of an oncotic cell death pathway and/or formation of toxin oligomers that bind to GPI-anchored proteins and concentrate on regions of the cell membrane called lipid rafts. Accumulation of toxin oligomers results in toxin insertion in the membrane, pore formation, osmotic cell shock, and ultimately insect death (Adams *et al.,* 1996).

How does the *Bt* toxin work to control insects?

1. Caterpillars ingest Bt protein while feeding.
2. The protein gets activated into toxin in the caterpillar's midgut.
3. Activated toxin binds to receptors on the midgut epithelium.
4. Pores form in the midgut causing the haemolymph to leak into the gut lumen.
5. The digestive system gets paralyzed.
6. The caterpillar stops feeding, and finally dies.

3.2 Baculoviruses (Baculoviridae)

Baculoviruses are pathogens that attack insects and other arthropods. Like some human viruses, they are usually extremely small (less than a thousandth of a millimeter across), and are composed primarily of double-stranded DNA that codes for genes needed for virus establishment and reproduction. Because this genetic material is easily destroyed by exposure to sunlight or by conditions

Ingestion of the protein by an insect larva

⇓ midgut alkaline pH (~9.0)

Solubilization/Dissolving of crystal proteins and release of
protoxins(proteins) in the insect midgut

⇓ Proteolytic enzymes

Activation of protoxins by cleavage of amino acid chains

⇓

Binding of the activated protein to receptors on
the cell membrane in the midgut

Aminopeptidases N Cadherins ⇓ Alkaline phosphatases

Insertion of the protein into cell membrane and
formation of a pore into the body cavity

⇓

Loss of ionic homeostasis across cell membrane
and osmotic lysis of midgut cell

⇓

Starvation, destruction of cell tissue and septicaemia
cause the death of the insect larvae

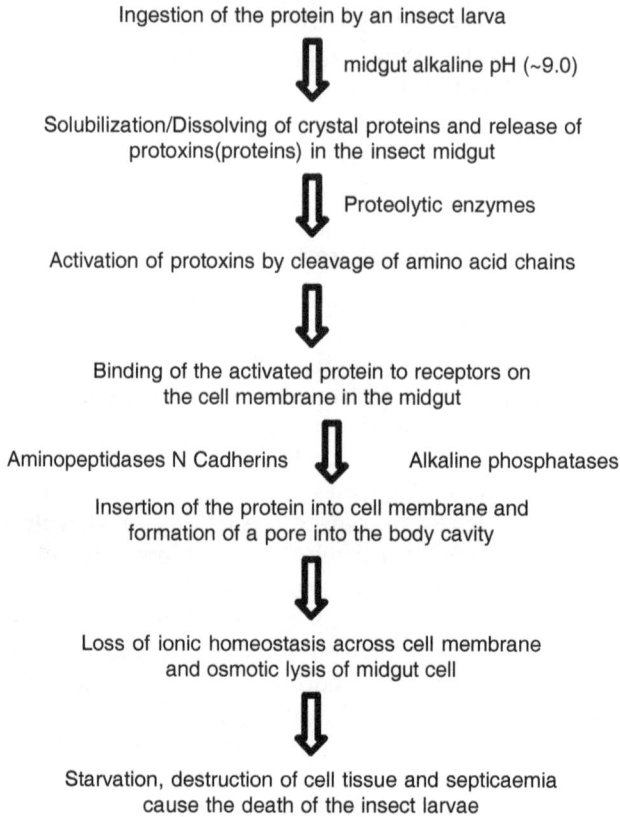

Mode of Action of *Bacillus thuringiensis*

in the host's gut, an infective baculovirus particle (*virion*) is protected by protein coat called a *polyhedron*.Most insect baculoviruses must be eaten by the host to produce an infection, which is typically fatal to the insect. The majority of baculoviruses used as biological control agents are in the genus *Nucleopolyhedrovirus*, so "baculovirus" or "virus" will hereafter refer to **nucleopolyhedro viruses.** These viruses are excellent candidates for species-specific, narrow spectrum insecticidal applications. They have been shown to have no negative impacts on plants, mammals, birds, fish, or even on non-target insects. This is especially desirable when beneficial insects are being conserved to aid in an overall IPM program, or when an ecologically sensitive area is being treated (Miller, 1998; Battu *et al.,* 2001).

3.2.1 Life Cycle

Viruses are unable to reproduce without a host–they are *obligate parasites.* Baculoviruses are no exception. The cells of the host's body are taken over by the genetic message carried within each virion, and forced to produce more virus particles until the cell, and ultimately the insect, dies. Most baculoviruses cause the host insect to die in a way that will maximize the chance that other insects will come in contact with the virus and become infected in turn. Infection by baculovirus begins when an insect eats virus particles on a plant–perhaps from a sprayed treatment. The infected insect dies and "melts" or falls apart on foliage, releasing more virus. This additional infective material can infect more insects, continuing the cycle (Hawtin, 1993).

Baculoviruses have a distinctive rod shaped nucleocapsids that are 30-60nm in diameter and 250-300nm in length. GVs are occluded in ovicylindrical bodies with dimensions of about 0.3µm x 0.5µm. The occluded form of NPVs are polyhedral in shaped and are approximately 0.15-15µm in size. The occluded form of both the GVs and the NPVs can clearly be seen using a light microscope. The occlusion derived virus (ODV) is the form of the virus which is produced in the latter stages of viral infection and is enclosed in a proteinaceous occlusion body. They allow for horizontal spread of the virus from insect to insect and allow the virus to persist for long periods in the environment (King *et al.*, 1994)

3.2.2 Relative Effectiveness

It is widely acknowledged that baculoviruses can be as effective as chemical pesticides in controlling specific insect pests. However, the expense of treating a hectare of land with a baculovirus product invariably costs more than an equally efficacious chemical treament. This difference in price is due primarily to the labor intensive nature of baculovirus production. Some viruses can be produced *in vitro* (within cell cultures in the laboratory, not requiring whole, living insects). These are less expensive than those that can only be produced *in vivo*, that is, inside of living insects. The cost of rearing live hosts adds greatly to the final cost of the product. It is to be hoped that insect cell culture systems currently being developed for other uses may ultimately make viral pesticides more cost-effective.

3.2.3 Appearance

Insects killed by baculoviruses have a characteristic shiny-oily appearance, and are often seen hanging limply from vegetation. They are extremely fragile to the touch, rupturing to release fluid filled with infective virus particles. This tendency to remain attached to foliage and then rupture is an important aspect of the virus life-cycle. As discussed above, infection of other insects will only occur if they eat foliage that has been contaminated by virus-killed larvae. It is interesting to note that most baculoviruses, unlike many other viruses, can be seen with a light microscope. The polyhedra of many viruses look like clear, irregular crystals of salt or sand when viewed at 400x or 1000x. The fluid inside a dead insect is composed largely of virus polyhedra - many billions are produced inside of one cadaver.

3.2.4 Pesticide Compatibility

Viruses particles *per se* are generally unaffected by pesticides, although some chlorine compounds should be expected to damage or destroy viruses if applied at the same time. Baculovirus *efficacy*, however, can be altered in many ways by the effects of chemical pesticides on the host insect.

3.2.5 Mode of Action

Once the virions have entered the midgut epithelia the nucleocapsids migrate to the nucleus of the cell. This migration may be in association with the cellular actin. Once the nucleocapsids reach the nucleus the DNA is uncoated into the nucleus. Phosphorylation of the DNA binding protein causes the DNA to unwind allowing expression and replication of the viral genome. This process is slightly different for the granuloviruses. These viruses release their DNA into the nucleus through the nuclear pore.

Table 4: Current Use of Baculoviruses as Biological Insecticides.

Commodity	Insect Pest	Virus Used	Virus Product
Apple, pear, walnut	Codling moth	Codling moth GV	Cyd-Xe
Cabbage, tomatoes, cotton	Cabbage moth, American bollworm, diamondback moth	Cabbage army worm NPV	Mamestrin
Cotton, corn, tomatoes	*Spodoptera littoralis*	*Spodoptera littoralis* NPV	Spodopterin
Cotton and vegetables	Tobacco budworm *Helicoverpa zea*	*Helicoverpa zea* NPV	Gemstar LC, Biotrol, Elcar
Vegetable crops, flowers	*(Spodoptera exigua)*	*Spodoptera exigua* NPV	Spod-X
Vegetables	Celery looper *(Anagrapha falcifera)*	*Anagrapha falcifera* NPV	None at present
Alfalfa and other crops	Alfalfa looper *(Autographa californica)*	*Autographa californica* NPV	Gusano Biological Pesticide
Forest Habitat, Lumber	Douglas fir tussock moth *(Orgyia psuedotsugata)*	*Orgyia psuedotsugata* NPV	TM Biocontrol
Forest Habitat, Lumber	Gypsy moth *(Lymantria dispar)*	*Lymantria dispar* NPV	Gypchek

3.3 Fungal Pathogens

Fungi is divided into 2 divisions, *Myxomycota* and *Eumycota*. Entomopathogenic fungi are found in division *Eumycota* in the following sub divisions : *Mastigomycotina, Zygomycotina, Ascomycotina, Basidiomycotina* and *Deuteromycotina* (Tanada and Kaya,1993). Fungi may be associated with insects as *ectoparasites* and *endoparasites*. The fungi gain entrance into the insect mainly through the integument and in some cases thro' natural body openings. Fungi infect individuals in all orders of insects. Most common are *Hemiptera, Diptera, Coleoptera, Lepidoptera, Orthoptera* and *Hymenoptera*. In some insect orders, the immature (nymphal or larval) stages are more often infected than the mature or adult stage; in others, reverse may be the case. Host specificity varies considerably, some fungi infect a broad range of hosts and

Pre starve 4[th]instar larva- over night

⇩

Prepare virus suspension containing 10^8 POB/ml in water containing 0.1 per cent teepol

⇩

Dip clean castor leaves in virus suspension and shade dry

⇩

Allow the caterpillar to feed for 48hr and subsequently on untreated leaves

⇩

Collect the diseased larvae in distilled water

⇩

Macerate it in Blender and Filter through Muslin cloth

⇩

Centrifuge for 1 min at 500 RPM → Discard pellet (only tissue)

⇩

Supernatant containing POB s

⇩

Centrifuge at 2500 RPM for 15 min → Discard supernatant

⇩

Collect pellet (POB s)

⇩

Re suspend in distilled water

⇩

Repeat differential centrifugation

⇩

Pure POB s

⇩

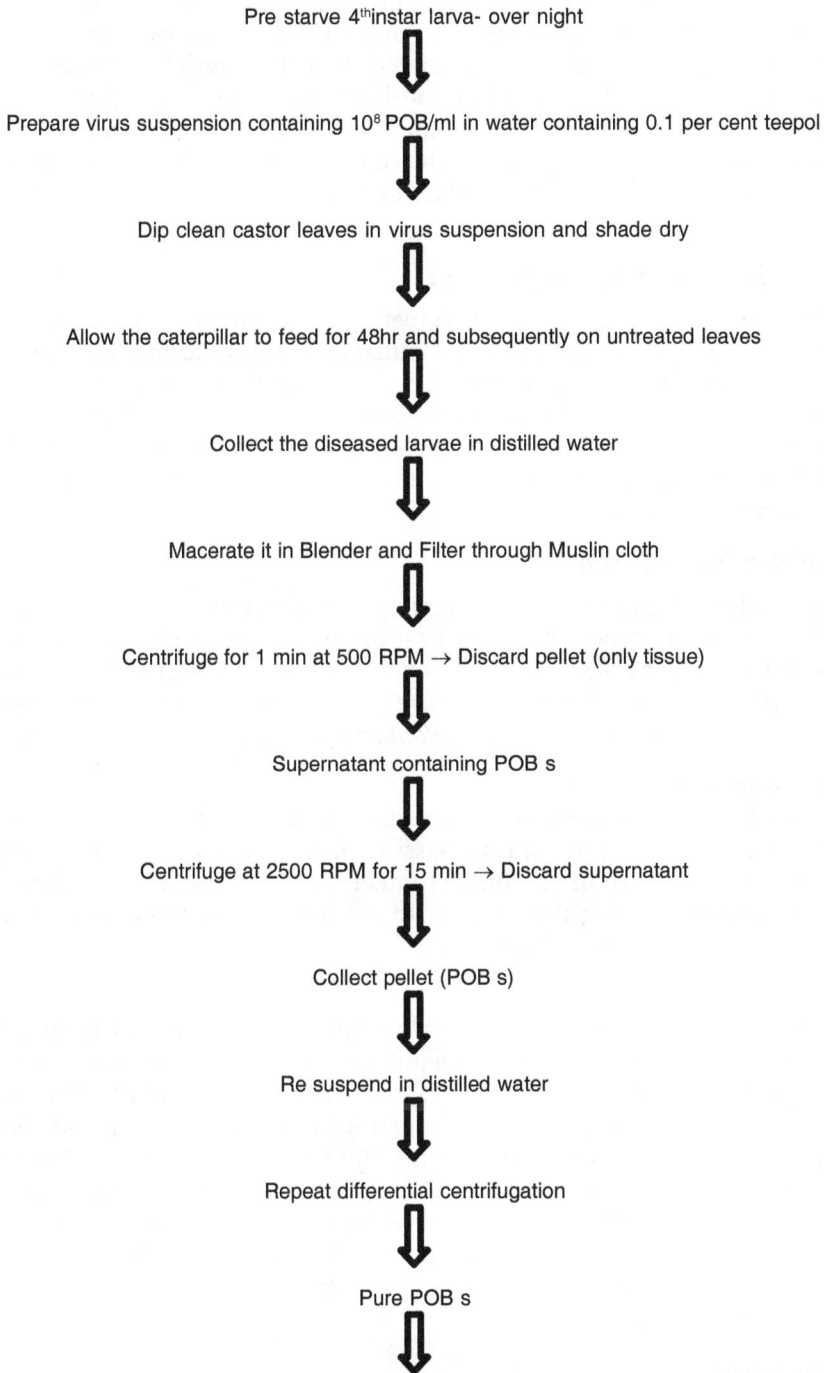

The dose of virus is expressed as larval equivalent (LE) and one LE is 6 x 10^9 POB.
One LE can be had from three fully grown up and virus infected larvae.

Mass Production of NPV of *Spodoptera litura.*

others are restricted to a few or a single insect species. *Beauveria bassiana* and *Metarhizium anisopliae* infect over 100 different insect species in several insect orders. Host specificity may be associated with the physiological state of the host system (*i.e.* insect maturation and host plant) the properties of insect's integument and/or with nutritional requirement of the fungus and the cellular defence of the host. Three types of defence reactions have been observed in the insect's haemocoel : Phagocytosis, cellular encapsulation and humoral encapsulation.

Sub Division Deuteromycotina

The Deuteromycotina or fungi Imperfecti are imperfect in the sense that most of them are recognized by their conidial (anamorph) forms since the sexual forms occur rarely or are unknown. The entomopathogenic Deuteromycotina are found in two classes, Hyphomycetes and Coelomycetes. Many are highly virulent pathogens and have been applied for control of insect pests. Unlike many Entomophthorales the imperfect fungi are easily cultured on most media used to cultivate fungi.

Class Hyphomycetes

A number of fungi in class Hyphomycetes cause muscardine diseases in insects. The term muscardine was first applied to the white muscardine of silkworm caused by *Beauveria bassiana*. Other muscardines are Green muscardine: *Metarhizium anisopliae, Aspergillus flavus, Paecilomyces farinosus, Sorosporella uvella, Paecilomyces fumosa.*

3.3.1. *Beauveria* Species

The white muscardine was the first disease in animals shown to be caused by a fungus or any microorganism. Bassi de Lodi demonstrated the contagious and pathogenic nature of the fungus infecting silkworm and also developed measures for controlling the disease. In his honour, Balsamo described and named the fungus *Botrytis bassiana.*

Host

Bb occurs worldwide. It has one of the largest host list among the imperfect fungi and also occurs in soils as a ubiquitous saprophyte. The hosts are mainly *Lepidoptera, Coleoptera* and *Hemiptera* but others occur in *Diptera* and *Hymenoptera.* This fungus has been found infecting the lungs of wild rodents, nasal passages of humans, horses and giant tortoises. Some of the major economic insect pests that are susceptible to this fungus are the European Corn borer, *O. nubilalis,* the codling moth *Laspeyresia* (Cydia) *Pomonella,* the Japanese beetle, *Popillia japonica,* the Colarado potato beetle, *Leptinotarsa decimlineata,* the chinch bug, *Blissus leucopterus* and the European cabbage worm, *Pieris brassicae.*

Pathogenicity

Infection generally takes place thro' the integument. *B.b.* produces detectable mycotoxins in culture media. One of the mycotoxins is beauvericin, an antibiotic that is highly toxic to the brine shrimp and moderately inhibiting to mosquito larvae. The cydodepsipeptide, bassianolide produced by *B.b* and *V.l* is toxic to silkworm when inoculated into the haemocoel or when fed to larvae. The secondary metabolites, which are anti-biotics may prevent bacterial putrefaction and thus permit fungal mummification of the host insect. The composition of nutrient media influences the production of mycotoxins by *B.b.* The best media for the production of the toxic proteolytic complex contain maize meal, yeast extract and leaf extract. The presence of fructose in the medium enhances the virulence of *B.b.*

3.3.2. *Metarhizium* Species

The green muscardine fungus, *M. anisopliae* is also widely distributed with a wide host range as that of *B.b.* In 1879, Metchnikoff isolated the fungus from the beetle *Anisoplia austriaca* and suggested its use as microbial agent against insect pests. *M. anisopliae* has tow types, the short-spored form, *M. anisopliae* var. *anisopliae* (conidia 3.5 – 9.00 µm) and the long spored, *M. anisopliae* var *major* (Conidia 9.0 – 18.0 µm). These species differ in virulence and may also infect different hosts. *Oryctes* spp are susceptible mainly to the major form. This strain has promise in MC because it has early dense sporulation, rapid *in vitro* spore germination and enhanced production of mycotoxins (Zimmermann, 1993).

Mycotoxins

A number of secondary metabolites act as *mycotoxins* and are produced by entomopathogenic fungi. Cultures of *M.a.* contain the *Cydodepsipeptides,* destruxins A, B, C, D and E and desmethyl destruxin B. Destruxins have been considered as new generation insecticides. They cause tetanic paralysis when inoculated into larvae of *G.m.* They are also produced in fungus infected larvae and are important in the development of symptoms. The rapid production of destruxins in the larvae causes death. The destruxins are toxic to insects only by ingestion and not thro' the integument

3.3.3. *Paecilomyces* Species

It is commonly found in nature and has a wide range of hosts mainly lepidopteran larvae. Entomopathogenic fungi in this genera are: *P. farinosus, P. fumosa – roseus, P. javanicus, P. ramosus.*

3.3.4. *Nomuraea* Species

N. rileyi infects a number of *lepidopteran* larvae, such as *H. zea, H. virescens* and *Trichoplusia ni.* Exposure of the infected larvae to light is essential for maximal conidial production. Also produces a toxin. Epizootics by this fungus often develop late in the growing season of crops. On soybean the source of infection is spores that persist on the leaves.

3.3.5. *Verticillium*

The genus *Cephalosporium* has been placed in synonymy with *Verticillium. V. lecanii* is a common pathogen of scale insects in tropical and semi tropical environment. It is known as the white-halo fungus. It also infects insects other than scales. The mycelium of *V.l.* produces the cyclodepsipeptide toxin, bassianolide which is also produced by *B.b.* In addition this fungus forms other insecticidal toxins such as dipicolinic acid and C_{25} compounds. Strains of *V.l.* are reported to be pathogenic for the rust fungi, which cause plant diseases. Such strains offer the interesting possibility of using a single fungal species to control both the pest insects and pest fungi of plants artificial media for its mass multiplication has been developed by DOR, hyderabad. (Vimala Devi and Prasad, 1997).

3.3.6. *Hirsutella* Species

The genus *Hirsutella* includes over 30 species that infect nearly all systematic groups of insects and certain mites. In the infection of some *Hirsutella* species, the amount of mycelium that covers the cadaver is very sparse and can be easily over looked by the inexperienced worker. Some of these fungi are highly virulent (eg) *H. thompsoni* on the citrus rust mite, *Phyllocoptruta oleivora*. They are promising candidates for MC but their growth and sporulation on solid media are alow as compared with other hyphomycetes.

3.3.7. *Aspergillus* Species

There are a number of entomopathogenic spp. of *Aspergillus* such as *A. flavus, A. parasiticus, A. tamari, A. ochraceus, A. fumigatus, A. repens* and *A. uersicolor*. These fungi are mainly saphrophytic but may infect a wide range of insect species. *A. flavus* has been studied most widely. In the honey bee *A. flavus* infect the brood and adult bees. The infection in the brood is called 'Stone brood'. In general, stone brood of honey bee is rare and is of minor importance to bee keepers. Several toxins have been isolated from cultures of *A. flavus* such as kojic acid which is toxic to housefly maggot. The virulence of strains of *A. flavus* to the silkworm is related to the production of kojic acid and the resistance of the strains to formalin. The most widely known and investigated toxins of *A. flavus* and *A. parasiticus* are the afla toxins a group of at least 12 highly substituted polycyclic compounds. The afla toxins are very potent carcinogens and produce tumors primarily in the liver of vertebrates including humans.

Sub Division Deuteromycotina class. Coelomycetes

Has two genera, *Tetranacrium* and *Aschersonia* whose members are entomopathogens, *Aschersonia aleyrodis* infects greenhouse whitefly *Trialeurodes vaporariorum*.

3.3.8 The Process of Pathogenesis Begins with

1. Adhesion of fungal infective units or conidium to the insect epicuticle.

2. Germination of infective units on cuticle.

3. Penetration of the cuticle.

4. Multiplication in the haemolymph.

5. Death of the host (Nutritional deficiency, destruction of tissues and releasing toxins).

6. Mycelial growth with invasion of all host organs.

7. Penetration of hyphae from the interior through the cuticle to exterior of the insect.

8. Production of infective conidia on the exterior of the insect.

☆ **Spore germination:** The development of mycosis can be separated into three phases: 1. adhesion and germination of the spore on the insects cuticle, 2. penetration into the haemocoel and 3. development of the fungus, which generally results in the death of an insect. Spore germination largely depends on the environmental humidity and temperature and to a lesser extent light conditions and the nutritional environment.

☆ **Penetration:** The process of penetration thro' the insect's instigates by a hypha germinating from a spore involves chemical (enzymatic) and physical forces. The enzymes detected on germ tubes are proteases, aminopeptidases, lipase, esterase and N-acetyl glucosamidase (chitinase).

☆ **Penetration thro' body openings:** Fungi can infect insects thro' buccal cavity, spiracles and other external openings. Since moisture is not a problem in the alimentary tract, the spore may germinate readily in this environment, on the other hand the digestive fluids may destroy the spore or germinating hypha. In mosquito larvae, the fungus *Culicinomyces* sp. invades thro the fore- and hindguts. Fungi can infect insects thro' spiracles and other body openings.

☆ **Pathogenicity:** Fungi usually cause insect mortality by one or more of the following: nutritional deficiency, invasion and destruction of tissues, and release of toxins. Some fungi are obligate pathogens (*e.g. Coelomomyces,* certain *Entomophthora*) and their complete life cycles have not been cultured outside the living host. Most entomopathogenic fungi, however, are facultative pathogens, capable of growing without an insect host. Successive transmission within a host may result in the enhancement of virulence or the isolation of a more virulent strain. On the other hand, the *in vitro* culture often results in diminished virulence.

☆ **Effect of Environmental conditions:** Environ. and particularly humidity and temperature and to a lesser extent light and air movement are very important in infection and sporulation of entomopathogenic fungi. Optimum temperatures for development, pathogenicity and survival of the fungi generally fall between 20 to 30°C. Resting spores may tolerate high temperatures from 80 to 100° for 5 to 60 min. and low temperatures of 7 to 10°C. Very high humidity (above 90 per cent RH) is required for spore germination and sporulation outside the host. Although high humidity is essential for sporulation the release of the conidia of *B. bassiana* and *M. anisopliae* from conidiophores is stimulated by low humidity (less than 50 per cent), darkness and vibration. The effect of light in fungal infection is not completely known. Since light often interacts with temperature and humidity, its action is difficult to evaluate separately. Extended exposure to far ultraviolet light generally destroys microorganisms but *Hirsutella thompsoni* survives such exposure by photo reactivation.

3.3.29 Mass Production of Green Muscardine Fungus

Metarhizium anisopliae infects coconut rhinoceros beetle/grub, sugarcane *Pyrilla*

Take 40 g of carrot bits in 250 ml conical flask with 65 ml of water

⇓

Autoclave at 20 Psi for 30 min

⇓

Cool and inoculate with the fungus

⇓

Fungus can be applied to manure pit after a fortnight

3.4 Protozoa

The protozoa are minute unicellular organisms. Because of their minute size, remained unobserved until the development of the microscopes. Van Leeuwenhoek is generally recognized as the father of protozoology for his observations on numerous protozoa. During the early part of the nineteenth century, silkworm industry in Europe was devastated by the disease pebrine in the silkworm and a number of investigators, particularly Pasteur became involved in the study of this disease. The organism causing pebrine was described by Naegeli in 1857 and named *Nosema bombycis.* The use of microsporidia as biocontrol agents was recommended as early as 1905 by Perez against green crab, *Carcinus maenas,* a serious predator of oysters, but no steps were taken in this direction. Increasing number of attempts to control insect pests were made with protozoa, in particular the microsporidia (Battu *et al.,* 1993).

Relation of Protozoa to Insects

There are about 1200 species of protozoa, out of about 15,000 described species, that are associated with insects. The entomogenous protozoa are commonly found in the digestive tracts of insects as commensals or they are in a mutalistic association with insects. Some insects serve as vectors of protozoan diseases of vertebrates and plants. In many of these cases the protozoa multiply in the insect vectors and may even cause harm to some vectors. A great number of protozoa are pathogenic to insects. The majority of the highly pathogenic forms occur in Phyla Apicomplexa and Microspora, particularly those that invade the haemocoel and develop intracellularly.

Portals of Entry

The majority of protozoa enter the insects by way of the mouth and digestive tract. The infective stage is generally a spore or cyst, but it may be the vegetative or non-cysted reproductive form (*e.g.* In flagellates and ciliates). Those protozoa that remain in the lumen of the digestive tract are attached to the epithelium, or enter apaendages associated with the digestive tract and generally cause no obvious pathology. These forms are mainly ciliates, flagellates and gregarines. Others penetrate into the haemocoel and exist extracellularly in the haemolymph intracellularly within the cells of various tissues and organs and cause pathologies. They are mainly the apicomplexans and microsporidia. Vertical transmission from parent to off-spring occurs in many protozoa, especially the microsporidia. The transmission is by transovum and transovarian modes.

Pathogensis, Signs and Symptoms

Most entomopathogenic protozoa have low virulence and cause a chronic infection that often does not kill an insect. Such a chronically infected insect frequently doest not exhibit marked external signs and symptoms *e.g.* color changes and abnormal movement or behaviour. Some protozoa are however are highly virulent and depending on the tissues attacked, the infection may be acute and fatal. In some cases the infected insects become chlorotic or whitish, are reduced in size and remain in immature stages much longer than the uninfected individuals. The enormous numbers of protozoan spores in the fat, midgut or haemolymph may cause these structures to turn miky white. The integument of dead insects (mainly larvae) generally remains firm and does not readily disintegrate. The gregarines of the order *Neogregarinida* are called neogregarines or schizogregarines. A number of entomopathogenic neogregarives produce lethal infections in important insect pests in the orders *Diptera, Coleoptera* and *Hemiptera.* Most neogregarines infect a single or a few insect species, but some such as *Farinocystis tribolii* and *Matesia grandis* attack a large number of insect species. *Farinocystis tribolii* infects flour beetle, *Tribolium castaneum,* causes a juvenile hormonal effect and produces larval-pupal and pupal-adult intermediate forms. A similar effect is produced when ether extracts of spores (oocysts) are applied to pupae of *T. castaneum* and to final instar nymphs or to bug, *Dysdercus cingulatus. Mattesia grandis* infects perorally the larva and adult of the cotton boll weevil, *Anthonomas grandis.*

Phylum Microspora, Class Microsporea

Members are called microsporidia. The disease they cause is called microsporidiosis. Microsporidia ranks among the smallest of eukaryotes and have an obligate intracellular habit. They posses unicellular spores, containing uninucleate or binucleate sporoplasm and an extrusion apparatus always with a polar filament and polar cap. They are the most promising protozoa for use in microbial control.

Signs and Symptoms

An insect infected with a microsporidium generally is altered in colour, size, form and activity, depending on the tissues and organs infected. A translucent larva turns increasingly opaque or dull, milky white as the infection progresses. The integument usually remains intact throughout infection. The activity of an infected larva is often reduced at an advanced stage of infection. *Nosema locustae* microsporidium registered to control insects in US. *N.l.* can provide long-term control against grasshoppers and crickets on rangeland and pasture. *Nosema fumiferanae* introduced into the populations of the spruce budworm. *Nosema pyrausta*, a naturally occurring microsporidium of the European corn borer when applied as a foliar spray, reduced the number of larvae per plant. *Mattesia grandis* against boll weevil *Vairimorpha necatrix* against tobacco budworm and corn earworm.

3.5 Entomopathogenic Nematodes (EPNs)

A plethora of nematode species in more than 30 families is associated with insects and other invertebrates. The major focus of research and development has been on nematode species in 7 families, Mermithidae, Tetradonematidae, Allantonematidae, Phaenopsitylenchidae, Sphaerulariidae, Steinernematidae, and Heterorhabditidae, because of their potential as biological control agents of insects (Kaya and Stock, 1997). The mermithid *Romanomermis culicivorax* had been used as an inundative agent for mosquito larval suppression (Petersen, 1985). It recycles in certain host habitats and produces high levels of infection in selected target mosquito species, but it is intolerant to polluted, organically enriched, or high salinity habitats and high temperatures. Furthermore, it could be produced only *in vivo* and could not compete commercially with another biological agent, *B. Thuringiensis* subsp. *israelensis.*

Currently, the steinernematids and heterorhabditids are receiving the most attention as microbial control agents of soil insects. After *B. thuringiensis,* these nematodes are next in commercial sales at US$2–3 million annually (Georgis, 1997). The complex known as DD-136 with the nematode *Neoaplectana carpocapsae* (also known as Dutky nematode) and the bacterium *Achromobacter nematophilus.* The nematode serves as a vector for the bacterium, which produces a septicemia (sporulation in blood; Milky disease) in the insect body. The bacteria are retained in the nematode intestine as the latter does not feed during its free-living existence. When such bacteria possessing nematodes invade fresh insect hosts, the latter are killed. Though a few nematodes can kill the host, sufficient number of them should invade the host. In India entampapthogenic nematodes were tried against rice and sugarcane borers.

The entomopathogenic steinernematid and heterorhabditid nematode species possess many attributes of parasitoids and pathogens. They are analogous to parasitoids because they have chemoreceptors and can actively search for their hosts. They are similar to pathogens because of their association with mutualistic bacteria in the genera *Xenorhabdus,* for steinernematids, and *Photorhabdus,* for heterorhabditids. The nematode/bacterial complex is highly virulent, killing its host within 48 h through the action of the mutualistic bacteria. These nematodes can be cultured *in vitro,* have a high reproductive potential, and have a numerical, but no functional, response. Moreover, they infect a number of insect pest species, yet pose no threat to plants, vertebrates, and many invertebrates. They can be mass produced, formulated, and easily applied as biopesticides, have been exempt from registration in many countries, are compatible with many pesticides, and are amenable to genetic selection.

The third-stage infective nematode (dauer stage or infective juvenile) of steinernematids and heterorhabditids has been likened to a guided missile because it carries the "warheads" of the mutualistic bacterial cells in its intestine (Akhurst, 1993). Each species of nematode is associated with a specific bacterium, but some bacterial species are associated with more than one nematode species. The link between the matching of the appropriate phase of these bacteria (the bacterium occurs as phase I or phase II with phase I more suitable for nematode production) and the successful production of efficacious entomopathogenic nematodes is essential. The large-scale production of nematodes on solid, monoxenic artificial medium on foam (Bedding, 1984) or particularly on liquid monoxenic media (Ehlers, 1996) expands the commercial possibilities. However, *in vivo* production using the wax moth, *Galleria mellonella* (L.), or another suitable insect host is still used by some in the cottage industry. Formulation of the infective juveniles in a wettable dispersible granule has permitted storage capability of 6 months at room temperature and has increased the options for their application.

The entomopathogenic activity of steinernematid and heterorhabditid species has been documented against a broad range of insect pests in a variety of habitats. These nematodes are especially efficacious against insects in soil and cryptic habitats. They have been used inundatively in a number of highvalue cropping systems. For example, the citrus root weevil, *D. abbreviatus,* in citrus, the black vine weevil, *Otiorhynchus sulcatus* (F.), in nurseries and cranberries, the black cutworm, *Agrotis ipsilon* (Hufnagel), and mole crickets, *Scapteriscus* spp., in turfgrass, and the peach borer moth, *Carposina niponensis* Walsingham, in apples have been successfully controlled. When an entomopathogenic nematode species is used against a pest insect, it is critical to match the right nematode species against the insect pest. Some nematode species are more efficacious against a particular insect group than against another insect group. For example, *Steinernema kushidai* Mamiya is effective against scarab grubs and less so against lepidopteran larvae, and *Steinernema scapterisci* Nguyen and Smart is effective against mole crickets and house crickets, but not effective against other insect groups. In addition, foraging behaviour of entomopathogenic nematodes can affect their efficacy.

Table 5: Steinernematid and Heterorhabditid Nematodes as Microbial Control Agents of Insects

Nematode Families and Species	Targeted Groups	Selected References
Heterorhabditidae		
Heterorhabditis bacteriophora	Lepidoptera, Coleoptera	Begley (1990)
Heterorhabditis megidis	Coleoptera	Klein (1990)
Heterorhabditis marelatus	Coleoptera, Lepidoptera	Berry *et al.* (1997)
Steinernematidae		
Steinernema carpocapsae	Lepidoptera, Coleoptera, Siphonaptera	Georgis and Manweiler (1994)
Steinernema feltiae	Diptera (Sciaridae)	Klein (1990)
Steinernema glaseri	Coleoptera (Scarabaeidae)	Klein (1990)
Steinernema kushidai	Coleoptera (Scarabaeidae)	Ogura (1993)
Steinernema riobrave	Lepidoptera, Orthoptera	Cabanillas *et al.* (1994)
Steinernema scapterisci	Orthoptera (mole crickets)	Parkman *et al.* (1993)

Some species are ambushers *(e.g., Steinernema carpocapsae* (Weiser) and *S. scapterisci)* that tend to remain near the soil surface and attach to and infect mobile hosts at the soil–litter interface (Lewis *et al.,* 1993). Other species *(e.g., S. glaseri* (Steiner) and *Heterorhabditis bacteriophora* Poinar) are cruisers that have an active searching strategy and are more effective against less mobile insects in the soil (Gaugler, 1997). Although they are used primarily as biopesticides, some species of nematodes persist and recycle in the host habitat, bringing about sustained suppression of some insect pests.

Genetic improvements in entomopathogenic nematodes may expand their potential as biocontrol agents by increasing search capacity, virulence, and resistance to environmental extremes, among other attributes. Gaugler *et al.* (1997), using molecular techniques, have inserted a heat-shock protein into *H. bacteriophora,* resulting in transgenic nematodes that were 18 times better than the wild types at surviving high-temperature stress. Field release of the transgenic and wild-type nematodes showed no differences in their abilities to persist. Significant advances have been made with these entomopathogenic nematodes, but the high costs associated with production and formulation in comparison to those costs of chemical pesticides and other biological *(i.e., B. thuringiensis)* will restrict their use to highvalue niche markets and sensitive areas where chemicals cannot be used. However, advances have also been made with insecticides *(e.g.,* imidacloprid) that are more environmentally friendly than organophosphates, carbamates, or chlorinated hydrocarbons. Insects will probably become resistant to these new chemical pesticides, and entomopathogenic nematodes and other entomopathogens may play a more important role in IPM. For example, the combination of imidacloprid and entomopathogenic nematodes has shown synergistic activity against 3rd instar

Table 6: *Bacillus thuringiensis* (*Bt*) Products Available in India

Bt Variety	Trade Name	Target Pest	Crop	Manufacturer
Bt kurstaki	Halt	Diamondback moth, Lepidopterous caterpillars	Cabbage	Wockhardt Ltd., Bombay
Bt kurstaki	Biolep	Cotton bollworms	Cotton	Biotech International Ltd., New Delhi
Bt kurstaki	Bioasap	Cotton bollworms	Cotton	Biotech International Ltd., New Delhi
Bt kurstaki	Delfin WG	Cotton bollworms, tobacco caterpillar	Cotton, castor	M/s Sandoz India Ltd., Bombay
Bt kurstaki	Dipel 8L	Cotton bollworms, fruit borers	Cotton, tomato, brinjal, okra	Lupin \agrochemicals Pvt.Ltd., Bombay
Bt kurstaki	Spicturin	American bollworm, leaf-folder, tobacco caterpillar, diamondback moth	Cotton, rice, cabbage, cauliflower, chillies	Tuticorin Alkali Chemicals and Fertilizers Ltd., Chennai
Bt kurstaki	Biobit	Lepidopterous caterpillars	Several	Rallis India Ltd., Bangalore

Source: Dhaliwal and Arora (2004).

scarabs (Koppenho ¨fer *et al.,* 2000). Imidacloprid is most efficacious against 1st- and 2nd-instar scarabs, but most damage is done by the 3rd instar. Accordingly, the combination of these two agents may be useful in the management of scarab pests in turf.

Table 7: Examples of Commercial *Bacillus thuringiensis* Insecticides

Trade Name	Manufacturer	B. thuringiensis Strain	Target Insect Pest
Lepidoptera			
Agree	Ciba Geigy	Krustaki HD – 1 + aizawai	Vegetable/Crop pests, *Spodoptera* spp.
Biobit	Novo-Nordisk	Kurstaki HD – 1 + aizawai	Vegetable/Crop pests, *Galleria mellonella*
Certan	Sandoz	Kurstaki HD – 1 aizawai	Forest Pests
Condor	Ecogen	Kurstaki HD – 1	Vegetable/Crop pests
Cutlass	Ecogen	Kurstaki HD – 1	Vegetable/Crop pests
Cybout	Cynamid	Kurstaki HD – 1	Vegetable/Crop pests
Dipel	Abbot	Kurstaki HD – 1	Forest Pests
Foray	Novo-Nordisk	Kurstaki HD – 1	Spodoptera spp.
Javelin	Sandoz	Kurstaki NRD –12	Vegetable pests
MVP	Mycogen	Kurstaki HD – 1 in *Pseudomonas*	
Thuricide	Sandoz	Kurstaki HD – 1	Vegetable/Crop pests
Coleoptera			
Di Terra	Abbot	Tenebrionis	Vegetable pests
Foil	Ecogen	tenebrionis	Ostrinia nubilalis (Hubner)
M-One	Mycogen	tenebrionis	Leptinotarsa decemlineata (Say)
M-Trak	Mycogen	tenebrionis	Vegetable Pests
Novodor	Novo-Nordisk	tenebrionis	Vegetable Pests
Trident	Sandoz	tenebrionis	Vegetable Pests
Trident II	Sandoz	tenebrionis	Vegetable Pests
Diptera			
Acrobe	Cyanamid	israelensis	Medical pests
Skeetal	Sandoz	israelensis	Medical pests
Teknar	Zoecon/Sandoz	israelensis	Medical pests
Vectobac	Abbot	israelensis	Medical pests

Source. Lee *et al.* (1998).

Steinernematids and heterorhabditids have been used successfully against a number of soil-inhabiting insect pests. However, this realm of insect nematology is a very young discipline with major contributions being made since the mid-1980s. However, there is a need for more in-depth basic information on their biology, including the ecology, behavior, and genetics of

Table 8: Guidelines and Indian Standards for Commercialization for Microbial Biopesticides Under Section 9[3b] and 9[3] of the Insecticide Act,1968; [w.e.f.1st Jan, 2011]

Sl.No.	Microbial Biopesticides	Composition
1.	**Baculoviruses**	Viral Units
	Nuclear Polyhedrosis Virus [NPV]	HaNPV and SLNPV: 1 X 10^9 POB/ML
	Granulosis Virus [GV]	PxGV, CiGV and AjGV: 5 X 10^9 Capsules/ML
	pH	7–8
	Pathological Contaminants	*Salmonella, Shigella, Vibrio etc* should be absent
	Other Microbial Contaminants	Not exceeding: 1 X 10^4/ml or g
	Bioassay for Determing LC$_{50}$	*H. armigera* NPV: < 0.5 POB/mm^2
		S.litura NPV: < 20.0 POB/mm^2
		P. xylostella GV: 0.15 OB/mm^2
		Chilo infuscatellus GV. 1 X 10^3 OB/mm^2
		Acheae janata GV. < 4 OB/mm^2
2.	**Entomopathogenic bacteria**	Minimum of 2 X 10^8 CFU/ml or gm
	Delta endotoxin content	Minimum of 2.0 per cent
	pH	7-8
	Moisture content	12 per cent
	Pathological Contaminants	*Salmonella, Shigella, Vibrio etc* should be absent
	Other Microbial Contaminants	Not exceeding: 1 X 10^4/ml or g
3.	**Entomopathogenic Fungi**	Minimum of 1 X 10^8 CFU/ml or gm
	pH	7.5
	Moisture content	8 per cent
	Pathological Contaminants	*Salmonella, Shigella, Vibrio etc* should be absent
	Other Microbial Contaminants	Not exceeding: 1 X 10^4/ml or g
	Bioassay for Determing LC$_{50}$	*S. litura.* not more than 2 x 10^6 spores/ml [3 x 10^3 spores/mm^2]
		H. armigera. not more than 4 x 10^6 spores/ml [6 x 10^3 spores/mm^2]
		P. xylostella. not more than 3 x 10^9 cfu/g

these nematodes, to help understand the underlying reasons for their successes and failures as biological control agents. Armed with this information, innovative approaches through genetic engineering and combinations with other control agents offer promise in insect suppression. More traditional approaches ofclassical biological control or augmentation with new or previously described species of nematodes may provide population reduction through inoculative releases.

Chapter 4

Host Plant Resistance

The IPM concept stresses the need to use multiple tactics to maintain insect pest abundance and damage below levels of economic significance. Thus, a major advantage to the use of insect-resistant crop varieties as a component of IPM arises from the ecological compatibility and compatibility with other direct control tactics. Insect-resistant cultivars synergize the effects of natural, biological, and cultural insect pest-suppression tactics. The "built-in" protection of resistant plants from insect pests functions at a very basic level, disrupting the normal association of the insect pest with its host plant. The compatible, complementary role plant resistance to insect pests plays with other direct control tactics is, in theory and practice, in concert with the objectives of IPM. All crop cultivars should contain resistance to insect pests. Plant resistance to insect pests has advantages over other direct control tactics. For example, plant resistance to insects is compatible with insecticide use, while biological control is not. Plant resistance to insects is not density dependent, whereas biological control is. Plant resistance is specific, only affecting the target pest. Often effects of use of insect-resistant cultivars are cumulative over time. Usually the effectiveness of resistant cultivars is long-lasting.

4.1 Definition

Painter (1951) described plant resistance as the relative amount of heritable qualities possessed by the plant which influence the ultimate degree of damage done by the insect.Lesser damage than average damage is taken as resistance while more damage than average damage constitutes susceptibility. A resistant variety produces higher yield than susceptible variety when both are subjected to the same extent of infestation by same insect at the same time. Resistance is a relative term only compared with less resistance or susceptibility. Absolute resistance or Immunity refers to the inability of a specific pest to consume or injure a particular variety under any known-conditions. Immune varieties are rare.

4.2 Advantages to the Use of Insect-Resistant Crop Varieties

Use of insect-resistant crop varieties is economically, ecologically, and environmentally advantageous. Economic benefits occur because crop yields are saved from loss to insect pests and money is saved by not applying insecticides that would have been applied to susceptible varieties. In most cases, seed of insect-resistant cultivars costs no more, or little more, than for susceptible cultivars. Ecological and environmental benefits arise from increases in species diversity in the agroecosystem, in part because of reduced use of insecticides.

4.3 Different Theories Regarding the Fundamental Basis of Host Plant Resistance

1. **Non-Preference and Antibiosis theory**—R.H. Painter [1951]—1. Non Preference: The plant characters that discourage its use for oviposition, food and shelter. 2. Antibiosis: Plant metabolites that interrupt the normal insect life/biology. 3. Tolrance: Plants ability to sustain insect population.

2. **Dual discrimination Theory**—Kenedy and Booth [1951]—Host selection is based on insect response to two type of stimuli, the "flavour stimuli" and "nutrient stimuli".

3. **Chemosensory Theory**—Dethier [1953]—Act of host selection is regulated by chemical senses in 5 phases *viz.*, Orientation of food, biting response, Continued feeding, cessation of feeding, dispersal

4. **Token stimulus Theory**—Frankel [1959]—Host selection must be based on some "odd" or "secondary biochemicals" within the plant tissues like essential oils, alkaloids, glucosides etc which provide stimulus to which insect responds. These are called " token stimuli"

5. **Thornsteinsons Theory**—Thornsteinson [1960]—Host selection is based on summation effect of several stimuli of the same chemotactic like, optimal feding response, feeding inhibitors, deterrents, chemotactic stimulants, sapid nutrients, piquant stimulant.

6. **Nutrient imbalance Theory**—House [1966]—Nutrient imbalance affects the intake and the insect selects only those food materials which are most suited to their requirement.

4.4 Host Plant Selection Process by an Insect

Host plant selection is a process by which an insect detects a resource providing plant within an environment of large population of diversified plant species. The process of host plant selection involves a sequence of five steps:

1. **Host-habitat finding**: The adult population of any species arrives at general host habitat by phototaxis or anemotaxis and geotaxis. Temperature and humidity play important role. Normally crop pests stay within general area where crops are planted and hence, this becomes less important in host plant selection.

⇓

2. **Host finding**: After locating habitat the insect pest makes a purposeful search to locate its appropriate host plant for its establishment. The essential visual or olfactory mechanisms help the contact. Once the pest reaches or contacts the host plants, tactile and olfactory sensory organs arrest further movement causing the insects to remain on the plant.

⇓

3. **Host recognition**: Although larvae are with sensorial receptors for host recognition, this phase is usually taken care of by ovipositing female adult. It is usually done with the help of specific volatile from the plants. *e.g.,* -Onion maggots, *Delia sp* attracted to its host by the odour of propryl disulphide. Cabbage maggot fly, *Delia brassica* get attracted by crucifer due to presence of few glucocyanolides.

⇓

4. **Host acceptance**: Various chemicals present in the host species actually govern the feeding process of insects. These chemicals responsible for initial biting, swallowing and continuation feeding. *e.g.,* Presence of phagostimulants like *morin* in mulberry *Morus alba* is key in continuation of feeding of silkworm *Bombyx mori.*

⇓

5. **Host suitability**: The nutritional value in terms of sugars, proteins, lipids and vitamins or absence of deleterious toxic compounds determines the suitability of the host for the pest in relation to the development of larvae, longevity and feeding.

Secondary Metabolites that Influence the Host Selection are Catogorised as:

A. Allomones

Gives adaptive advantage to producer

1. Repellents: Orient insects away from the plant
2. Suppresants: Inhibit biting or piercing

3. Deterrants: Prevents maintenance of feeding or oviposition

4. Antibiotics: Disrupt normal larval growth and reduce fecundity

B. Kairomones

Gives adaptive advantage to reciever

1. Attractants: Orients insects towards the host

2. Arrestants: Slow or stops the movement

3. Excitants: Elicit biting, piercing or oviposition

4.5 Mechanisms of Host Plant Resistance

R. H. Painter (1951) has grouped the mechanisms of host plant resistance into three main categories as 1. Non-preference (Antixenosis) 2. Antibiosis and 3. Tolerance

Though various workers have attempted to classify the mechanisms of resistance, the terms defined by Painter (1951)–non preference, antibiosis and tolerance were widely accepted. However, Kogan and Ortman (1978) proposed that the term non preference should be replaced by antixenosis because the former describes a pest reaction and not a plant characteristic. The three types of resistance are described in the context of the functional relationships between the plant and the insect.

4.5.1 Non-Preference or Antixenosis

The term 'Non-preference'referes to the responce of the insect to the charecteristics of the host plant, which make is unattractive to the insect for feeding, oviposition or shelter. Kogan and Ortman (1978) proposed the term 'Antigenosis', as the term 'Non-preference' pertains to the insect and not to the host plant. Some plants are not choosen by insects for food shelter or oviposition because of either the absence of desirable characters in that plant like texture, hairyness taste, flavour, or presence of undesirable characters. Such plants are less damaged by that pest and the phenomenon is called non preference *e.g.* Hairy varieties of soybean and cotton are not preffered by leafhoppers for oviposition Open panicle of sorghum supports less *Helicoverpa armigera* Wax bloom on crucifers deter diamondback moth *Plutella xylostella*.

4.5.2 Antibiosis

Antibiosis refers to the adverse effect of host plant on the insect due to the presence of some toxic substances or absence of required nutritional components. Such plants are said to exhibit antibiosis and hence do not suffer as much damage as normal plants. The adverse effects may be reduced fecundity, decreased size, long life cycle, failure of larva to pupate or failure of adult emergence, and increased mortality. Indirectly, antibiosis may result in an increased exposure of the insect to its natural enemies. *e.g.*: The most classical example ofhost plant resistance is DIMBOA (2,4 Di hydroxy -7- methoxy – 1,4 benzaxin – 3) content in maize which imparts chemical defense against the

European corn borer *Ostrinia nubilalis*. Nutrionally related antibiotic effect in rice variety Mudgo which is resistant to BPH.

4.5.3 Tolerance

A few plants withstand the injure caused by the insect by producing more number of tillers,roots, leaves etc in the place of damaged plant parts such plants are said to be tolerant to that particular pest.Tolerance usually results from one or more of the following factors.1. General vigour of the plant, 2.Regrowth of the damaged tissues 3.Strength of stems and resistant to lodging 4. Production of additive branches 5. Efficient utilization of non vital plant parts by the insect and 6. Compensation by growth of neibhouring plants.

4.6 Plant Defense against Herbivory

Many plants produce secondary metabolites, known as allelochemicals, that influence the behavior, growth, or survival of herbivores. Plant defenses can be classified generally as constitutive or induced. Constitutive defenses are always present in the plant, while induced defenses are produced or mobilized to the site where a plant is injured. There is wide variation in the composition and concentration of constitutive defenses and these range from mechanical defenses to digestibility reducers and toxins. Many external mechanical defenses and large quantitative defenses are constitutive, as they require large amounts of resources to produce and difficult to mobilize. A variety of molecular and biochemical approaches are used to determine the mechanism of constitutive and induced plant defenses responses against herbivory. Induced defenses include secondary metabolic products, as well as morphological and physiological changes. An advantage of inducible, as opposed to constitutive defenses, is that they are only produced when needed, and are therefore potentially less costly, especially when herbivory is variable.

4.6.1 Chemical Defences

Persimmon, genus *Diospyros*, has a high tannin content which gives immature fruit, seen above, an astringent and bitter flavor. The evolution of chemical defences in plants is linked to the emergence of chemical substances that are not involved in the essential photosynthetic and metabolic activities. These substances, secondary metabolites, are organic compounds that are not directly involved in the normal growth, development or reproduction of organisms, and often produced as by-products during the synthesis of primary metabolic products. Although these secondary metabolites have been thought to play a major role in defenses against herbivores, a meta-analysis of recent relevant studies has suggested that they have either a more minimal (when compared to other non-secondary metabolites, such as primary chemistry and physiology) or more complex involvement in defense.

Secondary metabolites are often characterized as either *qualitative* or *quantitative*. Qualitative metabolites are defined as toxins that interfere with an herbivore's metabolism, often by blocking specific biochemical reactions.

Qualitative chemicals are present in plants in relatively low concentrations (often less than 2 per cent dry weight), and are not dosage dependent. They are usually small, water soluble molecules, and therefore can be rapidly synthesized, transported and stored with relatively little energy cost to the plant. Qualitative allelochemicals are usually effective against non-adapted specialists and generalist herbivores.

Quantitative chemicals are those that are present in high concentration in plants (5 – 40 per cent dry weight) and are equally effective against all specialists and generalist herbivores. Most quantitative metabolites are digestibility reducers that make plant cell walls indigestible to animals. The effects of quantitative metabolites are dosage dependent and the higher these chemicals' proportion in the herbivore's diet, the less nutrition the herbivore can gain from ingesting plant tissues. Because they are typically large molecules, these defenses are energetically expensive to produce and maintain, and often take longer to synthesize and transport.The geranium, for example, produces a unique chemical compound in its petals to defend itself from Japanese beetles. Within 30 minutes of ingestion the chemical paralyzes the herbivore. While the chemical usually wears off within a few hours, during this time the beetle is often consumed by its own predators.

A. Alkaloids

Alkaloids are derived from various amino acids. Over 3000 known alkaloids exist, examples include nicotine, caffeine, morphine, colchicine, ergolines, strychnine, and quinine. Alkaloids have pharmacological effects on humans and other animals. Some alkaloids can inhibit or activate enzymes, or alter carbohydrate and fat storage by inhibiting the formation phosphodiester bonds involved in their breakdown. Certain alkaloids bind to nucleic acids and can inhibit synthesis of proteins and affect DNA repair mechanisms. Alkaloids can also affect cell membrane and cytoskeletal structure causing the cells to weaken, collapse, or leak, and can affect nerve transmission. Although alkaloids act on a diversity of metabolic systems in humans and other animals, they almost uniformly invoke an aversively bitter taste.

B. Cyanogenic Glycosides

These are stored in inactive forms in plant vacuoles. They become toxic when herbivores eat the plant and break cell membranes allowing the glycosides to come into contact with enzymes in the cytoplasm releasing hydrogen cyanide which blocks cellular respiration. Glucosinolates are activated in much the same way as cyanogenic glucosides, and the products can cause gastroenteritis, salivation, diarrhea, and irritation of the mouth. Benzoxazinoids, secondary defence metabolites, which are characteristic for grasses (Poaceae), are also stored as inactive glucosides in the plant vacuole. Upon tissue disruption they get into contact with β-glucosidases from the chloroplasts, which enzymatically release the toxic aglucones. Whereas some benzoxazinoids are constitutively present, others are only synthesised following herbivore infestation, and thus, considered inducible plant defenses against herbivory.

C. Terpenoids

The terpenoids, sometimes referred to as isoprenoids, are organic chemicals similar to terpenes, derived from five-carbon isoprene units. There are over 10,000 known types of terpenoids. Most are multicyclic structures which differ from one another in both functional groups, and in basic carbon skeletons. Monoterpenoids, continuing 2 isoprene units, are volatile essential oils such as citronella, limonene, menthol, camphor, and pinene. Diterpenoids, 4 isoprene units, are widely distributed in latex and resins, and can be quite toxic. Diterpenes are responsible for making Rhododendron leaves poisonous. Plant steroids and sterols are also produced from terpenoid precursors, including vitamin D, glycosides (such as digitalis) and saponins (which lyse red blood cells of herbivores).

D. Phenolics Compounds

Phenols consist of an aromatic 6-carbon ring bonded to a hydroxy group. Some phenols have antiseptic properties, while others disrupt endocrine activity. Phenolics range from simple tannins to the more complex flavonoids that give plants much of their red, blue, yellow, and white pigments. Complex phenolics called polyphenols are capable of producing many different types of effects on humans, including antioxidant properties. Some examples of phenolics used for defense in plants are: lignin, silymarin and cannabinoids. Condensed tannins, polymers composed of 2 to 50 (or more) flavonoid molecules, inhibit herbivore digestion by binding to consumed plant proteins and making them more difficult for animals to digest, and by interfering with protein absorption and digestive enzymes. Silica and lignins, which are completely indigestible to animals, grind down insect mandibles.

In addition to the three larger groups of substances mentioned above, fatty acid derivates, amino acids and even peptides are also used as defense. The cholinergic toxine, cicutoxin of water hemlock, is a polyyne derived from the fatty acid metabolism. β-N-Oxalyl-L-α,β-diaminopropionic acid as simple amino acid is used by the sweet pea which leads also to intoxication in humans. The synthesis of fluoroacetate in several plants is an example of the use of small molecules to disrupt the metabolism of herbivores, in this case the citric acid cycle.

E. Toxins

Most food crops do not contain toxic substances to insects, and if they are naturally present, they probably exist in concentrations that would not significantly affect insect or man. However, certain components that maybe in toxic levels to insects and rendered harmless to man by preparation and cooking are known in legumes or pulses and in some rootcrops, notably cassava.

A] Protease Inhibitor (trypsin inhibitor)

Cowpeas (*Vigna unguiculata*) are generally very susceptible to attack by *C. maculatus* and cause substantial losses. The Grain Legume Improvement Programme of the International Institute of Tropical Agriculture (IITA) in Ibadan, Nigeria has the global mandate for improving cowpea production, and considerable effort has been devoted to developing resistant varieties in conjunction with the Tropical Development Research Institute (TDRI, formerly TPI) of the Overseas Development Administration. One variety form Northern Nigeria displayed resistance to a strain of *C. maculatus* emanating from Brazil, which was important if cowpea production was going to be substantially increased in that country. Females of *C. maculatus* readily laid eggs on the seed surface, and larvae began feeding on the underlying cotyledons, but growth was extremely slow, and eventually most if not all the larvae died. Eighty to ninety-five percent (80-95 per cent) of larvae completed their development on the control susceptible varieties, strongly suggesting the presence of a toxic or antibiotic compound(s). Saponins (triterpenoid glycosides of which 5 aglycones have been isolated in soybeans toxic to *Callosobruchus* spp.) and lectins (phytohaemag glutinins found in beans, *Phaseolus vulgaris* were not found, but the resistant variety was shown to contain twice the level of a protease inhibitor (trypsin inhibitor) than the susceptible varieties, and was found to be inheritable, but its stability was unknown.Trypsin is a peptidase enzyme splitting proteins at certain specific peptide links and is secreted as trypsinogen from the mammalian pancreas, occurring in the digestive juices of most animals. Peptides are compounds formed from two or more amino acids by the amino group (NH_2) of one joining the carboxyl (COOH) group of the next forming peptide bonds (-HNCO-) with the elimination of water. Peptones are large fragments of peptides formed by the splitting of proteins by peptidase or protease enzymes. Trypsinogen inhibitors are also known in maize starch.

Most proteinaceous inhibitors have been, orated from the seeds of different members of Leguminosae. Refined extraction methods have isolated three proteinaceous fractions from soybean which strongly inhibit the growth of *T. castaneum* larvae; and more inhibitors have been isolated from wheat and lima beans Certain cultivars of groundnuts and chickpeas possess higher concentrations of specific *J. castaneum* protease inhibitors, which would at least confer some partial resistance to damage in storage, and increasing their concentration via breeding programmes would not be detrimental to human and animal nutrition.

B] L-canavanine

Apparently, L-canavanine, which may contribute from 8-10 per cent of the seeds dry weight in some legumes of the family Lotoeideae, is toxic to insects and higher animals due to disruption of normal protein synthesis. L-canavanine indescriminately replaces Larginine in structural proteins which are physiologically incompetent.

C] Amylase Inhibitors

Amalyase (or diastase) is a group of enzymes which split starch or glycogen variously to dextrin, maltose and glucose and are widely distributed in both plants and animals (such as in malt, pancreatic juices and in microorganisms) and can be purified by Affinity Chromatography. A glycoprotein from red kidney bean (*Phaseolus vulgaris*) which inhibits mammalian amylase, has also been shown to inhibit amylase activity in insects such as the yellow mealworm *Tenebrio molitor*, the larvae of the mediterrancean flour moth, *Ephestia kuhniella*, and adults of the confused flour beetle, *Tribolium confusum* (Powers and Culbertson, 1982; 1983). The rate at which the bean glycoprotein inhibits insect amylase and therefore their ability to breakdown starch to glucose, is dependent on pH, temperature and ionic strength. Glycoprotein amylase inhibitors from *Phaseolus vulgaris* beans have been purified by either conventional techniques or affinity chromotography. Bean amylase inhibitors apparently inhibit mammalian but not microbial or plant amylases.

4.6.2 Mechanical Defenses

The thorns on the stem of this raspberry plant, serve as a mechanical defense against herbivory. Plants have many external structural defenses that discourage herbivory. Depending on the herbivore's physical characteristics (*i.e.* size and defensive armor), plant structural defenses on stems and leaves can deter, injure, or kill the pest. Some defensive compounds are produced internally but are released onto the plant's surface; for example, resins, lignins, silica, and wax cover the epidermis of terrestrial plants and alter the texture of the plant tissue. The leaves of holly plants, for instance, are very smooth and slippery making feeding difficult. Some plants produce gummosis or sap that traps insects.

A plant's leaves and stem may be covered with sharp prickles, spines, thorns, or trichomes- hairs on the leaf often with barbs, sometimes containing irritants or poisons. Plant structural features like spines and thorns reduce feeding by large ungulate herbivores by restricting the herbivores' feeding rate, or by wearing down the mothparts. Raphides are sharp needles of calcium oxalate or calcium carbonate in plant tissues, making ingestion painful, damaging a herbivore's mouth and gullet and causing more efficient delivery of the plant's toxins. The structure of a plant, its branching and leaf arrangement may also be evolved to reduce herbivore impact. Coconut palms protect their fruit by surrounding it with multiple layers of armor. Trees such as coconut and other palms, may protect their fruit by multiple layers of armor, needing efficient tools to break through to the seed contents, and special skills to climb the tall and relatively smooth trunk.

4.3.3 Thigmonasty

Thigmonastic movements, those that occur in response to touch, are used as a defense in some plants. The leaves of the sensitive plant, *Mimosa pudica*, close up rapidly in response to direct touch, vibration, or even electrical and thermal stimuli. The proximate cause of this mechanical response is an abrupt change in the turgor pressure in the pulvini at the base of leaves resulting from osmotic phenomena. This is then spread via both electrical and chemical means through the plant; only a single leaflet need be disturbed.This response lowers the surface area available to herbivores, which are presented with the underside of each leaflet, and results in a wilted appearance. It may also physically dislodge small herbivores, such as insects.

4.3.4 Mimicry and Camouflage

Some plants mimic the presence of insect eggs on their leaves, dissuading insect species from laying their eggs there. Because female butterflies are less likely to lay their eggs on plants that already have butterfly eggs, some species of neotropical vines of the genus *Passiflora* (Passion flowers) contain physical structures resembling the yellow eggs of *Heliconius* butterflies on their leaves, which discourage oviposition by butterflies.

4.3.5 Indirect Defenses

The large thorn-like stipules of *Acacia collinsii* are hollow and offer shelter for ants, which in return protect the plant against herbivores.

Another category of plant defenses are those features that indirectly protect the plant by enhancing the probability of attracting the natural enemies of herbivores. Such an arrangement is known as mutualism, in this case of the "enemy of my enemy" variety. One such feature are semiochemicals, given off by plants. Semiochemicals are a group of volatile organic compounds involved in interactions between organisms. One group of semiochemicals are allelochemicals; consisting of allomones, which play a defensive role in interspecies communication, and kairomones, which are used by members of higher trophic levels to locate food sources. When a plant is attacked it releases allelochemics containing an abnormal ratio of volatiles. Predators sense these volatiles as food cues, attracting them to the damaged plant, and to feeding herbivores. The subsequent reduction in the number of herbivores confers a fitness benefit to the plant and demonstrates the indirect defensive capabilities of semiochemicals. Induced volatiles also have drawbacks, however; some studies have suggested that these volatiles also attract herbivores.

Plant use of endophytic fungi in defense is a very common phenomenon. Most plants have endophytes, microbial organisms that live within them. While some cause disease, others protect plants from herbivores and pathogenic microbes. Endophytes can help the plant by producing toxins harmful to other organisms that would attack the plant, such as alkaloid producing fungi which are common in grasses such as tall fescue (*Festuca arundinacea*).

4.3.6 Leaf Shedding and Colour

Leaf shedding may be a response that provides protection against diseases and certain kinds of pests such as leaf miners and gall forming insects. Other responses such as the change of leaf colours prior to fall have also been suggested as adaptations that may help undermine the camouflage of herbivores. Autumn leaf color has also been suggested to act as an honest warning signal of defensive commitment towards insect pests that migrate to the trees in autumn.[

4.3.7 Carbon: Nutrient Balance Hypothesis

The carbon:nutrient balance hypothesis, also known as the *environmental constraint hypothesis* or *Carbon Nutrient Balance Model* (CNBM), states that the various types of plant defenses are responses to variations in the levels of nutrients in the environment. This hypothesis predicts the Carbon/Nitrogen ratio in plants determines which secondary metabolites will be synthesized. For example, plants growing in nitrogen-poor soils will use carbon-based defenses (mostly digestibility reducers), while those growing in low-carbon environments (such as shady conditions) are more likely to produce nitrogen-based toxins. The hypothesis further predicts that plants can change their defenses in response to changes in nutrients. For example, if plants are grown in low-nitrogen conditions, then these plants will implement a defensive strategy composed of constitutive carbon-based defenses. If nutrient levels subsequently increase, by for example the addition of fertilizers, these carbon-based defenses will decrease.

4.7 Classification of Plant Resistance to Insects

A. Ecological Resistance or Pseudo Resistance or Apparent Resistance

Ecological resistance relies more on environmental conditions than on genetics. Certain crop varieties may overcome the most susceptible stage rapidly and thus avoid insect damage. Early maturing crop cultivars have been used in agriculture as an effective pest management strategy. However, plants that evade insect attack by this mechanism are likely to be damaged if the pest populations build-up early.

Pseudoresistance may be one or combination of the following:

1. **Host evasion**: Under some conditions, a host plant may pass through the most susceptible stage quickly or at time when insects are less in number. *e.g.,* Early planting of paddy in *kharif* minimize the infestation of stem borer *Scirpophaga incertulas* Sowing of sorghum soon after onset of monsoon in June helps to overcome shoot fly infestation

2. **Induced resistance**: It is a form of temporarily increased resistance as resulting from some conditions of plant or its environment such as changes in the amount of nutrients or water applied to the crop. *e.g.,* Application of potassium fertilizers (Kogan, 1994).

3. **Host escape**: It refers to lack of infestation or injury to the host plant because of transitory circumstances like incomplete infestation, thus finding of uninfested plant in a susceptible population does not necessarily mean that it is resistant.

B. Genetic Resistance

The factors that determine the resistance of host plant to insect establishment include the presence of structural barriers, allelochemicals and nutritional imbalance. These resistance qualities are heritable and operate in a concerted manner, and tend to render the plant unsuitable for insect utilization.

Genetic Resistance may be Grouped Based on:

A. Number of Genes

1. **Monogenic resistance**: Resistance is controlled by a single gene

2. **Oligogenic resistance**: Resistance is governed by a few genes

3. **Polygenic resistance**: Resistance is governed by many genes. This is also termed as horizontal resistance

B. Major or Minor Genes

1. **Major gene resistance**: The resistance is controlled by one or few major genes. Major genes have a strong effect and these can be identified easily. This is also called Vertical resistance.

2. **Minor gene resistance**: The resistance is controlled by a number of minor genes, each contributing a small effect. It is called minor gene resistance. This is also referred to as horizontal resistance.

C. Biotype Reaction

1. **Vertical resistance**: If a series of different cultivars of a crop show different reactions when infested with different insect biotypes, resistance is vertical. It is also referred to also as a qualitative or biotypespecific resistance. Vertical resistance is generally, but not always, of a high level and is controlled by a major genes or oligogenes. It is considered less stable.

2. **Horizontal resistance**: Horizontal resistance describes the situation where a series of different cultivars' of a crop show no differential interaction when infested with different biotypes of an insect. Generally, horizontal resistance is controlled by several poly genes or minor genes, each with a small contribution to the resistance trait. Horizontal resistance is moderate, does not exert a high selection pressure on the insect, and is thus more durable or stable.

4.8 Intensity of Resistance

Interactions between host plants and insects are spread over a wide spectrum of intensity. In terms of the host plant, lesser the population of the insect and or lesser the damage they cause to the plant, more resistant the plant is likely to be. On the other hand, from the point of view of the insect, interaction varies from totally unsuitable host to completely suitable for growth and development of insect. Therefore, intensity of resistance is a relative term and should be discussed in relation to a susceptible cultivar of the same species. Painter (1951) used the following scale to classify degree of resistance based on intensity:

a. **Immunity**: An immune variety is one which a specific insect will never consume or injure under any known conditions. There are thus, few, if any, cultivars immune to the attack of specific insects which are otherwise known to attack cultivars of the same species.

b. **High resistance**: A variety with high resistance is one which possesses qualities resulting in small damage by a specific insect under a given set of conditions.

Table 9: Function of Allelochemicals in Insect-Plant Interactions

Allelochemical	Insect Reaction	
	Stimulant	Deterrent
Cucurbitacin	*Diabrotica undecimpunctata*	*Epilachna tredecimnotata*
Cucurbitacin E and I	*Diabroticites*	*Phyllotreta nemorum* (Linnaeus)
Cyanogenic glycoside	*Epilachna varivestris* Mulsant	Many phytophagous insects
Furanocoumarins	*Papilio polyxenes* Fabricius	*Spodoptera litura*
Glucosinolate	*Pieris rapae* (Linnaeus)	*P. polyxenes*
Gosspypol	*Anthonomus grandis* Boheman	*Helicoverpa zea* (Boddie)
Iridoid	*Euphydryas editha*	*Locusta migratoria* (Linnaeus)
Lignin	*Bootettix argentalus*	*Ligurotettix coequilletti*
Lupanin	*Macrosiphon albifons* Essig	*Acyrthosiphon pisum* (Harris)
Tannin	*Anacridium melanorhodon*	*H. zea*
Tomatine	*Pieris brasasicae* (Linnaeus)	*Leptinotarsa decemlineata* (Say)

Source: Panda and Kush (1995).

Table 10: Feeding Deterrents in Host Plants

Plant	Feeding Deterrent(s)	Insect(s)
Cruciferae	Sinigrin	*Pieris brassicae* (Linnaeus)
Cucumis sativus	Cucurbitacin	*Tetranychus urticae* Koch, *Phyllotreta nemarum*
Gossypium spp.	Gossypol	*Heliothis virescens* (Fabricius), *Spodoptera littoralis* (Boisduval), *Eariasinsulana* (Boisduval)
	Isoquercitrin, quercitrin, quercetin	*Helicoverpa zea* (Boddie), *Pectinophora gossypiella* (Saunders)
Hordeum spp.	Gramine	*Schizaphis graminum* (Rondani)
Lycopersicon esculentum	Rutin, chlorogenic acid, a-tomatine	*H. zea*
Melilotus spp.	Coumarin	*Listroderes costirustris*
Momordica charantia	Momordicine II	*Aulacophora foveicollis* (Lucas)
Solanaceae (potato, tomato)	Demissine, solacauline, tomatine, leptine I and II	*Leptinotarsa decemlineata* (Say), *Manduca sexta* (Johannsen)
Solanum spp.	Tomatine, solanidine, a-chaconine	*Choristoneura fumiferana* (Clemens)
Solanum tuberosum	Tomatine	*Empoasca fabae* (Harris)
Sorghum bicolor	p-Hydroxybenzaldehyde, dhurrin, procyanidin	*S. graminum*
Triticum aestivum	Hydroxamic acid	*Metopolophism drhodum*
Zea mays	DIMBOA	*Ostrinia nubilalis* (Hubner)

Source: Panda and Kush (1995).

c. **Low resistance**: A low level of resistance indicates the possession of qualities which cause a variety to show lesser damage or infestation by an insect than the average for the crop under consideration.

d. **Susceptibility**: A susceptible variety is one which shows average or more than average damage caused by an insect.

e. **High susceptibility**: A variety shows high susceptibility when much more than average damage is done by the insect under consideration.

These terms are relevant to express resistance vis-a-vis screening of varieties under field conditions and have nothing to do with the mechanism of resistance. An intermediate level of resistance is, sometimes, referred to as **moderate resistance**.

Table 11: Examples of Insect Biotypes Involved in Host Plant Resistance to Insects

Insect Species	Common Name	Crop	Number of Biotypes
Acyrthosiphon pisum	Pea aphid	Alfa alfa	4
Amphorophora idaei	Raspberry aphid	Raspberry	4
Eriosoma lanigerum	Wooly apple aphid	Apple	3
Mayetiola destructor	Hessian fly	Wheat	11
Nephotettix virescens	Green leafhopper	Rice	3
Nilaparvata lugens	Brown planthopper	Rice	4
Orseolia oryzae	Gall midge	Rice	4
Rhopalosiphum maidis	Corn leaf aphid	Corn	5
Schizaphis graminum	Greenbug	SorghumWheat	57
Therioaphis maculate	Spotted alfalfa aphid	Alfalfa	6

Source: Panda and Kush (1995).

Chapter 5

Application of Biotechnology in Pest Management

Definition

Biotechnology can be broadly defined as "using living organisms or their products for commercial purposes." As such, biotechnology has been practiced by human society since the beginning of recorded history in such activities as baking bread, brewing alcoholic beverages, or breeding food crops or domestic animals. A narrower and more specific definition of biotechnology is "the commercial application of living organisms or their products, which involves the deliberate manipulation of their DNA molecules."

5.1 Transgenic Plants (Genetically Modified or GM Crops)

Transgenic technology can be utilized to develop plants with various beneficial traits such as a) Crop protection traits which include resistance to pests, diseases and herbicides. b) Abiotic stressin the form of tolerance to drought, heat, cold or salinity, thus enabling plants to be grown in inhospitable habitats, adding more land for cultivation; and c) Quality traitsleading to enhanced nutrition; prolonged shelf-life or improved taste, colour or fragrance of fruits, vegetables and flowers; and increased crop yield India made its long-awaited entry into commercial agricultural biotechnology when the Genetic Engineering Approval Committee (GEAC), Ministry of Environment and Forests, Govt of India, at its 32nd meeting held in New Delhi on 26th March 2002 approved three *Bt*-cotton hybrids for commercial cultivation. This is a historic decision as *Bt*-cotton became the first transgenic crop to receive such an approval in

India. These transgenic hybrids were developed by MAHYCO (Maharashtra Hybrid Seed Company Limited) in collaboration with Monsanto.

Combining DNA from different existing organisms (plants, animals, insects, bacteria, etc.) results in modified organisms with a combination of traits from the parents. The sharing of DNA information takes place naturally through sexual reproduction and has been exploited in plant and animal breeding programs for many years. However, sexual reproduction can occur only between individuals of the same species. What's new since 1972 is that scientists have been able to identify the specific DNA genes for many desirable traits and transfer only those genes, usually carried on a plasmid or virus, into another organism. This process is called genetic engineering and the transfer of DNA is accomplished using either direct injection or the *Agrobacterium*, electroporation, or particle gun transformation techniques. It provides a method to transfer DNA between any living cells (plant, animal, insect, bacterial, etc.). Virtually any desirable trait found in nature can, in principle, be transferred into any chosen organism. An organism modified by genetic engineering is called transgenic. A transgenic plant is simply a normal plant with one or more additional genes from diverse sources. Scientists have bred crops for resistance against insect pests for a long time. These transgenic plants produce insecticidal or antifeedant proteins continuously in the plants under field conditions. Before taking up any attempt to produce transgenic plants to counter insect attack, the following requirements and priorities need to be identified.

The factors for resistance should be controlled by single genes:

1. Standardization of methods for transfer of such genes can easily be accomplished.
2. Expression of transferred gene should occur in the desired tissues at the appropriate time.
3. The transgenic plant should be safe far consumption.
4. Inheritance of the gene in the successive generations should be very stable.
5. There should be no penalty for yield in terms of other quantitative characters.

5.1.1 *Bacillus thuringiensis* and Genetic Engineering

Plant Genetic Systems, a Belgian Biotechnology Company, in July 1987 were the first to report development of transgenic plants of tobacco containing delta-enddtoxin in enough amount to kill the first instar larvae of *Manduca sexta* (Johannsen) and *Heliothis virescens* (Fabricius) within 3 days. Plants infested with tobacco hornworm showed very limited feeding damage compared to non-transgenic control plants, which suffered heavy feeding damage and were almost completely defoliated within two weeks.

Transgenic plants carrying *Bt* genes have now been produced in a wide range of crop species including tobacco, tomato, potato, cotton, maize, rice, broccoli, oilseed rape, soybean, walnut, larch, poplar, sugarcane, apple, peanut, chickpea and alfalfa with different crystal protein genes. Many field trials of crops expressing these modified *Bt* genes have been carried out and most of these have produced impressive results, *e.g. CryIA* cotton against *H. zea* and *Pectinophora gossypiella* (Saunders), *Cry3A* potato against *Leptinotarsa decemlineata* (Say) and *CryIA* maize against O. *nubinalis.* Through extensive gene synthesis, promoter optimization and protein targeting, researchers have now improved expression of nuclear *cry* genes such that tobacco leaves can accumulate up to 0.8 per cent of their soluble protein as *Bt* toxin in about 10 per cent of the plants while the remaining transformants produced between 0.001-0.06 per cent of their soluble protein as toxin. Recently, it was reported that amplicization of an unmodified *CryIA (c)* coding sequence in chloroplasts up to 10000 copies per cell resulted in the accumulation of an unprecedented 3-5 per cent of the soluble protein in tobacco leaves as protoxin. The technique known as *plasmid transfonnation* resulted in highest level of toxin accumulation presently reported in plants. The plants were extremely toxic to larvae of *Heliothis virescens (Fabricius), Helicovepa zea* (Boddie) and *Spodoptera exigua* (Hubner) Also, the isolation of toxin genes from *B. thuringiensis* strains with specificity toward other classes is being pursued.

Table 12: Transgenic Plants Carrying *Bt* Genes for Insect Resistance

Plant	Gene(s)	Target Insect Pest (s)
Cotton	Cry 1A (b), Cry1A (c)	*Helicoverpa zea* (Boddie), *Pectinophora gossypiella* (Saunders), *Spodoptera exigua* (Hubner), *Trichoplusiani* (Hubner)
Egg plant	Cry (III)b	*Leptinotarsa decemlineata* (Say)
Maize	Cry1A (b)	*Ostrinia nubilalis* (Hubner)
Poplar	Cry 1 A	*Lymantria dispar* (Linnaeus)
Potato	Cry 3 A, Cry1A (b)	*L. decemlineata, Phthorimaea operculella* (Zeller)
Rice	Cry 1A (b), Cry1A (c)	*Chilo suppressalis* (Walker), *Cnaphalocrocis medinalis* (Guenee), *Scirpophaga incertulas* (Walker)
Sugarcane	Cry1A (b)	*Diatraea saccharalis* (Fabricius)
Tobacco	Cry1A (b)	*Heliothis virescens* (Fabricius), *H. zea, M. sexta*
Tomato	Cry1A (b) Bt (k)	*M. sexta, H. zea, M. sexta, Keifera lycopersicella*

Source: Lee *et al.,* 1998.

5.2 Protease Inhibitors

Proteinase inhibitors (PIs) occur frequently within the tissues of most plant families and have been ascribed several putative functions including that of conferring protection against phytophagous insect attack. Through specifically binding to proteolytic enzymes within the gut, forming highly stable complexes, they interfere with the digestive enzymes of pest insects causing reduced growth, increased levels of pest mortality and reduced plant damage (Ryan, 1990). In recent years, a large quantity of research has concentrated on producing genetically modified crops expressing genes encoding plant PIs for resistance to pest insects (Gatehouse and Gatehouse, 1999). To date, a number of PIs have been demonstrated to have potential for pest control when expressed in transgenic plants. For example, the serine proteinase inhibitor cowpea trypsin inhibitor (CpTI) has been shown to have a detrimental effect against lepidopteran and coleopteran pests when expressed in transgenic plants. Similarly, soybean Kunitz trypsin inhibitor (SKTI) has also been shown to be effective against a range of pests (Gatehouse *et al.,* 1999). Oryzacystatin (OC-1), an inhibitor of cysteine proteinases derived from rice, has similarly been demonstrated to have insecticidal effects against aphids, coleopteran insects and nematodes when expressed in crop plants, including potato and oilseed rape. Proteinase inhibitors expressed in transgenic plants, however, have the potential to adversely affect beneficial insects that feed on the target pest.

The possible role of protease inhibitors (PIs) in plant protection was investigated as early as 1947 when, Mickel and Standish observed that the larvae of certain insects were unable to develop normally on soybean products. Subsequently the trypsin inhibitors present in soybean were shown to be toxic to the larvae of flour beetle, *Tribolium confusum* (Lipke *et al.,* 1954). Following these early studies, there have been many examples of protease inhibitors active against certain insect species, both in *in vitro* assays against insect gut proteases. The term "protease includes both endopeptidases" and "exopeptidases" whereas, the term "proteinase" is used to describe only endopeptidases. Several non-homologous families of proteinase inhibitors are recognized among the animal, microorganisms and plant kingdom. Majority of proteinase inhibitors studied in plant kingdom originates from three main families namely leguminosae, solanaceae and gramineae.

These protease inhibitor genes have practical advantages over genes encoding for complex pathways *i.e.*by transferring single defensive gene from one plant species to another and expressing them from their own wound inducible or constitutive promoters thereby imparting resistance against insect pests. This was first demonstrated by Hilder *et al.,* 1987 by transferring trypsin inhibitor gene from *Vigna unguiculata* to tobacco, which conferred resistance to wide range of insect pests including lepidopterans, such as *Heliothis* and *Spodoptera*, coleopterans such as *Diabrotica, Anthonomnous* and orthoptera such as *Locusts*. Further, there is no evidence that it had toxic or deleterious effects on mammals. Many of these protease inhibitors are rich in cysteine and lysine, contributing to better and enhanced nutritional quality (Ryan, 1989). Protease inhibitors also exhibit a very broad spectrum of activity including suppression of pathogenic nematodes like *Globodera tabaccum, G. pallida,* and *Meloidogyne incognita* by CpTi (Williamson and Hussey, 1996), inhibition of spore germination and mycelium growth of *Alternaria alternata* by buckwheat trypsin/chymotrypsin (Dunaevskii *et al.,* 1997) and cysteine PIs from pearl millet inhibit growth of many pathogenic fungi including *Trichoderma reesei* (Joshi *et al.,* 1998). These advantages make protease inhibitors an ideal choice to be used in developing transgenic crops resistant to insect pests. Further, transformation of plant genomes with PI-encoding cDNA clones appears attractive not only for the control of plant pests and pathogens, but also as a means to produce PIs, useful in alternative systems and the use of plants as factories for the production of heterologous proteins (Sardana *et al.,* 1998). These inhibitor families that have been found are specific for each of the four mechanistic classes of proteolytic enzymes, and based on the active amino acid in their "reaction center" (Koiwa *et al.,* 1997), are classified as serine, cysteine, aspartic and metallo-proteases.

5.3 Serine Proteinase Inhibitors

The role of serine PIs as defensive compounds against predators is particularly well established, since the major proteinases present in plants, used for processes such as protein mobilization in storage tissues, contain a cysteine residue as the catalytically active nucleophile in the enzyme active site. Serine proteinases are not used by plants in processes involving large scale protein digestion, and hence the presence of significant quantities of inhibitors with specificity towards these enzymes in plants cannot be used for the purposes of regulating endogenous proteinase activity (Reeck *et al.,* 1997). In contrast, a major role for serine PIs in animals is to block the activity of endogenous proteinases in tissues where this activity would be harmful, as in case of pancreatic trypsin inhibitors found in mammals. The serine class of proteinases such as trypsin, chymotrypsin and elastase, which belong to a common protein superfamily, are responsible for the initial digestion of proteins in the gut of most higher animals (GarciaOlmedo *et al.,* 1987). *In vivo* they are used to cleave long, essentially intact polypeptide chains into short peptides which are then acted upon by exopeptidases to generate amino acids, the end products of protein digestion. These three types of digestive serine proteinases are distinguished based on their specificity, trypsin specifically cleaving the C-terminal to residues carrying a basic side chain (Lys, Arg), chymotrypsin showing a preference for cleaving C-terminal to residues carrying a large hydrophobic side chain (Phe, Tyr, Leu), and elastase showing a preference for cleaving C-terminal to residues carrying a small neutral side chain (Ala, Gly) (Ryan, 1990). Inhibitors of these serine proteinases have been described in many plant species, and are universal throughout the plant kingdom, with trypsin inhibitors being the most common type. At least, part of this bias can be accounted for by the fact that (mammalian) trypsin is readily available and is the easiest of all the proteinases to assay using synthetic substrates, and hence is used in screening procedures. Because of these reasons the members of the serine class of proteinases have been the subject of intense research than any other class of proteinase inhibitors. Such studies have provided a basic understanding of the mechanism of action (Huber and Carrell, 1989) that applies to most serine proteinase inhibitor families and probably to the cysteine and aspartyl proteinase inhibitor families as well. All serine inhibitor families from plants are competitive inhibitors and all of them inhibit proteinases with a similar standard mechanism (Laskowski and Kato, 1980).

Serine proteinases have been identified in extracts from the digestive tracts of insects from many families, particularly those of lepidoptera (Houseman *et al.,* 1989) and many of these enzymes are inhibited by proteinase inhibitors. The order lepidoptera, which includes a number of crop pests, the pH optima of the guts are in the alkaline range of 9-11 (Applebaum, 1985) where, serine proteinases and metallo-exopeptidases are most active. Additionally, serine proteinase inhibitors have anti-nutritional effects against several lepidopteran insect species (Shulke and Murdock, 1983; Applebaum, 1985). Purified Bowman-Birk trypsin inhibitor (Brovosky, 1986) at 5 per cent of the diet inhibited growth of these larvae but SBTI (Kunitz, 1945), another inhibitor of bovine trypsin, was less effective when fed at the same levels.

Broadway and Duffey (1986a) compared the effects of purified SBTI and potato inhibitor II (an inhibitor of both trypsin and chymotrypsin) on the growth and digestive physiology of larvae of *Heliothis zea* and *Spodoptera exigua* and demonstrated that growth of larvae was inhibited at levels of 10 per cent of the proteins in their diet. Trypsin inhibitors at 10 per cent of the diet were toxic to larvae of the *Callosobruchus maculatus* (Gatehouse and Boulter, 1983) and *Manduca sexta* (Shulke and Murdock, 1983).

Recent X-ray crystallography structure of winged bean, *Psophocarpus tetragonolobus* Kunitz-type double headed alpha-chymotrypsin shows 12 anti-parallel beta strands joined in a form of beta trefoil with two reactive site regions (Asn 38-Leu 43 and Gln 63-Phe 68) at the external loops (Ravichandaran *et al.,* 1999; Mukhopadhyay, 2000). Structural analysis of the Indian finger millet *(Eleusine coracana)* bifunctional inhibitor of alpha-amylase/trypsin with 122 amino acids has shown five disulphide bridges and a trypsin binding loop (Gourinath *et al.,* 2000). These structural analysis would greatly help in "enzyme engineering" of the native PIs to a potent form, against the target pest species than the native PIs.

5.4 Cysteine Proteinase Inhibitors

Isolation of the midgut proteinases from the larvae of cowpea weevil, *C. maculatus* (Kitch and Murdock, 1986; Campos *et al.,* 1989) and bruchid *Zabrotes subfaceatus* (Lemos *et al.,* 1987) confirmed the presence of cysteine mechanistic class of proteinase inhibitors. Similar proteinases have been isolated from midguts of the flour beetle *Tribolium castaneum*, Mexican beetle *Epilachna varivestis* (Murdock *et al.,* 1987) and the bean weevil *Ascanthoscelides obtectus* (Wieman and Nielsen, 1988). Cysteine proteinases isolated from insect larvae are inhibited by both synthetic and naturally occurring cysteine proteinase inhibitors (Wolfson and Murdock, 1987). In a study of the proteinases, from the midguts of several members of the order coleopteran, 10 of 11 beetle species representing 11 different families, had gut proteinases that were inhibited by p-chloromercuribenzene sulfonic acid (PCMBS), a potent sulphydryl reagent (Murdock *et al.,* 1988) indicating that the proteinases were of the cysteine mechanistic class. The optimum activity of cysteine proteinases is usually in the pH range of 5-7, which is the pH range of the insect gut that use cysteine proteinases (Murdock *et al.*, 1987). Another puzzling aspect of studies with *C. maculatus* is the apparent effects of certain members of Bowman-Birk trypsin inhibitor family on the growth and development of these larvae. Although cysteine proteinase is primarily responsible for protein digestion in *C. maculatus*, it is not clear, how the cowpea and soybean Bowman-Birk inhibitors are exert their anti-nutritional effects on this organism. Advances in enzymology has revealed the existence of a variety of cysteine proteinases resulting in their classification into several families namely papain, calpin and asparagines specific processing enzyme (Turk and Bode, 1991). Cystanins have also been characterized from potato (Waldron *et al.*, 1993), ragweed (Rogers *et al.*, 1993), cowpea (Fernandes *et al.*, 1993) papaya (Song *et al.*, 1995) and avacado (Kimura *et al.*, 1995).

The rice cysteine proteinase inhibitors are the most studied of all the cysteine PIs which is proteinaceous in nature (Abe and Arai, 1985) and highly heat stable (Abe *et al.*, 1987). Recent three dimensional structure analysis of oryzacystatin OC-I by Tanokura's group (Nagata *et al.*, 2000), using NMR has showed a well defined main body consisting of amino acids from Glu 13–Asp 97 and an alpha helix with five stranded anti parallel beta-sheet, while the N terminus (Ser 2-Val 12) and C terminus (Ala 98-Ala 102) are less defined. Further, analysis has demonstrated OC-I to be similar to chicken cystatin which belongs to type-2 animal cystatin. Another rice cystatin named as OC-II, with a putative target binding motif gln-x-val-x-gly shares similar motif with OC-I but has a different inhibition constant (Ki) value (Arai *et al.*, 1991).

5.5 Aspartic and Metallo-Proteinase Inhibitors

Knowledge on the role of aspartic proteinases in insect digestion is limited than that of cysteine proteinases. In species of six families of the order hemiptera, aspartic proteinases (cathepsin D-like proteinases) were found along with cysteine proteinases (Houseman and Downe, 1983). The low pH of midguts of many members of coleoptera and hemiptera provides more favourable environments for aspartic proteinases (pH optima ~ 3-5) than the high pH of most insect guts (pH optima ~ 8-11) where the aspartic and cysteine proteinases would not be active. No aspartic proteinases have been isolated from coleoptera but Wolfson and Murdock 1987 demonstrated that pepstatin, a powerful and specific inhibitor of aspartyl proteinases, strongly inhibited proteolysis of the midgut enzymes of Colorado potato beetle, *Leptinotarsa decemlineata* indicating that an aspartic proteinase was present in the midgut extracts. Potato tubers possess an aspartic proteinase inhibitor, cathepsin D (Mares *et al.*, 1989) that shares considerable amino acid sequence identity with the trypsin inhibitor SBTI from soybeans. Plants have also evolved at least two families of metallo-proteinase inhibitors, the metallo-carboxypeptidase inhibitor family in potato (Rancour and Ryan, 1968), and tomato plants and a cathepsin D inhibitor family in potatoes (Keilova and Tomasek, 1976).

The cathepsin D inhibitor (27 kDa) is unusual as it inhibits trypsin and chymotrypsin as well as cathepsin D, but does not inhibit aspartyl proteases such as pepsin, rennin or cathepsin E. The inhibitors of the metallo-carboxypeptidase from tissue of tomato and potato are polypeptides (4 kDa) that strongly and competitively inhibit a broad spectrum of carboxypeptidases from both animals and microorganisms, but not the serine carboxypeptidases from yeast and plants (Havkioja and Neuvonen, 1985). The inhibitor is found in tissues of potato tubers where it accumulates during tuber development along with potato inhibitor I and II families of serine proteinase inhibitor. The inhibitor also accumulates in potato leaf tissues along with inhibitor I and II proteins in response to wounding (Hollander-Czytko *et al.,* 1985). Thus, the inhibitors accumulated in the wounded leaf tissues of potato have the capacity to inhibit all the five major digestive enzymes *i.e.* trypsin, chymotrypsin, elastase, carboxypeptidase A and carboxypeptidase B of higher animals and many insects (Hollander-Czytko *et al.,* 1985). Aspartic PIs have been recently been isolated from sunflower (Park *et al.,* 2000), barley (Kervinen *et al.,* 1999) and cardoon *(Cyanara cardunculus)* flowers named as cardosin A (Frazao *et al.,* 1999).

The detailed structural analysis of prophytepsin, a zymogen of barley aspartic proteinase shows a pepsin like bilobe and a plant specific domain. The N terminal has 13 amino acids necessary for inactivation of the mature phytepsin (Kervinen *et al.,* 1999), and the aspartic PI cardosin A from cardoon shows regions of glycolylations at Asn-67 and Asn 257. The Arg-Gly-Asp sequences recogonizes the cardosin receptor which is found in a loop between two-beta strands on the molecular surface (Frazao *et al.,* 1999).

Mechanism of Toxicity

The mechanism of action of these proteinase inhibitors has been a subject of intense investigation (Barrett, 1986; MacPhalen and James, 1987; Greenblatt *et al.,* 1989). Knowledge on mechanisms of protease action and their regulation *in vitro,* and *in vivo,* in animals, plants, microorganisms and more recently in viruses have contributed to many practical applications for inhibitor proteins in medicine and agriculture.

Baker *et al.,* 1984 showed that the secretion of proteases in insect guts depends upon the midgut protein content rather than the food volume. The secretion of proteases has been attributed to two mechanisms, involving either a direct effect of food components (proteins) on the midgut epithelial cells, or a hormonal effect triggered by food consumption (Applebaum, 1985). Models for the synthesis and release of proteolytic enzymes in the midguts of insects proposed by Birk and Applebaum, 1960, Brovosky, 1986reveal that ingested food proteins trigger the synthesis and release of enzymes from the posterior midgut epithelial cells. The enzymes are released from membrane associated forms and sequestered in vesicles that are in turn associated with the cytoskeleton. The peptidases are secreted into the ectoperitrophic space between the epithelium, as a particulate complex (Eguchi *et al.,* 1982), from where the proteases move transversely into the lumen of the gut, where the food proteins are degraded. PIs inhibit the protease activity of these enzymes and reduce the quantity of proteins that can be digested, and also cause hyper-production of the digestive enzymes which enhances the loss of sulfur amino acids (Shulke and Murdock, 1983) as a result of which, the insects become weak with stunted growth and ultimately die.

The digestive proteolytic enzymes in the different orders of commercially important insect pests belong to one of the major classes of proteinases predominantly. Coleopteran and hemipteran species tend to utilize cysteine proteinases (Murdock *et al.,* 1987) while lepidopteran, hymenopteran, orthopteran and dipteran species mainly use serine proteinases (Ryan, 1990; Wolfson and Murdock, 1990). Examples from both of these classes of proteinases have been shown to be inhibited by their cognate proregions (Taylor *et al.,* 1995). The effect of class specific inhibitors on the pest digestive enzymes is not always a simple inhibition of proteolytic activity, but recent studies have indicated the reverse may happen. It would appear that there are often two populations of digestive enzymes in target pests, those that are susceptible to inhibition and those that are resistant (Bown *et al.,* 1997), and some insects respond to ingestion of plant PIs such as soybean trypsin inhibitor (Broadway and Duffey, 1986b) and oryzacystatin by hyper-producing inhibitor-resistant enzymes.

The mechanism of binding of the plant protease inhibitors to the insect proteases appear to be similar with all the four classes of inhibitors. The inhibitor binds to the active site on the enzyme to form a complex with a very low dissociation constant (10^7 to 10^{14} M at neutral pH values), thus effectively blocking the active site. A binding loop on the inhibitor, usually "locked" into conformation by a disulphide bond, projects from the surface of the molecule and contains a peptide bond (reactive site) cleavable by the enzyme (Terra *et al.,* 1996; Walker *et al.,* 1998). This peptide bond may be cleaved in the enzyme inhibitor complex, but cleavage does not affect the interaction, so that a hydrolyzed inhibitor molecule is bound similar to an unhydrolyzed one. The inhibitor thus directly mimics a normal substrate for the enzyme, but does not allow the normal enzyme mechanism of peptide bond cleavage to proceed to completion *i.e.,* dissociation of the product (Walker *et al.,* 1998). It would also appear that insect digestive trypsins do not fall into the classification of peptidase hydrolases, as defined by inhibition spectra. It has been shown, notably, that the trypsin like digestive proteases of several lepidopteran species are inhibited by (l-3-trans carboxiran-2-carbonyl)-l-leu-agmatin (E-64) (Novillo *et al.,* 1997) an inhibitor generally considered to be specific to cysteine proteinases (Dunn, 1989). Thus, true interactions will become clear only when we have protein crystals and X-ray diffraction data for the structure of insect enzyme/inhibitor complexes. Further, specificity of the inhibitor enzyme interaction is primarily determined by the specificity of proteolysis determined by the enzyme (Blancolabra *et al.,* 1996).

5.6 Lectins

Lectins have deleterious effects against larvae, developing stages and mature forms of insects from orders Coleoptera, Diptera, Hemiptera, Homoptera,

Hymenoptera, Isoptera, Lepidoptera and Neuroptera. Insecticide activity of lectin is generally evaluated by bioassays that incorporate the lectin into artificial diets offered to insects, with insects dying from nutritional deprivation. It has been shown that lectins are resistant to proteases present in the insect gut, a property responsible for their active presence in the digestive tract, eventually with insecticide effects. The precise mechanisms of Insecticidal action of lectins remain unknown, though it has been suggested that this entomotoxic activity seems to depend upon the carbohydrate recognition property they exhibit. Plant lectins with affinity for N-acetylglucosamine and chitin-binding property are able to bind chitin and glycosylated proteins of the peritrophic matrix, interfering in the digestion and absorption of nutrients. The peritrophic matrix constitutes a membrane found in the midgut that separates the contents of the gut lumen from the digestive epithelial cells. The matrix contains a network composed by chitin (polymer of N-acetylglucosamine) and glycoproteins such as peritrophins. The importance of the integrity of the peritrophic matrix lies in the protection it offers to midgut epithelial cells against microorganism infection and mechanical damage by abrasive food particles, as apart from the compartmentalization of digestive processes. Ultrastructural studies have shown abnormalities caused by Triticum vulgaris lectin in midgut of Ostrinia nubilalis and Drosophila such as hypersecretion of many disorganized layers of peritrophic matrix and morphological changes of microvilli. Lectin may also cross the midgut epithelial barrier by transcytosis, entering the insect circulatory system and resulting in a toxic action against endogenous lectins involved in haemolymph self-defense mechanisms. Lectin may be internalized by endocytotic vesicles into the epithelial cells, blocking nuclear localization and nuclear sequence-dependent protein import, thus inhibiting cell proliferation.

5.7 Trypsin Modulating Oostatic Factor

Trypsin Modulating Oostatic Factor (TMOF) is a new, proprietary approach to controlling pest insect populations without the undesired side effects of current synthetic chemical insecticides or the insect resistance problems anticipated for *Bacillus thurengensis* based insecticides. TMOF is a ten amino acid peptide first isolated from mosquito ovaries. The normal function of TMOF is to signal the epithelial cells lining the midgut, to cease making digestive proteases, when sufficient essential amino acids have been borne to the ovaries to complete egg protein synthesis. When administered to mosquitoes and other insects the resultant lack of trypsin-like digestive enzymes results in no protein digestion, leading to reproductive failure, starvation and death. TMOF is active at low concentrations (nanograms per insect). Because of its unique mode of action, development of resistance by the target insect is expected to be an extremely low frequency event. Egg development and blood digestion in the mid gut of female mosquito are two inter related processes regulated by complex series of hormonal interactions. Following blood feeding, JH and an ovarian factor corpus cardiacum stimulating factor (CCSF) are released. CCSF releases egg development neurosecretory hormone (EDNH), which stimulates the ovary to synthesize and release ecdysone. Ecdysone is converted into 20-OH-ecdysone by the mosquito fat body an din concert with JH, induces the fat body to synthesis vitellogenin. At the same time, the blood meal induces a *de novo* Synthesis of trypsin like enzyme in the mid gut of the female mosquito Thus, stopping any of these processes may cause arrest of egg development causing sterility.

5.8 Tissue Culture of Insecticidal Plants

The various application of *Azadirachtin*, neem based triterpenoid in the pharmaceutical and agricultural industries needs a large production of high azadirachtin containing seeds, due to numerous applications of neem derived products in various cosmetics, insecticidal, nematicidal products increases the demand of neem seed production. It is therefore necessary to have large scale neem plantation, which can be achieved through tissue culture. Tissue Culture would help in establishing a better productive plantation with uniform, genetically superior, high azadirachtin containing planting material.

5.9 Challenging of Cell Lines–*In vitro* Replication of Insect Viruses

In vitro production of. insect virus in cell lines has got several advantages. Cell cultured viruses are free from undesirable viral or microbial contamination, secondary biotypes and insect debris. In insect cell line, virus production can be monitored easily. It is easy to maintain cell lines continuously for virus production. Large scale production is possible using fermentorslbioreactors. Baculovirus replication in insect cell line has been reported for several nuclear polyhedrosis viruses.

Preparation of Virus Inoculum

Fourth instar larvae are inoculated with the homologous NPV by allowing them to feed on diet surface contaminated with 106 polyhedra per bottle. After three days of inoculation, the larvae are washed in sterile distilled water and thl'J haemolymph is drawn by cutting the prolegs and collected in 10ml of TNM-FH medium containing cysteine to prevent melanisation. The medium is then centrifuged at 780 g for 10 min to sediment haemocytes and passed through a 0.45 mm millipore membrane filter under a laminar hood.

Challenging of Cell Lines

At log phase of the cell line, the spent medium is removed and 2.5 ml of medium containing virus inoculum is added to the cells and an adsorption time of two hours is allowed with periodical rocking of flasks at an interval of 10 min. The medium is then removed. The cells washed twice gently with sterile Hanks balanced salt solution (HBSS). Finally, after removing the HBSS completely, five ml of TNM-FH medium containing 10 per cent foetal bovine serum is added and incubated at 27.5°C. The cells were observed daily for virus infection under a microscope. Virus can be harvested after five to six days.

Identification of *in vitro*–Cultured Viruses

The cell cultured virus can be subjected to DNA analysis using restriction endonucleases. The methodology is described elsewhere. The DNA profiles of cell cultured virus and *in vivo* produced virus can be compared. Similarity in the DNA profiles of both viruses will indicate that in cell line the virus has not undergone any recombination or mutation.

Biological Activity of *in vitro* Cultured Viruses

The efficacy of cell cultured viruses has to be tested by conducting bioassays. The virus infected cells were harvested 10 days after inoculation and stored at -20°C. Before processing, the cells were thawed out and sonicated in ice bath using a Branson Sonifier 450 to disrupt the cells and release the polyhedra. The cell debris were removed by centrifugation at 200 x g for one min. The supernatant was again centrifuged at 780 x g for five min. to pellet the virus. The polyhedra were suspended in distilled water and the concentration assessed with the help of a haemocytometer under phase contrast microscope.

The virus harvested from the cell line can be bioassayed to compare their biological activity with that of the *in vivo* produced virus. A standard diet surface contamination method is followed starting with a dose of 3.5×10^5 polyhedral occlusion bodies (POB) ml^{-1}. Seven-fold serial dilutions are made to give five doses each. Chickpea seed based-sem: -syhthetic diet is dispensed into sterile five ml glass vials and allowed to cool down to room temperature under a laminar hood. After 2-3h, an aliquot of 10 ml of the different viral dilutions are applied on the diet and spread uniformly over the entire diet surface using a four mm glass rod with rounded tip. After 15 min., one larva of second instar is introduced into each vial and plugged with. sterile cotton. Thirty five to forty larvae are used in each dose. A similar set of larvae without virus treatment is maintained as control. Observations on the larval mortality is recorded daily for ten days and data are subjected to probit analysis. LC50 values of the two viruses can be compared.

5.10 Application of PCR Techniques in Entomological Research

The evolution of Polymerase Chain Reaction (PCR), dates back to 1985 just a decade after the invention of gene cloning. Since its conception by Kary B. Mullis, this novel approach has been put into use to study the biological processes in the field of Entomology and allied sciences. In Insect Pathology, PCR is used for the molecular characterisation of insect pathogens such as viruses, bacteria, fungi and nematodes. Following are some applications of PCR to different groups of insect pathogens.

1. **Insect viruses:** PCR based methods facilitate investigation of phylogeny and grouping of insect viruses by allowing the determination of the nucleotide sequence of a specific gene from any virus isolate. These studies help in the understanding of biological adaptations such as virus host specificity (Bulach *et al.,* 1999). The DNA polymerase and polyhedrin gene are the target sequences for

amplification. PCR amplification of specific gene of virus aids in detection of viral DNA in infected insects at early stages of infection, cross infectivity, latent infection and ecological studies. PCR is used for identification of expression of rer.ombinant baculoviruses in insect cell lines.

2. **Bacteria :** PCR is widely used in the characterisation of *Bacillus thuringiensis (B.t.)*. Brousseau *et al.* (1993) developed a methodology of arbitrary primers (AP) PCRto distinguish commercial strains and to differentiate *B.t* from *B. cereus.* The purity of commercial *B.t.* products can be checked using AP-PCR. A modification of PCR namely Exclusive PCR or E-PCR has been developed to detect novel cry genes by using mixtures of degenerate and specific primers. Electrophoretic patterns of PCR products using specific primers of cry genes predicts the insecticidal activity of *B.t.* strains.Besides, PCR techniques are used in the detection of *B.t.* strains in the environment and its expression in transgenic plants.

3. **Entomopathogenic fungi:** PCR offers objective and reliable methods for identification of fungal isolates which overcome the limitations of RFLP and RAPD analysis in the classification of entomopathogenic fungi. Combination of PCR with RAPD and SSCP (Single Strain DNA Confirmation Polymorphism) characterise and identify geographic isolate variation of *Beauveria* spp. (Hegedus and Khachatourians, 1996). Geonmic DNA, mitochondrial DNA (mtDNA) and ribosomal DNA (rDNA) are used for synthesis of primer sequence for amplification. Reverse Transcriptase- Differential Display (RT-DD-PCR) technique was used to detect and characterise Prl gene and chit1 gene of *Metarhizium anisopliae* responsible for the pathogenicity of the fungi. Besides, PCR techniques are used for early detection of fungal disease, development and cloning of pathogenesis related gene (Pri gene) from virulent strain in non-virulent strain of fungi.

Chapter 6

Chemical Control

6.1 Trends in Pesticide Consumption

Synthetic pesticides have been extensively used in developing countries mostly after the adoption of green revolution and the control of vector borne diseases. By the early 1980's the developing countries were thought to use 10 – 25 per cent of the world pesticide production. However, about 1/3rd of the crops were still lost to pests each year and malaria alone affected 100 million people annually. By 1990, the third world countries used 26 per cent of the world pesticide production. Around 55 per cent of agricultural land situated in these countries is related to much lower consumption of pesticides than developed countries. Taiwan tops the list using 17 kg a.i./ha followed by Japan (12 kg a.i./ha) and Africa, the least with 0.13 kg a.i./ha, while India used 0.57 kg a.i./ha in the year 1998. Many developing countries including India, China, Bangladesh and Indonesia are participating in the global expansion of agricultural output. The value of pesticide imports to Asia increased three fold between 1970s to 1980s. The fastest growing pesticide markets are India, Brazil, China and Spain. Of the total world pesticide production 24 per cent reach the developing countries, 12 per cent goes to Asia, 8 per cent to Latin America and 4 per cent to Africa.

The total amount of pesticides used in India increased from 2.35 thousand metric tones in 1950-51 to nearly 85,000 metric tones in 1993-94. Earlier projections had put the pesticide demand by 2000 at nearly 1-lakh metric tones. But in view of the high priority being accorded to IPM, the pesticide consumption has shown a decreasing trend in the recent years. In Tamil Nadu the synthetic pesticide consumption has decreased by more than 50 per cent during last 7 years, a decreasing trend has also been recorded in Andhra Pradesh and Karnataka. In contrast pesticide consumption continues to rise rapidly in Punjab and Rajasthan. In 1992, the world consumption of herbicides was 44 per cent, insecticides 30 per cent, fungicides 20 per cent and others 6 per cent of the total pesticide consumption compared to 77 per cent insecticides, 12 per cent herbicides, 8 per cent fungicides and 3 per cent others in India. When cotton utilized 54 per cent, rice 17 per cent, cereals and millets 6 per cent, and others 235 of total pesticide consumption in 1979 in India it was 39 per cent in cotton, 35 per cent in rice, 17 per cent in cereals and millets and 9 per cent in others in the year 1988. In contrast to this, 27 per cent of the total pesticides were used in horticulture, 17 per cent in rice, 24 per cent in cotton, 7 per cent in maize and 25 per cent in others in the world in 1992. The world market on pesticides is estimated to grow @ 4.5 per cent each year with the largest growth occurring in herbicides. The average growth rate in Asia Pacific region is approximately 5–7 per cent, but in Indonesia and Pakistan, the market is expanding @ 20-30 per cent per annum. Along with the increase in the amount of pesticide consumption, there is a change in the potency of some new chemicals observed in recent past. DDT was applied @ 1-2kg a.i./ha for the control of different pests, the organophosphates in general are effective @ 250- 500 g a.i./ha, synthetic pyrethroids @ 12.5- 100 g a.i./ha. Some recently developed chemicals like nitroguanidines are effective @ 25 g a.i./ha. Thus there has been more than 100-fold increase in the potency of new insecticides.

6.2 Pesticide Definition

Pesticide is defined as any substance, or mixture of substances, or micro-organisms including viruses, intended for repelling, destroying or controlling any pest, including vectors of human or animal disease, nuisance pests, unwanted species of plants, or animals causing harm during or otherwise interfering with the production, processing, storage, transport, or marketing of food, agricultural commodities, wood and wood products or animal feeding stuffs, or which may be administered to animals for the control of insects, arachnids or other pests in or on their bodies. Control of insects with chemicals is known is chemical control.The term includes substances intended for use as insect or plant growth regulators; defoliants; desiccants; agents for setting, thinning or preventing the premature fall of fruit; and substances applied to crops either before or after harvest to protect the commodity from deterioration during storage and transport. The term also includes pesticide synergists and safeners, where they are integral to the satisfactory performance of the pesticide.

Properties of an Ideal Insecticide or Pesticide

1. It should be freely available in the market under different formulations.

2. It should be toxic and kill the pest required to be controlled.

3. It should not be phytotoxic to the crops on which it is used.

4. It should not be toxic to non target species like animals, fish, natural enemies etc.

5. It should be less harmful to human beings and other animals.

6. Should not leave residues in crops like vegetables.

7. It should have wide range of compatibility.

8. Should possess quick known down effect.

9. Should be stable on application.

10. Should not possess tainting effects and hould be free from offensive odour.

11. Should be cheaper.

6.3 Different Classifications of Insecticides

Insecticides are classified in several ways taking into consideration their origin, mode of entry, mode of action and the chemical nature of the toxicant.

6.3.1 Based on the Origin and Source of Supply

A. **Inorganic Insecticides**: Comprise compounds of mineral origin and elemental sulphur. This group includes arsenate and fluorine compounds as insecticides. Sulphur as acaricides and zinc phosphide as rodenticides.

B. **Organic Insecticides**: 1. Insecticides of animal origin: Nereistoxin isolated from marine annelids, fish oil rosin soap from fishes.

2. Plant Origin insecticides or Botanical insecticides: Nicotinoids, pyrethroids, Rotenoids etc.

3. Synthetic organic insecticides: Organochlorines, Organophosphorous, Carbamate insecticides etc.

4. Hydrocarbon oils etc.

6.3.2 Based on the Mode of Entry of the Insecticides into the Body of the Insect

A. **Contact poisons**: These insecticides are capable of gaining entry into the insect body either through spiracles and trachea or through the cuticle itself. Hence, these poisons can kill the insects by mere coming in contact with the body of the insects. *e.g.* DDT and HCH.

B. **Stomach poisons**: The insecticides applied on the leaves and other parts of plants when ingested act on the digestive system of the insect and bring aboutthe kill of the insect. *e.g.*, Calcium arsenate, lead arsenate

C. **Fumigants**: A fumigant is a chemical substance which is volatile at ordinary temperatures and sufficiently toxic to the insects. Fumigation is the process of subjecting the infested material to the toxic fumes or vapours of chemicals or gases which have insecticidal properties. Chemical used in the fumigant and a reasonably airtight container or room is known as fumigation chamber or "Fumigatorium". Fumigants mostly gain entry into the body of the insect through spiracles in the trachea.

Commonly used Fumigants and their doses

1. Aluminium phosphide, marketed as Celphos tablets used against field rats, groundnut bruchids etc.
2. Carbon disulphide 8-20 lbs/1000 ft³ of food grains.
3. EDCT (Ethylene Dichloride Carbon Tetrachloride) 20-30 lbs/1000cft of food grains
4. EDB Ethylene dibromide 1 lb/1000 ft³ of food grains.
5. SO2: By burning sulphur in godowns SO$_2$ fumes are released.

D. **Systemic insecticides**: Chemicals that are capable of moving through the vascular systems of plants irrespective of site of application and poisoning insects that feed on the plants. Ex: Methyl demeton, Phosphamidon, Acephate 'Non systemic insecticides' are not possessing systemic action are called non systemic insecticides. Somenon systemic insecticides, however, have ability to move from one surface leaf to the other. They are called as 'trans laminar insecticides'. *e.g.* Malathion, Diazinon, spinosad etc.

6.3.3 Based on Mode of Action

1. **Physical poisons**: Bring about the kill of insects by exerting a physical effect. *e.g.,* Heavy oils, tar oils etc. which cause death by asphyxiation. Inert dusts effect loss of body moisture by their abrasiveness as in aluminium oxide or absorb moisture from the body as in charcoal.
2. **Protoplasmic poisons**: A toxicant responsible for precipitation of protein especially destruction of cellular protoplasm of midgut epithelium. *e.g.* Arsenical compounds.
3. **Respiratory poisons**: Chemicals which block cellular respiration as in hydrogen cyanide (HCN), carbon monoxide etc.
4. **Nerve poisons**: Chemicals which block Acetyl cholinesterase (AChE) and effect the nervous system. *e.g.* Organophosphorous, carbamates.
5. **Chitin Synthesis inhibitors**: Chitin inhibitors interfere with process of synthesis of chitin due to which normal moulting and development is disrupted. *e.g.* Novaluron, Diflubenzuran, Lufenuron, Buprofezin
6. **General Poisons**: Compounds which include neurotoxic symptoms after some period and do not belong to the above categories. *e.g.* Chlordane, Toxaphene, aldrin

6.3.4 Based on Stage Specificity

1. Ovicides, 2. Larvicides, 3. Pupicides, 4. Adulticides

6.3.5 Based on Toxicity

Sl.No.	Classification	Symbol and Colour	Oral LD50	DermalLD50
1.	Extremely toxic	Skull and Pioson, Red	1-50	1-200
2.	Highly toxic	Danger and Pioson, Yellow	51-500	201-2000
3.	Moderately toxic	Caution and Blue	501-5000	2001-20,000
4.	Less toxic	Green	>5000	>20,000

6.3.6 Generation wise Classification Pesticides

Sl.No.	Generation	Pesticides
1.	First generation	Inorganic pesticides like Lead arsenate
2.	Second generation	Chlorinated hydrocarbons like DDT
3.	Third generation	Juvenile hormones
4.	Fourth generation	Anti Juvenile hormones

6.4 Classification of Insecticides based on Chemical nature

6.4.1 Inorganic Insecticides

1. **Arsenic compounds**: In an arsenical compound, the total arsenic content and the water soluble arsenic content are of importance, the water solubility of arsenic may result in entering the foliage and causing burning injury to plants, and hence water insoluble compounds are preferred for insect control. Arsenates are more stable and safe for application on plants then arsenites. Arsenites are mainly used in poison baits since they are phytotoxic. However arsenates are less toxic to insects then arsenites. In insects arsenates cause regurgitation, torpor (sluggishness) and quiescence. Disintegration of epithelial cells of the midgut and clumping of the chromatin of the nuclei are the effects noticed in poisoned insects. Slow decrease in oxygen consumption is also evident and kill of the insect is primarily due to the inhibition of respiratory enzymes. Water soluble arsenic causes wilting followed by browning and shriveling of the tissue.

   a. **Calcium arsenate**: It was first used by about 1906 as an insecticide. It is a white flocculent power, formulated as a dust of 25 to 30 per cent metallic arsenic equivalent. Dosage – Calcium arsenate at 0.675 to 1.350 kg with equal quantity of slaked lime in 450 litres of water. LD_{50} for mammals oral – 35 to 100. Being a stomach poison it was mainly used for control of leaf eating insects.

   b. **Lead arsenate**: It was first used as in insecticide in 1892 for the control of gypsy moth. It is a stomach poison with little contact action LD_{50} for rat oral 10-100, dermal 2400 mg/kg. It is rarely used as dust. 450 g to 1800 g of load arsenate is diluted with 200-240 litres of water.

    c. **Arsenite Paris green**: It is a double salt of copper acetate and copper arsenite. It was first used in 1867 for the control of Colorado potato beetle, *Leptinotarsa decemlineata*. It is now used as bait for the control of slugs. LD_{50} for rat oral- 22 mg/kg. Very good against termites

2. **Flourine Compounds**: These compounds were used since 1890. They are principally stomach poisons and to a limited extent contact poisons. The kill is more rapid than that of arsenicals. Their insecticidal properties are related to the fluorine content and solubility in the digestive juices of insect. Flouride poisoning produces spasms, regurgitation, flaccid paralysis and death

    a. **Sodium fluoride**: It is a white power. Available in 93 to 99 per cent purity in commercial products. It is highly phytotoxic and used in poison baits used exclusively against cockroaches, earwigs, cutworms, grasshoppers etc.

3. **Sulphur**: It is primarily fungicide and acaricide. Formulated as fine dust (90 to 95 per cent a.i with 10 per cent inert material. It is also formulated as wettable powders. Effectiveness increases with fineness of sulphur particles.

Lime sulphur: Aqueous solution of calcium polysulphide. It is prepared by sulphur solution in calcium hydroxide suspension, preferably under pressure in the absence of air and is used against scales, mites, aphids besides powdery mildew.

Properties: 1. Affect nervous system causing excitement at lower doses and paralysis at higher concentration.2. Not phytoloxic.3. Leave no harmful side effects.4. Mighly toxic to mammals.5. Disappear rapidly from the treated surface. So can be used safely before harvest of the produce.

6.4.2 Organic Insecticides/Natural Occurring Insecticids

6.4.2.1 Botanical Insecticides

The insecticides of plant origin extracted from seeds, flowers, leaves, stem and roots, are termed as botanical insecticides. Insecticides of plant origin unlike synthetic organic insecticides are safer to use but since they are expensive and lack residual toxicity, their use has been limited in the country.

1. Neem

The neem tree*, Azadirachta indica is a* perennial tree distributed in tropical, subtropical, semi-arid and arid zones. It posses medicinal, insecticidal, insect repellent, antifeedant, growth regulatory, nematicidal and antifungal properties. Neem seed extract and oil contains a number of components such as azadirachtin, salannin, nimbin, epinimbin nimbidin that gives insecticidal, insect repellent, ovicidal, antifeedant and growth regulator characters. Azadiractitin disrupts moulting by antagonizing the insect hormone ecdysone. Azadiractitin Acute oral LD_{50} for rat is 5000mg/kg, Acute dermal for rabbit is >2000mg/kg.

Azadirachtin, a chemical compound belonging to the limonoid group, is a secondary metabolite present in neem seeds. It is a highly oxidized tetranortriterpenoid which boasts a plethora of oxygen functionality, comprising an enol ether, acetal, hemiacetal, and tetra-substituted oxirane as well as a variety of carboxylic esters. Azadirachtin has a complex molecular structure and the first total synthesis was completed by Steven Ley in 2007. Both secondary and tertiary hydroxyl groups and tetrahydrofuran ether are present and the molecular structure reveals 16 stereogenic centres, 7 of which are tetrasubstituted. These characteristics explain the great difficulty encountered when trying to produce it by a synthetic approach. It was initially found to be active as a feeding inhibitor towards the desert locust (*Schistocerca gregaria*), it is now known to affect over 200 species of insect, by acting mainly as an antifeedant and growth disruptor, and as such it possesses considerable toxicity toward insects (LD_{50}(*S. littoralis*): 15 µg/g). It fulfills many of the criteria needed for a natural insecticide if it is to replace synthetic compounds. Azadirachtin is biodegradable (it degrades within 100 hours when exposed to light and water) and shows very low toxicity to mammals (theLD_{50} in rats is > 3,540 mg/kg making it practically non-toxic).

Azadirachtin is found in the seeds (0.2 to 0.8 percent by weight) of the neem tree, *Azadirachta indica* (hence the prefix aza does not imply an aza compound, but refers to the scientific species name). Many more compounds, related to azadirachtin, are present in the seeds as well as in the leaves and the bark of the neem tree which also show strong biological activities among various insect pests. Effects of these preparations on beneficial arthropods are generally considered to be minimal. Some laboratory and field studies have found neem extracts to be compatible with biological control.

Preparation of Neem Seed Kernel Extract (NSKE 5 per cent)

Take 50 g of powdered neem seed kernels soak it in one litre of water for 8 hours and stir the contents often. Squeeze the soaked material repeatedly for better extraction of the azadirachtin in the aquous suspension. Filter the contents through muslin cloth. Make the filtrate to one litre. Add 1ml teepol or triton or sandovit or soap water (2 per cent) and spray. Commercial formulations of neem are available in 10000 ppm, 1500 ppm and 300 ppm the market. Some of the neem formulations are Margosan, Neemark,Neemrich, Achook, Bioneem, Neemazal, Neemax, Nimbicidine,Vepacide, Margocide,Neemgold etc.

2. Nicotine

Nicotine is a simple alkolide derived from tobacco, *Nicotiana tabacum*, and other Nicotiana species Insecticidal formulations generally contain nicotine in the form of 40 per cent nicotine sulfate and are currently imported in small quantities from India. In both insects and mammals, nicotine is an extremely fast-acting nerve toxin. It competes with acetylcholine, the major neurotransmitter, by bonding to acetylcholine receptors at nerve synapses and causing uncontrolled nerve firing. This disruption of normal nerve impulse activity results in rapid failure of those body systems that depend on nervous input for proper functioning. In insects, the action of nicotine is fairly selective, and only certain types of insects are affected.

Nicotine is found in the leaves of *Nicotiana tabacum* and *N.rustica* from 2 per cent to 14 per cent. Nicotine sulphate has been mainly used as a contact insecticide with marked fumigant action in the control of sucking insect's*viz.*, aphids, thrips, psyllids, leafminers and jassids. Nicotine sulphate is more stable and less volatile. It is a nerve poison beinghighly toxic when absorbed through the cuticle taken in through the tracheae or when ingested. It affects the ganglian blocking conduction at higher levels. Tobacco decoction, useful for controlling aphids, Thrips etc. Can be prepared by boiling 1kg of tobacco waste in 10lts of water for 30 minutes or steep itin cold water for a day. Then make it up to 30 litres and add about 90gm of soap. $LD_{50}$for rat oral- 50-60 mg/kg.

3. Rotenone

Rotenone is insecticidal compound that occurs in the roots of *Lonchocarpus* species in South America, Derris species in Asia, and several other related tropical legumes. Commercial rotenone was at one time produced from Malaysian Derris. Currently the main commercial source of rotenone is Peruvian *Lonchocarpus*, which often is referred to as cube root. Rotenone is a powerful inhibitor of cellular respiration, the process that converts nutrient compounds into energy at the cellular level. In insects rotenone exerts its toxic effects primarily on nerve and muscle cells, causing rapid cessation of feeding. Death occurs several hours to a few days after exposure. Rotenone is extremely toxic to fish, and is often used as a fish poison (piscicide) in water management programs. It is effectively synergized by PBO. LD_{50} to white rat oral-130 to 1500. Dust or spray containing 0.5 to 1.0 per cent and 0.001 to 0.002 percent rotenone are used commercially.

4. Plumbagin

Plumbagin is naturally occurring napthoquinone of plant origin from the roots of *Plumbago europea* L. (Plumbaginaceae) and named so in 1828 by Bulong d' Astafort. Plumbagin is known for its medicinal, antifertility, antimicrobial, molluscicidal, nematicidal and other pharmacological properties on diverse fauna. The yield of plumbagin ranges between 0.5-3.0percent on dry weight basis. The cold alcoholic extract (5 per cent) of roots of *P. zeylanica* L. was toxic to *Euproctis fraterna* larvae as contact spray. Contact toxicity of 5 per cent petroleum ether extracts of *P. zeylanica* root against *Spodoptera litura* Fab., *Dystercus koenigii* Fab., *Dipaphis erysimi* Kalt, *Dactynops carthami* H.R.L, *Coccinella septumpunctata* L. was also reported.

5. Pyrethrum

It is extracted from dried flower heads of *Chrysanthemum cinerariaefolium* (Asteraceae). The actual chemical ingredients having insecticidal action are identified as five esters. They are: Pyrethrin I, Pyrethrin II, cinerins-I and cinerin-II and Jasmoline, which are predominately found in achenes of flowers from 0.7 to 3 per cent. The esters are derived from : 2 acids and 3 alcohols.

Two acids – Chrysanthemic acid and Pyrethric acid

Three alcohols – Pyretholone, Cinerolone and Jasmolone

Active Principles/Esters
1. Pyrethrin I = Pyrethrolone + Chrysanthemc acid.
2. Pyrethrin II = Pyrethrolone + Pyrethric acid.
3. Cinerin I = Cynerolone + Chrysanthemc acid.
4. Cinerin II = Cynerolone + Pyrethric acid.
5. Jasmolin II = Jasmolone + Pyrethric acid.

Pyrethrins are powerful contact insecticides but appear to be poor stomach poisons. A characteristic action of Pyrethroid is the rapid paralysis or 'knock down' effect and substantial recovery that follow it. This recovery is due to rapid enzymatic detoxification in the insect. To bring about mortality equivalent to knock down effect three times increase in dosage may be required. Compounds such as piperonyl butoxide, propyl isome and sulfoxide are known to inhibit the detoxication enzyme and increase the toxicities of pyrethroids. These synergists are used at 10 parts to 100 part of pyrethroid. LD_{50} for white rat oral-200 dermal for rat-1800. Pyrocon E 2/22 (1 part of pyrethrin + 10 parts of piperonyl butoxide) is used for the control of coconut red palm weevil. In household sprays and as a repellent against external parasites of livestock pyrethrum is useful. It is also mixed with grains in storage to protect from stored grain pests. Its use alone or in combination with piperonyl butoxide as food packages has been permitted by the food and Drug Administration and no other chemical has been approved.

Properties of Pyrethrum
1. Highly unstable in light, moisture and air.
2. Have no residual effect.
3. Paralyse by more contact.
4. Gains entry through spiracle and cuticle.
5. Act on central nervous system.
6. Having rapid knock down effect.
7. Practically no mammalian toxicity.
8. Good insecticides against household and cattle pests.

6. Sabadilla (veratrine alkaloids)
Sabadilla is derived from the ripe seeds os *Schoenocaulon officinale*, a tropical lily plant which grows in Central and South America. Sabadilla is also sometimes known as cevadilla or caustic barley. When sabadilla seeds are aged,

heated, or treated with alkali, several insecticidal alkaloids are formed or activated. Alkaloids are physiologically active compounds that occur naturally in many plants. In chemical terms they are a heterogeneous class of cyclic compounds that contain nitrogen in their ring structures. Caffeine, nicotine, cocaine, quinine, and strychnine are some of the more familiar alkaloids. The alkaloids in sabadilla are known collectively as veratrine or as the veratrine alkaloids. They constitute 3-6 per cent of aged, ripe sabadilla seeds. Of these alkaloids, cevadine and veratridine are the most active insecticidally. European white hellebore (*Veratrum album*) also contains veratridine in its roots. In insects, sabadilla's toxic alkaloids affect nerve cell membrane action, causing loss of nerve cell membrane action, causing loss of nerve function, paralysis and death. Sabadilla kills insects of some species immediately, while others may survive in a state of paralysis for several days before dying. Sabadilla is effectively synergized by PBO.

7. Ryania

Ryania comes from the woody stems of *Ryania speciosa*, a South American shrub. Powdered Ryania stem wood is combined with carriers to produce a dust or is extracted to produce a liquid concentrate. The most active compound in ryania is the alkaloid ryanodine, which constitutes approximately 0.2 per cent of the dry weight of stem wood.

Ryania is a slow-acting stomach poison. Although it does not produce rapid knockdown paralysis, it does cause insects to stop feeding soon after ingesting it. Little has been published concerning its exact mode of action in insect systems. Ryania is effectively synergized by PBO and is reported to be most effective in hot weather

8. Chinaberry

Melia azedarach Linn, also known as Chinaberry, China tree, Chinaball tree, Persian lilac, pride of India, Indian Lilac, bead tree, Texas umbrella tree. It is a small to medium-sized shrub or tree in the mahogany family (Meliaceae) native to north western India and has long been recognized for its insecticidal properties (Emmanuel *et al.*, 2008).

Meliaceae plants are rich source of limnoids, and a typical plant, *Melia azedarach* Linn which is also known as Chinaberry is of particular interest, because it contains several types of limnoids possessing insect antifeeding activity. It is a small to medium-sized shub or tree and distinguished from other members of the Meliaceae by the nature of its compound leaves, and by its drooping, persistent clusters of yellowish fruits. It is native to northwestern India. Strong antifeedant, insect growth regulatory and toxic activity of *M. azedarach* seeds extracts have been reported by Chiu and Zhang (1984), Chiu *et al.* (1987), Anwar *et al.* (1992) and Dilawari *et al.* (1994), Juan *et al.* (2000) and Kaur and Singh (2003) against *Spodoptera litura, Mythimna separata, Sesamia nonagrioides, Plutella xylostella, Spilosoma rhodophila* and *Earias vittella*. The antifeedant, IGR and toxic activity of *M. azedarach* extractives is attributed to a large number of chemically diverse tetranor triterpines such as meliacarpins, meliacarpinins, salanin and other meliacins (Bohnenstengel *et al.*1999).

9. Caesalpinia crista

Caesalpinia crista Linn., known as karanjwa in India, is a perennial shrub branches finely-downy, leaflets membranous, elliptic-oblong, obtuse; petiolules very short; hooked spines. Flowers in dense (usually spicate) long-peduncled terminal and superaxillary racemose dense at the top, lax downward. Pods shortly stalked, oblong, densely armed on the faces with prickles. Distributed thoughout India, generally tropically and in Asia, Australasia, Indian Ocean Islands, and Pacific Ocean Islands. Akhtar *et al.* (1985) has reported antihelminthic property of *C. crista* seeds extracts. Seed kernel extract of *C. crista* contains four cassane type furano diterpenes, which have been characterized as caesalpinins and its analogues. Some other compounds identified from the seeds include norcaesalpinin-E, caesalmin-C and its desacetyl and desacetoxy derivatives, caesaldekarin and its acetoxy derivatives (Kalauni *et al.*, 2004).

Melia azedarach seeds and leaves extracts when applied by the leaf dip and artificial diet incorporation method adversely effected the feeding, survival, growth and development of *H. armigera* larvae. The descending order of antifeedant and IGR activity of various extracts was methanol > hexane > ethyl acetate > aqueous > butanol extract. Amongst the various extracts, methanol extract showed maximum larval and pupal duration, larval pupal intermediates, pupal mortality and malformed adult at the highest concentration tested [Emmanuel, *et al.*, 2008.]. *M. azedarch* seeds and leaves extracts *viz.*, hexane, methanol, ethyl acetate, aqueous and butanol when incorporated into the diet was 4.59, 5.85, 3.25, 2.81, 9.70; 16.21, 18.32, 28.38, 17.50, 28.62 times more effective than when applied by leaf dip method in causing 50.00 per cent antifeedance. Also, the larval mortality was exhibited at lower dose when the various extracts of *M. azedarach* seeds and leaves were incorporated into the diet than when applied by the leaf dip method. Further, the application of these extracts by leaf dip method required higher dose for the inhibition of normal adult emergence as compared to artificial diet incorporation method [Emmanuel Nathala and Swaran Dhingra, 2005 a].

The hexane and methanol extract of *M. azedarach* and *Caesalpinnia crista* seeds when applied on the leaf disc showed 1.13, 1.23; 1.04, 2.05 times better antifeedant activity than the neem oil [Emmanuel Nathala and Swaran Dhingra, 2005 b]. Interestingly, hexane and methanol extract of *M. azedarach* seeds proved to be 1.61 and 1.94 times more effective than the neem oil in inhibiting 50 per cent normal adult emergence. As compared to the aqueous extract of neem seeds kernel, the aqueous extract of *M. azedarach* and *C. crista* seeds when applied by leaf dip method were found to be 2.24 and 1.95 times more effective than NSKE in causing 50 per cent antifeedance [Emmanuel Nathala and Swaran Dhingra, 2006 a]. Aqueous extract of *M. azedarach* seeds and leaves when applied on the leaf disc showed 2.97 and 16.36 times lower cumulative

Insect Growth Regulatory Effects by Extracts of
Melia azedarach in *Helicoverpa armigera.*

[*Source*: N. Emmanuel. 2005. Antifeedant and growth inhibitory effects of *Melia azedarach* and *Caesalpinia crista* extracts against *Helicoverpa armigera* (Hubner). PhD Thesis, IARI, New Delhi].

larval toxicity than the NSKE. Also, the aqueous extract of *M. azedarach* seeds was 9.42 times more active in inhibiting normal adult emergence than aqueous neem seeds extract [Emmanuel Nathala and Swaran Dhingra, 2006 b]. Neem oil and NSKE when incorporated in diet showed better antifeedant, larval toxicity and IGR effect than the hexane, methanol and aqueous extract of *M. azedarach* and *C. crista* seeds and leaves when applied by leaf dip method. Thus, the chronic effect of neem oil and NSKE is more as compared to the *M. azedarach* and *C. crista* seeds and leaves extracts. C. *septumpunctata* were safe feeding on the aphids treated with various extracts of seeds and leaves of *M. azedarach* and *C.crista*. Except for hexane extract of *M. azedarach* seeds, all the extracts were safer to *Apanteles sp* [Emmanuel Nathala and Swaran Dhingra, 2006 a].

Effect of Neem Seed Kernel Extract and Neem Oil on Insect Pests

The effect of neem seed kernel extract (NSKE) and neem oil on different insect pests reported by various workers is presented in table 18. The data available on antifeedant, IGR and larval toxicity highlights the effectiveness of the NSKE and neem oil obtained from *A. indica*.

Table 13: Biological Effects of NSKE and Neem Oil on various Insect Pests.

Pradhan *et al.* (1962)	First to prove the antifeedant property of neem and showed that 0.001 per cent aqueous suspension of crushed neem kernel when sprayed on cabbage plant totally stopped the feeding of desert locust, *Schistocerca gregaria* on treated foliage.
Mane (1968)	Aqueous extract of neem seed kernels gave considerable protection to various crops against larvae of *Euproctis lunata*, *S. litura, Utetheisa pulchella.*
Gill and Lewis (1971)	*Pieris* sp larvae fed on foliage treated with neem kernel extract failed to develop to maturity and most of them died while moulting.
Sandhu and Singh (1975)	1 per cent neem kernel extract reduced the damage by *P. brassicae.*
Attri (1975)	Neem oil extractive was found to be 40 times less effective as antifeedant than water extract to *S. gregaria.*
Ladd and Jacobson, (1980)	1 per cent neem kernel extracts as spray on soybean deterred the feeding of *Popillia japonica.*
Saxena (1982)	Neem epell extract showed per cent antifeedance of 95.1 to 98.4 against *Amsacta moorie.*
Joshi *et al.* (1984)	Spray application of 0.5, 0.75 and 0.1 per cent suspension of neem seed kernel protected tobacco plants from feeding by *S. litura*
Saxena and Khan (1984)	Phloem feeding by *N. lugens* on rice plants treated with neem oil was significantly less than on untreated plants.
Chari and Muralidharan (1985)	2 per cent kernel suspension completely inhibited the feeding by *Achaea janata* larvae.
Singh and Sharma (1986)	Cabbage and cauliflower crops sprayed with various concentrations of neem kernel suspension showed strong epellent and antifeedant property against aphid *Brevicoryne brassicae.*
Prabhakar *et al.* (1986)	Neem kernel extract caused larval mortality, prolonged larval duration and inhibited pupation of *Trichoplusia ni* and *S. exigua.*

Contd...

Table 13–*Contd...*

Miesner *et al.* (1986)	Aqueous extract of neem significantly affected the larval weight and pupation of *Ostrinia nubilalis*.
Singh *et al.* (1988)	Neem oil and ethanolic extract at 0.5 per cent exhibited 48.00 and 9.00 per cent mortality of *Lipaphis erysimi* while the aqueous extract caused no mortality at the same concentration.
Gujar and Mehotra (1988)	Effective concentration for 50 per cent antifeedant activity of neem oil against *Aulacophora foveicollis* was 0.4 per cent.
Kareem *et al.* (1998)	3 per cent NSKE protected the mung bean crop from *Heliothis* sp, *Etiella* sp and *Maruca* sp.
Mikolajczak *et al.* (1989)	Neem seed extracts reduced the larval growth rate and increased the time of pupation of *S. frugiperda*.
Singh *et al.* (1990)	Neem oil was at par with monocrotophos in reducing the *Scirpophaga incertulas* incidence in rice.
Thakar *et al.* (1992)	Aqueous neem extract (10 per cent) was comparable to quinalphos (0.04 per cent) in damage reduction of gram *(Vigna)* by *H. armigera*.
Patel *et al.* (1993)	Five per cent suspension sprays of neem seeds found to be effective against 2nd instar larvae of *Amsacta moorie*.
Sarode and Gabhane (1994)	5 per cent NSKE reduced the infestation of okra fruit borer.
Dhingra (1996)	Neem oil provided 14 and 12 fold synergism of cypermethin at 1: 1 and 1: 5 ratios against *Mylabris pustulata*.
Behera and Satpathy (1996)	NSKE caused 100 per cent mortality of 4th instar larvae of *S. litura* at 10 DAT.
Rao and Dhingra (1997)	Synergistic activity of neem oil in mixed formulation with cypermthin (1: 5) was pronounced against 4 and 9 day old larvae of *S. litura*.
Jeyakumar and Gupta (1999)	NSKE (10 per cent) showed oviposition deterrency and ovicidal effect against *H. armigera*.
Dhingra *et al.* (2002)	Neem oil microemulsion reduced the larval and pupal weight gain, and adult emergence, and increased the larval mortality, larval-pupal intermediates and the pupal deformity in the 3rd instar treated-larvae of *H. armigera*. Its relative effectiveness in inhibiting adult emergence compared to the macroemulsion (1.0) was 1.68.
Bajpai and Sehgal (2003)	Neem oil caused high level pupal mortality of *S. litura*.
Simmonds *et al.* (2004)	Amongst the important limonoids of neem seeds, isosalanninolide showed antifeedant activity greater than that of nimbin or salannin and comparable to azadirachtin against *S. littoralis*, *S. frugiperda* and *H. armigera* larvae.
Emmanuel Nathala and Swaran Dhingra [2006 b]	Evaluated the relative bioactivity of various extracts of seeds and leaves of *Melia azedarach* and *Caesalpinia crista* extracts against *Helicoverpa armigera* by leaf dip and artificial diet incorporation bioassay techniques and compared with neem oil and NSKE. The hexane and methanol extract of *M. azedarach* and *C. crista* seeds when applied on the leaf disc showed 1.13, 1.23; 1.04, 2.05 times better antifeedant activity than the neem oil. All the extracts required significantly higher dose than neem oil to cause 50.00 per cent cumulative larval mortality. Interestingly, hexane and methanol extract of *M. azedarach* seeds proved to be 1.61 and 1.94 times more effective than the neem oil in inhibiting 50 per cent normal adult emergence.

Contd...

Table 13–_Contd..._

Kraiss and Cullen (2008).	Studied the Insect growth regulator effects of azadirachtin and neem oil on survivorship, development and fecundity of *Aphis glycines* (Homoptera: Aphididae) and its predator, *Harmonia axyridis* (Coleoptera: Coccinellidae).
Kulkarn, *i et al.*, 2008.	Stated that Azadirachtin 1 per cent and 5 per cent formulations along with chemical insecticides and bio-pesticides was better alternative for minimizing pesticide residues in grapes and in combination of neem formulations +bio-pesticides + chemical pesticides the yield was 12.50 kg/vine against 7.80 kg/vine in untreated check.
Misra, H. P. 2009.	Reported that the per cent damaged heads due to *Spodoptera litura* F., remained significantly low with Neemarin 1500 ppm and 10, 000 ppm both @ 5 and 6 ml/lit of water that was on par with cartap hydrochloride with 37.19-41.36 per cent reduction in head damage at 10 DAS. Significant suppression of diamondback moth (DBM) larvae/plant was observed with Neemarin 1500 ppm and 10, 000 ppm @ 4-6 ml/lit and cartap hydrochloride.
Rafiq, *et al.*, 2012.	Reported the Efficacy of neem *(Azadirachta indica* A. Juss) callus and cells suspension extracts against three lepidopteron insects of cotton. The larvae showed a repellent behavior, decrease in weight, and negligible leaf area damage. The preparation T3 (1: 1000 v/v Extract: D.H$_2$O) showed 76 to 84 per cent mortality, T4 (1: 10000 v/v Extract: D.H$_2$O) 28 to 72 per cent, and T5 (1: 10000 v/v Extract: D.H$_2$O) 12 to 40 per cent mortality after the 5th day of incubation.

6.4.3 Synthetic Organic Insecticides

1. Chlorinated Hydrocarbons (Organo-chlorines)

The plant protection across the world and India owe its growth to the chemicals under this group which have modernized the pest control tactics. The properties which have lead to their extensive use are high insecticidal efficacy, long residual action, wide range of insect susceptibility, cheapness per unit area and available in different formulations. They are also known as chlorinated synthetics or chlorinated organics or chlorinated hydrocarbons. The important organochlorines are:

Dichloro Diphenyl Trichloroethane [DDT]

Othmar Zeidler in 1874 synthesized DDT by the reaction of chloral with chlorobenzene in the presence of sulfuric acid and its insecticidal properties were discovered in 1939 by a Swiss chemist, Paul Hermann Müller. Pure DDT is a colourless, crystalline solid that melts at 109° C (228° F); the commercial product, which is usually 65 to 80 percent active compound, along with related substances, is an amorphous powder that has a lower melting point. DDT is applied as a dust or by spraying its aqueous suspension. This discovery brought the 'Nobel Prize' for medicine to Paul Muller in 1948 for the life saving discovery. Dichloro Diphenyl Trichloroethane (DDT) is stomach and contact insecticide. It has got long residual action. It is also non-phytotoxic except to cucurbits. It is not much effective against phytophagous mites. Due to low cost of DDT and effectiveness against a variety of insects particularly against house flies and mosquitoes, it is much popularized but due to long residual life and accumulation, it is banned in several countries. The acute oral LD_{50} for rats is 113-118 mg/kg. It does affect the nervous system preventing normal transmission of nerve impulses. DDT causes a violent excitatory neurotoxic system in most insects which are having uncoordinated movement and DDT Jitters (tremor of the entire body).

Hexa Chloro Cyclohexane (HCH)

Hexachlorocyclohexane or the abbreviation HCH was Synthesised by Michael Faraday in the year 1825.The gamma-isomer of BHC has the insecticidal activity. BHC is a stomach and contact insecticide. It has got slight fumigant action. It is persistent insecticide. It is non-phytotoxic except cucurbits. HCH was first used to combat the Colorado beetle. It has been extensively used as soil insecticides particularly to control termites, white grubs and cutworms. Highly purified product containing 99 per cent of gamma isomer of HCH is known as lindane, this name was proposed in 1949 after Vander Linden, a German chemist who isolated this isomer in 1912. Lindane is more acute neurotoxicant than DDT results in tremors, ataxia, convolutions, falling prostration and ultimately leading to death.

2. Cyclodines

Cyclodienes also act as neurotoxicants which disturb the balance of sodium and potassium ions within the neuron resulting into tremors, convulsions, prostration and ultimately the death. The outstanding characteristic of the cyclodienes is their longer stability in the soil, resulting in more control of soil inhabiting insect pests. Some of the compound belonging to this group are chlordane (1945), aldrin and dieldrin (1948), heptachlor (1949), endrin (1951), mirex (1954), endosulfan (1956) and chlordecone 1958). Among them aldrin, chlordane and heptachlor were often in use for termite control as they are most effective, long lasting and economical insecticides but now banned by Govt.of India.

Aldrin

Aldrin is a broad-spectrum insecticide used in particular to combat soil and cotton pests as well as locusts. Aldrin is an insecticide that enters the insect's body (and in the environment) and quickly converts todieldrin.

It is persistent and non-systemic soil insecticide. It is usually recommended for the control of termites throughout India. Two German chemists Otto Paul Hermann Diels and Kurt Alder first documented the Diels-Aldernovel reaction in 1928 for which they were awarded the Nobel Prize in Chemistry in 1950 for their work. Formulations: EC 30 per cent, Granule 5 per cent and Dusts 5 per cent Trade names: Octalene, Aldrex, Aldrosol, and Aldrite. LD_{50} value: 67 mg/kg.

Dieldrin

Dieldrin is a chlorinated hydrocarbon originally produced in 1948 by J. Hyman and Co, Denver, as an insecticide. Dieldrin is closely related to aldrin, which reacts further to form dieldrin. Aldrin is not toxic to insects; it is oxidized in the insect to form dieldrin which is the active compound. Both dieldrin and aldrin are named after the Diels-Alder reaction which is used to form aldrin from a mixture of norbornadiene and hexachlorocyclopentadiene. Originally developed in the 1940s as an alternative to DDT, dieldrin proved to be a highly effective insecticide and was very widely used during the 1950s to early 1970s. Endrin is a stereoisomer of dieldrin. However, it is an extremely persistent organic pollutant; it does not easily break down. Furthermore it tends to biomagnify as it is passed along the food chain. Long-term exposure has proven toxic to a very wide range of animals including humans, far greater than to the original insect targets. For this reason it is now banned in most of the world. It has been linked to health problems such as Parkinson's, breast cancer, and immune, reproductive, and nervous system damage. It can also adversely affect testicular descent in the fetus if a pregnant woman is exposed to the pesticide. It is persistent and non-systemic insecticide used for mainly soil inhabiting insect pests. It is also not phytotoxic in recommended doses. Formulations: Dust 2 per cent, Trade names: Quintox, Alvit, LD_{50} value: 46 mg/kg.

Heptachlor

Analogous to the synthesis of other cyclodienes, heptachlor is produced via the Diels-Alder reaction of hexa_chloro cyclo pentadiene and cyclopentadiene. The resulting adduct is brominated followed by treatment with hydrogen chloride in nitromethane in the presence of aluminum trichloride or with iodine monochloride. Compared to chlordane, it is about 3 - 5 times more active as an insecticide, but more inert chemically, being resistant to water and caustic alkalies. It is a non-systemic, contact poison with fumigant action. It is effective against termites, white grubs, grass hoppers etc Formulations: EC 20 per cent, and Dust 5 per cent.

Endosulfan

Endosulfan is a derivative of hexachlorocyclopentadiene, and is chemically similar to aldrin, chlordane, and heptachlor. Specifically, it is produced by the Diels-Alder reaction of hexachlorocyclopentadiene with *cis*-butene-1,4-diol and subsequent reaction of the adduct with thionyl chloride. Technical endosulfan is a 7:3 mixture of stereoisomers, designated α and β. α- and β-Endosulfan are conformational isomers arising from the pyramidal stereochemistry of sulfur. α-Endosulfan is the more thermodynamically stable of the two, thus β-endosulfan irreversibly converts to the α form, although the conversion is slow.

Endosulfan is an organochlorine insecticide and acaricide that is being phased out globally. The two isomers, endo and exo, are known popularly as I and II. Endosulfan sulfate is a product of oxidation containing one extra O atom attached to the S atom. Endosulfan became a highly controversial agrichemical due to its acute toxicity, potential for bioaccumulation, and role as an endocrine disruptor. Because of its threats to human health and the environment, a global ban on the manufacture and use of endosulfan was negotiated under the Stockholm Convention in April 2011. The ban was effected from mid-2012, with certain uses exempted for five additional years. It is still used extensively in India, China, and few other countries. It is produced byMakhteshim Agan and several manufacturers in India and China. Due to its unique mode of action, it is useful in resistance management; however, as it is not specific, it can negatively impact populations of beneficial insects. It is, however, considered to be moderately toxic to honey bees, and it is less toxic to bees than organophosphate insecticides.

History of Commercialization and Regulation of Endosulfan

☆ Early 1950s: Endosulfan was developed.

☆ 1954: Hoechst AG (now Bayer Crop Science) won USDA approval for the use of endosulfan in the United States.

☆ 2000: Home and garden use in the United States was terminated by agreement with the EPA.

☆ 2002: The U.S. Fish and Wildlife Service recommended that endosulfan registration should be cancelled, http://en.wikipedia.org/wiki/Endosulfan - cite_note-17 and the EPA determined that endosulfan residues on food and in water pose unacceptable risks. The agency allowed endosulfan to stay on the US market, but imposed restrictions on its agricultural uses.

☆ 2007: International steps were taken to restrict the use and trade of endosulfan. It is recommended for inclusion in the Rotterdam Convention on Prior Informed Consent, and the European Union proposed inclusion in the list of chemicals banned under the Stockholm Convention on Persistent Organic Pollutants. Such inclusion would ban all use and manufacture of endosulfan globally.Meanwhile, the Canadian government announced that endosulfan was under consideration for phase-out, and Bayer Crop Science voluntarily pulled its endosulfan products from the U.S. market but continues to sell the products elsewhere.

☆ 2008: In February, environmental, consumer, and farm labor groups including the Natural Resources Defense Council, Organic Consumers Association, and the United Farm Workers called on the U.S. EPA to ban endosulfan. In May, coalitions of scientists, environmental groups, and arctic tribes asked the EPA to cancel endosulfan, and in July a coalition of environmental and workers groups filed a lawsuit against the EPA challenging its 2002 decision to not ban it. In October, the Review Committee of the Stockholm Convention moved endosulfan along in the procedure for listing under the treaty, while India blocked its addition to the Rotterdam Convention.

☆ 2009: The Stockholm Convention's Persistent Organic Pollutants Review Committee (POPRC) agreed that endosulfan is a persistent organic pollutant and that "global action is warranted", setting the stage of a global ban. New Zealand banned endosulfan.

☆ 2010: The POPRC nominated endosulfan to be added to the Stockholm Convention at the Conference of Parties (COP) in April 2011, which would result in a global ban. The EPA announced that the registration of endosulfan in the U.S. will be cancelled· Australia banned the use of the chemical.

☆ 2011: The Supreme Court of India banned manufacture, sale, and use of toxic pesticide endosulfan in India. The apex court said the ban would remain effective for eight weeks during which an expert committee headed by DG, ICMR, will give an interim report to the court about the harmful effect of the widely used pesticide.

The Supreme Court of India passed interim order on May 13, 2011, in a Writ Petition filed by Democratic Youth Federation of India, (DYFI), a youth wing of Communist Party of India (Marxist) in the backdrop of the incidents reported in Kasargode, Kerala, and banned the production, distribution and use of endosulfan in India because the pesticide has debilitating effects on humans and the environment. A 2001 study by Centre for Science and Environment (CSE)CSE had established the linkages between the aerial spraying of the pesticide and the growing health disorders in Kasaragod. Over the years, other studies have confirmed these findings, and the health hazards associated with endosulfan are now widely known and accepted. However, in July 2012, the Government asked the Supreme Court to allow use of the pesticide in all states except Kerala and Karnataka, as these states are ready to use it for pest control.

It is a non-systemic, contact and stomach poison with slight fumigant action. It is highly toxic to fish. Formulations: EC 35 per cent, Granule 4 per cent and Dusts 4 per cent. Trade names: Thiodan, Endocel, Endodhan, Endotaf.LD_{50} value: 80-110 mg/kg.

3. Organophosphates

An organophosphate (sometimes abbreviated OP) or phosphate ester is the general name for esters of phosphoric acid. In health, agriculture, and government, the word "organophosphates" refers to a group of insecticides or nerve agents acting on the enzyme acetylcholine esterase.

Early pioneers in the field include Jean Louis Lssaigne (early 19th century) and Philippe de Clermont (1854). In 1932, German chemist Willy Lange and his graduate student, Gerde von Krueger, first described the cholinergic nervous system effects of organophosphates, noting a choking sensation and a dimming of vision after exposure. This discovery later inspired German chemist Gerhard Schrader at company IG Farben in the 1930s to experiment with these compounds as insecticides. Their potential use as chemical warfare agents soon became apparent, and the Nazi government put Schrader in charge of developing organophosphate (in the broader sense of the word) nerve gases. Schrader's laboratory discovered the G series of weapons, which included Sarin, Tabun, and Soman. The Nazis produced large quantities of these compounds, though did not use them during World War II.

British scientists experimented with a cholinergic organophosphate of their own, called diisopropylfluorophosphate (DFP), during the war. The British later produced VX nerve agent, which was many times more potent than the G series, in the early 1950s, almost 20 years after the Germans had discovered the G series.

After World War II, American companies gained access to some information from Schrader's laboratory, and began synthesizing organophosphate pesticides in large quantities. Parathion was among the first marketed, followed by malathion and azinphosmethyl. The popularity of these insecticides increased after many of the organochlorine insecticides like DDT, dieldrin, and heptachlorwere banned in the 1970s. Many *organophosphates* are potent nerve agents, functioning by inhibiting the action of acetylcholinesterase (AChE) in nerve cells.

Malathion

Malathion was used in the 1980s in California to combat the Mediterranean Fruit Fly. Malathion itself is of low toxicity; however, absorption or ingestion into the human body readily results in its metabolism to malaoxon, which is substantially more toxic. In studies of the effects of long-term exposure to oral ingestion of malaoxon in rats, malaoxon has been shown to be 61 times more toxic than malathion. It is a non systemic contact and stomach insecticideand acaricide of low mammalian toxicity. Hence it is recommended on fruits and vegetables till a few days prior to harvest. It is also recommended for storage insects and also for external application for parasites on animals. Formulations: EC 50 and Dusts 40. Trade names: Cythion and Himala LD_{50} value: 2800 mg/kg.

Methyl Parathion

Parathion was developed by Gerhard Schrader for the German trust IG Farben in the 1940s. After the war and the collapse of IG Farben due to the war crime trials, the Western allies seized the patent, and parathion was marketed worldwide by different companies and under different brand names. The most common German brand was E605 (banned in Germany after 2002); this was not a food-additive "E number" as used in the EU today. "E" stands for *Entwicklungsnummer* (German for "development number"). It is a contact and stomach poison with slight fumigant action. It is widely used in for sucking insects and foliage feeders. Formulations: EC 50 and Dusts 2. Trade names: Folidal, Metacid, Paratox, Dhanumar LD_{50} value: 13 mg/kg.

Diazinon

Diazinon was developed in 1952 by the Swiss company Ciba-Geigy as a replacement for the insecticide DDT. Diazinon became available for mass use in 1955, as DDT production tapered. Prior to 1970 diazinon had issues with contaminants in the solution. However, by the 1970s, alternative purification methods were utilized to reduce residual materials. After this, diazinon became an all-purpose indoor and outdoor commercial pest control product. It is a contact persistent insecticide with nematicidal properties. It is very much useful against household insects such as flies and cockroaches.It has contact, stomach poison and also fumigant action. Formulations: EC 20 and 5G Trade names: Basudin LD$_{50}$ value: 300-850 mg/kg.

Dichlorvos [DDVP]

Dichlorvos or 2,2-dichlorovinyl dimethyl phosphate is a highly volatile organophosphate, widely used as an organophosphorus insecticide to control household pests, in public health, and protecting stored product from insects. It is effective against mushroom flies, aphids, spider mites,caterpillars, thrips, and whiteflies in greenhouse, outdoor fruit, and vegetable crops. It is also used in the milling and grain handling industries and to treat a variety of parasitic worm infections in dogs, livestock, and humans. It is fed to livestock to control bot fly larvae in the manure. It acts against insects as both a contact and a stomach poison. It is available as an aerosol and soluble concentrate. It is also used in pet collars and "no-pest strips" as pesticide-impregnated plastic. In this form it has recently been labeled for use against bed bugs. It is contact poison but due to high vapour pressure it has got strong penetrating power.It is very effective against hidden insects due to its fumigation action. It is recommended for leaf miners and leaf webbers. It brings quick knock down effect. It does not leave toxicresidues. It is highly toxic to bees. It is acontact and stomach poison with fumigant action. Formulations: EC 76 and 5G. Trade names: nuvan, vapona, Doom, Divap LD$_{50}$ value: 56 – 108 mg/kg.

Quinolphos

Quinalphos is an organothiophosphate chemical chiefly used as a pesticide. It is a reddish brown liquid. The chemical is Ranked 'moderately hazardous' in World Health Organization's (WHO) Quinalphos, which is classified as a yellow label (highly toxic) pesticide in India, is widely used in the following crops: wheat, rice, coffee, sugarcane, andcotton. It is contact poison having good penetrating power and It is having acaricidal properties.It is widely used against caterpillars and borer on cotton, vegetables and other crops. Formulations: EC 25 and 5 G. Trade names: Ekalux, Shakthi Quick, Quinguard, Quinaltaf, Smash, Flash LD$_{50}$ value: 62–137 mg/kg.

Phosolone

Phosalone is an organophosphate chemical commonly used as an insecticide and acaricide. It is developed byRhône-Poulenc in France but EU eliminated it from pesticide registration on December 2006. It is weak acetylcholinesterase inhibitor. It is a non systemic contact insecticide and acaricide, effective against wide spectrum of species. Formulations: EC 35 Trade names: Zolone LD$_{50}$ value: 135 mg/kg.

Chlorpyriphos

Chlorpyrifos is a crystalline organophosphate insecticide. It was introduced in 1965 by Dow Chemical Company and is known by many trade names including

Dursban and Lorsban. It acts on the nervous system of insects by inhibiting acetylcholinesterase. Chlorpyrifos is moderately toxic to humans and chronic exposure has been linked to neurological effects, developmental disorders, and autoimmune disorders. Exposure during pregnancy retards the mental development of children, and most use in homes has been banned since 2001 in the U.S. In agriculture, it remains "one of the most widely used organophosphate insecticides". It is a non-systemic contact insecticide very effective against sucking and chewing insects. It is widely recommended as seed treatment chemical against white grub and termites. Formulations: EC 20 Trade names: Dursban, Chloroban, Durmet, Radar LD$_{50}$ value: 135-163 mg/kg.

Phosphomidon

Phosphamidon is an organophosphate insecticide first reported in 1960. Phosphamidon is very highly toxic to mammals and is listed as WHO Hazard Class Ia. A harvester developed symptoms of moderately severe poisoning after working in a field that had been sprayed with the chemical 2 weeks earlier. He collapsed and exhibited significant depression of serum cholinesterase, but recovered completely within 2 days after successful treatment with atropine. International trade of phosphamidon is covered by the Rotterdam Convention. It is a systemic insecticide having low contact action. It is very effective against sap sucking insect pests. On application it is absorbed in the plant tissues within 1-3 hours and is translocated more towards the top. It is less toxic to fish and more toxic to bees. Formulations: 40 SL Trade names: Demecron, Sumidon, Chemidan, Hydan, Phamidon LD$_{50}$ value: 17-30 mg/kg.

Monocrotophos

Monocrotophos is an organophosphorus (OP) insecticide, developed by Ciba-Geigy (now Novartis) and first registered in 1965. This non-specific systemic insecticide and acaricide, used to control common mites, ticks and spiders with contact and stomach action. The United Nations has asked India to ban the monocrotophos pesticide - the presence of which in mid-day meal led to the death of 23 school children in Bihar. However, the manufacturers of monocrotophos insecticide advocated that it was cheaper and more effective in checking pests when compared to other alternatives in. Use of Monocrotophos has accordingly been banned on vegetables. However, its use on crops like cotton, paddy, maize, pulses, sugarcane, coconut, coffee, etc is still allowed, keeping in view its bio-efficacy and cost effectiveness. It is a systemic insecticide and acaricide with contact action. It has wide range of susceptibility of insects. It is toxic to bees. Formulations: 36 SL Trade names: Monocil, Nuvacron, Monophos, Monochem, Monostar LD$_{50}$ value: 14-23 mg/kg.

Methyl Demeton

Demeton-S-methyl is used as an acaricide and insecticide; It is flammable. Demeton-s-methyl is a pale yellow oil that has a sulfur-like odor. It is incompatible with alkaline materials.It is hydrolyzed rapidly in alkaline media and more slowly in acidic and neutral aqueous media.It is non-corrosive. When heated to decomposition, demeton-s-methyl emits very toxic fumes. It is contact and systemic insecticide and acaricide.It is used against soft bodied insects, which suck the plant sap. Formulations: 25 EC Trade names: Metasystox and Dhanusyatax LD_{50} value: 57-106 mg/kg.

Dimethoate

Dimethoate is a widely used organophosphorus (OP) insecticide applied to kill mites and insects systemically and on contact. It was introduced in the 1950s, originally patented by American Cyanamid. Dimethoate is used against a broad range of insects such as thrips, aphids, mites, and whiteflie and on a number of crops including citrus, cotton, fruit, olives, potatoes, tea, tobacco and vegetables. It is also permitted for the control of the flies in livestock accommodation, home gardens and food storage. Formulations: 30 EC Trade names: Rogor, Celgor, Novogor, Tara 909, roxion LD_{50} value: 320-380 mg/kg.

Triazophos

Triazophos is an organophosphorous broad-spectrum insecticide and acaricide with contact and stomach action. It is non-systemic, but penetrates deeply into plant tissues through translaminar action. It controls aphids, thrips, midges, beetles, caterpillars, spider mites and whiteflies in field crops, vegetables, ornamentals and fruit trees. Formulations: 40 EC Trade names: Hostathion, Trizocel, Truzo, Suthation LD_{50} value: mg/kg.

Profenophos

Non-systemic insecticide and acaricide with contact and stomach action. Exhibits a translaminar effect. Has ovicidal properties. Control of insects (particularly Lepidoptera) and mites on cotton, maize, sugar beet, soya beans, potatoes, vegetables, tobacco, and other crops, at 250-1000 g/ha. (Rat): Oral LD_{50} 358 mg/kg. (Rabbit): Oral LD_{50} 700 mg/kg. Dermal LD_{50} 277 mg/kg. It is highly toxic to birds and fish. Formulations: 50 EC Trade names: Curacron, Celcron, Bolero, Proven.

Acephate

Acephate is an organophosphate foliar insecticide of moderate persistence with residual systemic activity of about 10–15 days at the recommended use rate. It is used primarily for control of aphids, including resistant species, in vegetables (*e.g.* potatoes, carrots, greenhouse tomatoes, and lettuce) and in horticulture (*e.g.* on roses and greenhouse ornamentals). It also controls leaf miners, caterpillars, sawflies and thrips in the previously stated crops as well as turf, and forestry. By direct application to mounds, it is effective in destroying imported fire ants. Acephate is sold as a soluble powder, as emulsifiable concentrates, as pressurized aerosol, and in tree injection systems and granular formulations. It is a systemic and contact poison.It has low toxicity and safe to environment. Formulations: 75 SP Trade names: Arthane, Starthane, Orthene. LD_{50} value: 866-945 mg/kg.

Phorate

It is a systemic granular insecticide and also possesses acaricidal properties.

Phorate is a systemic granular insecticide and acaricide. At normal conditions, it is a pale yellow mobile liquid poorly soluble in water but readily soluble in organic solvents. It is relatively stable and hydrolyses only at very acidic or basic conditions. It is very toxic both for target organisms and for mammals including human. Phorate is most commonly applied in granular form. It is non-biocumulative and has no residual action. But some metabolites may persist in soil. Phorate is absorbed readily through all ways. It is very effective against sucking insects and also against maize borers, cut worms, white grubs etc. Formulations: 10 G. Trade names: Thimet LD_{50} value: 1.6 – 3.7 mg/kg.

4. Carbamates

Carbamates are esters of N-methyl carbamic acid. Aldicarb, carbaryl, propoxur, oxamyl and terbucarb are carbamates. All carbamate insecticides are derivatives of carbamic acid. Many of the carbamic esters are insecticidal and a few are effective molluscicides Like organophosphates, the carbamate insecticides interfere in cholinergic transmission. The carbamate enters the synapse and inhibits the acetylcholine-esterase as a result the acetylcholine contains to depolarize the post synaptic membrane, causing prolonged stimulation resulting into the failure of the nerve or effector tissue. Carbamates have an analogous action, carbamylating rather than phosphorylating the enzyme and the ChE recovers more readily from carbamates than from organophosphates. Thus, unlike, organophosphates, they are known as reversible inhibitors.

Carbaryl

Union Carbide discovered carbaryl and introduced it commercially in 1958. Bayer purchased Aventis CropScience in 2002, a company that included Union Carbide pesticide operations. It Carbaryl is often produced using methyl isocyanate (MIC) as an intermediary. A leak of MIC used in the production of carbaryl caused the Bhopal disaster, the largest industrial accident in history. This accident caused around 11,000 deaths and over 500,000 injuries. Carbaryl is a contact and stomach insecticide. It is most popular insecticide because it is effective against a wide range of insects and possesses very low mammalian toxicity. It is compatible with many pesticides except Bordeaux mixture lime sulphur and urea. It is not effective against mites. Formulations: WP 50 per cent, Granule 4 per cent and Dusts 5 per cent Trade names: Sevin. LD_{50} value: 400 mg/kg.

Propoxur (Arprocarb)

Propoxur is a carbamate insecticide and was introduced in 1959. Propoxur is a non-systemic insecticide with a fast knockdown and long residual effect used against turf, forestry, and household pests and fleas. It is also used in pest control for other domestic animals, *Anopheles* mosquitoes, ants, gypsy moths, and other agricultural pests. It can also be used as a molluscicide. It has long residual action. Formulations: 20 per cent EC, 50 per cent WP Trade names: Baygon, Blattamen, Saphaer LD_{50} value: 90-128 mg/kg.

Carbofuran

Carbofuran is one of the most toxic carbamate pesticides. Since its inception in 1967, Furadan has become one of the most widely used pest control insecticides in the world. Used as both an at plant and foliar insecticide, Furadan provides control of both soil and above ground pests.It is a plant systemic broad spectrum and long residual insecticide, miticide and nematicide. It is recommended as soil insecticides against plant sap sucking and borer pests. Formulation: 3G, 48F Trade names: Furadan LD_{50} value: 8-14 mg/kg.

Carbosulfan

Carbosulfan is a brown viscose liquid. It is not very stable; itdecomposes slowlyat roomtemperature.Its solubility in water Carbosulfan has very low maximum residue limits for use in the EU and UK examples of this can be seen in apples and oranges, where it is 0.05 mg/kg.It is a systemic insecticide, and nematicide. It is recommended as seed dresser insecticide Formulation: 25 DS. Trade name: Marshal.

Thiodicarb

Thiodicarb is a carbamate insecticide and pesticide that consists of two methomyl groups linked by amino nitrogen through sulfur molecules. As an insecticide, thiodicarb is effective against eggs as well as larvae, although the latter must feed upon treated foliage in order to be controlled. Heavy infestations of larvae may require higher applications than the standard dose, but not to exceed 60 ounces (1.77 liters) per acre (4047 square meters) per season.It is a insecticide with ovicidal properties, and molluscicide. Formulation: 75 WP. Trade name: Larvin.

Aldicarb

Aldicarb is effective against sucking pests but is primarily used as a nematicide. Aldicarb is effective where resistance to organophosphate insecticides has developed. Aldicarb was one of the "dirty dozen" pesticides that the environmental group Pesticide Action Network North America targeted in 1985. EPA put a ban in place in 2010, requiring an end to distribution by 2017. For humans, it is the most toxic insecticide used on field crops.It is systemic pesticide usually applied in soil as seed furrow, band or broadcast treatments either pre-plant or at planting as well as post emergence side dress treatments. It has also possessing acaricidal property and toxic to higher animals Formulation: 10 G. Trade names: Temik LD_{50} value: 0.93 mg/kg.

Methomyl

Methomyl was introduced in 1966, but its use is restricted because of its high toxicity to humans. The EU and UK have imposed these restrictions by allowing a maximum pesticide residue limit of 0.02 mg/kg for apples and oranges.0.02 mg/kg is the limit of detection. It is a systemic with contact and stomach insecticide and nematicide. It is very effective against a wide variety of pests particularly army worms, cabbage semilooper, Okra stem fly, fruit borers, leaf defoliators, cotton boll worms, etc. Formulations: 90 WP,12.5 EC, 40 SP Trade names: Lannate, Dunnate LD_{50} value: 30 mg/kg.

5. Synthetic Pyrethroids

A **pyrethroid** is an organic compound similar to the natural pyrethrins produced by the flowers of pyrethrums (*Chrysanthemum cinerariaefolium* and *C. coccineum*). Pyrethroids were introduced in the late 1900s by a team of Rothamsted Research scientists following the elucidation of the structures of pyrethrin I and II by Hermann Staudinger and Leopold Ružièka in the 1920s. Their work consisted firstly of identifying the most active components of pyrethrum, extracted from East African chrysanthemum flowers and long known to have insecticidal properties. Pyrethrum rapidly knocks down flying insects but has negligible persistence — which is good for the environment but gives poor efficacy when applied in the field. Pyrethroids now constitute the majority of commercial household insecticides. In the concentrations used in such products, they may also have insect repellent properties and are generally harmless to human beings in low doses but can harm sensitive individuals. They are usually broken apart by sunlight and the atmosphere in one or two days. Pyrethroids are usually combined with piperonyl butoxide, a known inhibitor of key microsomal cytochrome P450 enzymes from metabolizing the pyrethroid, which would diminish its lethality.

Synthetic pyrethroids have the properties of plant derivative pyrethrum as insecticides but are considerably more stable in light and air. Allethrin was first synthetic analogue of pyrethroids. They act on tiny channel through which sodium is pumped to cause excitation of neurons and prevent the sodium channels from closing, resulting in continual nerve trans mission, tremors and eventually death. The synthetic pyrethroids have extremely high insecticidal activity at extremely low doses and are bio-degradable in nature. Their activity is most pronounced against lepidopterous pests and they are very effective against beetle, leaf miner and bugs. They are very effective against eggs, larval and adult stages of insects. They have antifeedant and repellent properties. They are not readily washed off from the plants by rain due to lipophilic characters. These synthetic pyrethroids are very less toxic to mammals and having a quick knock down activity to insects, the lower toxicity to mammals and increase safety for the user. Very low application rate of synthetic pyrethroids as compared to conventional insecticides brings reduced environmental pollution.

Limitations

1. These are generally not effective as soil insecticide.
2. Even at low dosages kill non target species.
3. Cause resurgence of several groups of insect pests especially whiteflies and aphids.
4. Rapid development of resistance to synthetic pyrethroids in many insect species.
5. This may be due to high selection pressure exerted by high mortality.
6. Synthetic pyrethroids are poor acaricides.

First Generation

First generation pyrethroids are considered to be of low toxicity to people and other mammals because they are rapidly broken down in the body. First generation pyrethroids decompose quickly in sunlight and air and thus pose little risk in the environment but all pyrethroids are toxic to aquatic animals.

Allethrin

The **allethrin** was the first pyrethroid synthesized by Milton S. Schechter in 1949 in the United States. The compounds have low toxicity for humans and birds, and are used in many household insecticides such as mosquito coils. They are, however, highly toxic to fish and bees. Insects subject to its exposure become paralyzed (nervous system effect) before dying. They are also highly toxic to cats because they either do not produce, or produce less of certain isoforms ofglucuronosyl transferase, which serve in hepatic detoxifying metabolism pathways. They are also used as an ultra-low volume spray for outdoor mosquito control. Trade name: Pynamin LD$_{50}$ value: rats 572-1100 mg/kg for rats and Dermal LD$_{50}$ >2000 mg/kg.

Second Generation

Second generation pyrethroids are not acutely toxic to people or other mammals.These pyrethroids decompose rapidly in sunlight. They thus pose little threat to the environment, but for the same reason they are not suitable for agricultural use.

Resmethrin

It is a mixture of four optical isomers which have different insecticidal activities. The 2-S *alpha* (or SS) configuration, known as esfenvalerate, is the most insecticidally active isomer. Fenvalerate consists of about 23 per cent of this isomer.Approximately 20 times more effective than pyrethrum in housefly knock down, and is not synergized to any appreciable extent with pyrethrum synergists. Trade name: NRDC – 104, SBP-1382, and FMC – 17370 LD$_{50}$ value Dermal LD$_{50}$ 2000-3000 mg/kg.

Bioresmethrin

It is stereoisomer of resmethrin. Appeared in 1967. It is 50 times more effective than pyrethrum against normal (susceptible to insects) houseflies, and also not synergized with pyrethrum synergists. Both resmethrin and Bioresmethrin decompose fairly rapidly on exposure to air and sunlight, so never developed for agricultural use. Trade name: NRDC-107,FMC −18739, and RU-1148 LD$_{50}$: 8,600 mg/kg (oral) and 10,000 mg/kg (dermal).

Bioallethrin

It is d-trans −allethrin, introduced in 1969. More potent than allethrin and readily synergized, but it is not as effective as resmethrin.

Third Generation

Third generation pyrethroids do not decompose in sunlight and contain some of the most powerful insecticides known. Third generation pyrethoids are not highly toxic to people or other mammals mainly because they decompose rapidly in the body.

Fenvalerate

Fenvalerate is an insecticide of moderate mammalian toxicity. In laboratory animals, central nervous system toxicity is observed following acute or short-term exposure. Fenvalerate has applications against a wide range of pests. Residue levels are minimized by low application rates. Fenvalerate is most toxic to bees and fish. It is found in some emulsifiable concentrates, ULV, wettable powders, slow release formulations, insecticidal fogs, and granules. It is most commonly used to control insects in food, feed, and cotton products, and for the control of flies and ticks in barns and stables. Fenvalerate does not affect plants, but is active for an extended period of time.Fenvalerate may irritate the skin and eyes on contact, and is also harmful if swallowed.It is contact insecticide and of broad spectrum in nature. It is stable in sunlight and has longer residual toxicity. Formulations: 20 EC Trade names: Fenvel, Bilfen, Belmark, Sumicidin, Pydrin LD_{50} value: 300-630 mg/kg.

Permethrin

Permethrin is widely used as an insecticide, acaricide, and insect repellent. In medicine, permethrin is a first-line treatment for scabies; a 5 per cent (w/w) cream is marketed by Johnson and Johnson under the name Lyclear. In Nordic countries and North America, it is marketed under trade name **Nix**, often available over the counter.

In agriculture, permethrin is mainly used on cotton, wheat, maize, and alfalfa crops. Its use is controversial because, as a broad-spectrum chemical, it kills indiscriminately; as well as the intended pests, it can harm beneficial insects including honey bees, and aquatic life.

Permethrin is used on humans to eradicate parasites such as head lice or mites responsible for scabies and as a pest-repellent clothing treatment. Permethrin is more effective in reducing itch persistence than crotamiton or lindane. The common prescription is a 5 per cent concentration of permethrin for scabies and a 1 per cent concentration for the over-the-counter (OTC) treatment for head lice or crabs. Pharmaceutical grade permethrin 99 per cent is differentiated from pesticide grade 94 per cent by a higher purity, well specified impurities, and lower content of the toxic CIS component at 25 per cent as opposed to 40 per cent in the pesticide grade. Permethrin is also used in industrial and domestic settings to control pests such as ants and termites. It may be incorporated in formulations of wood preservative. Formulations: 25 EC and 5 per cent smoke generation. Trade names: Ambush, pounce, pramex. LD_{50} value: Acute oral LD_{50}: 7000 mg/kg, Dermal LD_{50}: >5100 mg/kg.

Fourth Generation

Offer the most resistance to exposure to sunlight and air and, therefore, are more persistent. This is more toxic to people than other pyrethroids and therefore requires more care in use. More stable in the environment.

Cyhalothrin is an organic compound that is used as a pesticide It is a pyrethroid, a class of man-made insecticides that mimic the structure and insecticidal properties of the naturally-occurring insecticide pyrethrum which comes from the flowers of chrysanthemums. Synthetic pyrethroids, like lambda-cyhalothrin, are often preferred as an active ingredient in insecticides because they remain effective for longer periods of time. It is a colorless solid, although samples can appear beige, with a mild odor. It has a low water solubility and is nonvolatile. It is used to control insects in cotton crops.

λ -Cyhalothrin

Lambda-cyhalothrin is a mixture of isomers of cyhalothrin. Non-systemic insecticide with contact and stomach action, and repellent properties.gives rapid knockdown and long residual activity. It is an insecticide and acaricide used to control a wide range of pests. Formulations: 2.5 EC, 5 per cent EC Trade names: Kung-Fu,Reeva, Charge, Excaliber, Grenade, Hallmark, Karate, Matador, Samurai and Sentinel.

Cyfluthrin

Cyfluthrin is a synthetic pyrethroid insecticide and common household pesticide. It is a complex organic compound and the commercial product is sold as a mixture of isomers. Like most pyrethroids, it is highly toxic to fish, invertebrates, and insects, but it is far less toxic to humans. It is generally supplied as a 10-25 per cent liquid concentrate for commercial use and is diluted prior to spraying onto agricultural crops and outbuildings.

Excessive exposure can cause nausea, headache, muscle weakness, salivation, shortness of breath and seizures. In humans, it is deactivated by enzymatic hydrolysis to several carboxylic acid metabolites, whose urinary excretion half-lives are in a range of 5–7 hours. Worker exposure to the chemical can be monitored by measurement of the urinary metabolites, while severe overdosage may be confirmed by quantification of cyfluthrin in blood or plasma. It is a non-systemic contact and stomach poison,with rapid knock down effect. It is for control of chewing and sucking insects on crops. Cyfluthrin is also used in public health situations and for structural pest control. Formulations: 5 EC, 10 per cent EC Trade names: Contur, Laser, Responsar, Tempo LD_{50} value: 869 - 1271 mg/kg.

Cypermethrin

Cypermethrin acts as a stomach and contact insecticide. It has wide uses in cotton, cereals, vegetables and fruit, for food storage, in public health and in animal husbandry. Cypermethrin is is found in many household ant and cockroachkillers, including Raid and ant chalk.It was synthesised in 1974 and first marketed in 1977, by Shell (which has since sold their pesticide business to American Cyanamid. It is also used for impregnation of mosquito bed nets to prevent malaria, and extensively for indoor pests. Formulations: 10 EC, 25 EC Trade names: Cyper guard, Ripcord, Cymbush and Cyper kill. LD_{50} value: ha Oral LD_{50} 303-4123 mg/kg, dermal more than 2400 mg/kg.

Fenpropathrin

It is contact insecticide and of broad spectrum in nature.It is extremely toxic to fish,wildlife.and aquatic organisms. It have acaricidal and miticidal property. Formulations: 2.4 EC, 10 or 20 per cent EC. Trade names: Danitol, Rody and Meothrin LD_{50} value:54 mg/kg.

Deltamethrin

Deltamethrin plays key role in controlling malaria vectors, and is used in the manufacture of long-lasting insecticidal mosquito nets. It is used as one of a battery of pyrethroid insecticides in control of malarial vectors, particularly *Anopheles gambiae,* and whilst being the most employed pyrethroid insecticide, can be used in conjunction with, or as an alternative to, permethrin, cypermethrin and other organophosphate-based insecticides, such as malathion and fenthion. Resistance to deltamethrin (and its counterparts) is now extremely widespread and threatens the success of worldwide vector control programmes. It is more potent than any other insecticide. It has also proved effective even against insects resistant to conventional insecticides.It is contact and stomach insecticide. Formulations: 2.8 EC, 2.5 per cent WP Trade names: Decis, Decaguard, Deltex, LD_{50} value:135mg/kg.

Fluvalinate

Fluvalinate is a stable, non-volatile, fat-soluble compound. Fluvalinate commonly used to control varroa mites in honey beecolonies. It is a insecticide and acaricide with stomach and contact activity in target insects. It is used as a broad spectrum insecticide. Formulations: 25 EC. Trade names: Klartan, Mavrik, Mavrik Aqua Flow, Spur and Yardex LD_{50} value: 1,050 to 1,110 mg/kg.

Fenfluthrin

It is a very potent recent synthetic pyrethroid against a various groups of insects and mites. Highly toxic to *Daphnia* (Aquatic Invertebrate) Trade Names: Bayticol, Bayvarol, Baynac.

Tefluthrin

It was designed to be effective against soil pests particularly termites. With an LD_{50} for rats of at 29 mg/kg, Tefluthrin is one of the most toxic pyrethroids.

6. Botanicals as Insect Growth Regulators

Fixed oils are the less aromatic oils derived from plants. Oils generally work by clogging the respiratory openings of the insect, causing suffocation. Oils are usually emulsified with water for application.Ex: Neem oil, Citronella Oil,Garlic oil etc. Neem oil is extracted from the tropical neem tree, *Azadirachta indica*, contains insecticidal properties that are composed of a complex mixture of biologically active compounds. Its various active ingredients act as repellents, feeding inhibitors, egg laying deterrents,growth retardants, sterilants and direct toxins. Neem oil has very low toxicity to mammals. The advantages of oil applications are many, like they are inexpensive, usually result in good coverage, are simple to mix, and are safe to warm-blooded animals. Some disadvantages of use include phytotoxicity, instability in storage,and ineffectiveness against certain pests.

7. Novel Insecticides

A. Neonicotinoids

Neonicotinoids are a class of neuro-active insecticides chemically similar to nicotine. The development of this class of insecticides began with work in the 1980s by Shell and the 1990s by Bayer Neonicotinoids are the first new class of insecticides introduced in the last 50 years, and the neonicotinoid imidacloprid is currently the most widely used insecticide in the world. The neonicotinoids include acetamiprid, clothianidin, imidacloprid, nitenpyram, nithiazine, thiacloprid and thiamethoxam. Neonicotinoids, like nicotine, bind to nicotinic acetylcholine receptors of a cell and triggers a response by that cell and this receptor blockage causes paralysis and death. Most neonicotinoids are water-soluble and break down slowly in the environment, so they can be taken up by the plant and provide protection from insects as the plant grows. In addition, most seeds are also treated with a neonicotinoid insecticide, usually thiamethoxam.

Imidacloprid

Imidacloprid is a systemic chloronicotinyl pesticide, belonging to the class of neonicotinoid insecticides. It works by interfering with the transmission of nerve impulses in insects by binding irreversibly to specific insect nicotinic acetylcholine receptors. As a systemic pesticide, imidacloprid translocates or moves easily in the xylem of plants from the soil into the leaves, fruit, pollen, and nectar of a plant. Imidacloprid also exhibits excellent translaminar movement in plants and can penetrate the leaf cuticle and move readily into leaf tissue.

Imidacloprid is a systemic insecticide and It is sold under many names for many uses; it can be applied by soil injection, tree injection, application to the skin of the plant, broadcast foliar, ground application as a granular or liquid formulation, or as a pesticide-coated seed treatment. It is widely used for pest control in agriculture. Other uses include application to foundations to prevent termite damage, pest control for gardens and turf, treatment of domestic pets to control fleas, protection of trees from boring insects, and in preservative treatment of some types of lumber products.

Systemic insecticide with translaminar activity and with contact and stomach action.Used as a seed dressing, soil application and foliar application against sucking insects including leaf hoppers, plant hoppers, aphids, thrips and whitefly, also effective against soil insects, termites. It is highly toxic to birds. Formulations: 17.8 SL,70 WS Trade names: Confidor, Gaucho, Admire, Merit, Premier, Stalone. Tatamida, Maratho,Provado LD_{50} value: 450 mg/kg.

Acetamiprid

Acetamiprid is an odorless neonicotinoid insecticide. It is systemic and intended to control sucking insects on crops such as leafy vegetables, citrus fruits, pome fruits, grapes, cotton, cole crops, and ornamental plants. Used as a soil and foliar application against homoptera especially aphid and leafhoppers.Thysonaptera and Lepidoptera. Formulations: 20 SP. Trade names: Pride, Assail Intruder, Profil, Supreme, LD$_{50}$ value:>2000 mg/kg.

Thiomethoxam

Thiamethoxam has a broad spectrum of activity against many types of insects. Thiamethoxam is a systemic insecticide is effective against sucking inscts, stem borers and termites. Thiamethoxam is a moderately toxic substance. In normal use, there are no unacceptable risks involved. The substance is toxic to bees and harmful to aquatic and soil organisms, although the level of toxicity to bees is not yet clear. A metabolite of thiamethoxam in soil is clothianidin. Contact and stomach poison with translaminar and systemic movement used as a seed treatment and foliar application against sucking insects. It has very strong effect on viral transmitting insects. Formulations: 25 WG, 70 WS. Trade names: Actara, Cruiser, Crux, Flagship, Meridian, Adage, Rinova LD$_{50}$ value: 1563 mg/kg.

Clothianidin

Clothianidin is an alternative to organophosphate, carbamate, and pyrethroid pesticides. It has helped prevent insect pests build up resistance to organophosphate and pyrethroid pesticides. It is systemic and translaminar in action It shows inhibitory action on oviposition and feeding. Formulations: 50 WG. Trade names: Dantop, Celeso LD$_{50}$ value: >5000 mg/kg.

Thiacloprid

Thiacloprid, a new chloronicotinyl insecticide, is targeted to control sucking and biting insects in cotton, rice, vegetables, pome fruit, sugar beet, potatoes and ornamentals. Pests controlled include aphids, whitefly, beetles and lepidoptera such as leaf miners. It acts as acute contact and stomach poison, with systemic action. Thiacloprid is slightly mobile in soil and hence it has no potential for leaching into ground water. Formulations: 36 WG, 70 WG. Trade names: Calypso, Bariard, Alanto LD$_{50}$ value: 500mg/kg.

B. Phenyl Pyrazoles (Fiproles)

Fipronil

Fipronil is a broad-use insecticide that belongs to the phenylpyrazole chemical family. Fipronil is a broad-spectrum insecticide that disrupts the insect central nervous system by blocking GABA-gated chloride channels and glutamate-gated chloride (GluCl) channels, resulting in central nervous system toxicity. This causes hyperexcitation of contaminated insects' nerves and muscles. Specificity of fipronil on insects may come from a better efficacy on GABA receptor, but also because GluCl channels do not exist in mammals.

Fipronil is a slow acting poison. When used as bait, it allows the poisoned insect time to return to the colony or harborage. In cockroaches, the feces and carcass can contain sufficient residual pesticide to kill others in the same nesting site In ants, the sharing of the bait among colony members assists in the spreading of the poison throughout the colony With the cascading effect, the projected kill rate is about 95 per cent in three days for ants and cockroaches Fipronil serves as a good bait toxin not only because of its slow action, but also because most, if not all, of the target insects do not find it offensive or repulsive.

Fipronil is or has been used in these manners:

☆ Under the trade name Regent, it is used against major lepidopteran and orthopteran pests on a wide range of field and horticultural crops and against coleopteran larvae in soils.

☆ Under the trade names Goliath and Nexa, it is employed for cockroach and ant control.

☆ It has been used under the trade name Adonis for locust control in Madagascar and in Kazakhstan.

☆ Marketed under the names Termidor, Ultrathor, and Taurus in Africa and Australia, fipronil effectively controls termite pests, and was shown to be effective in field trials in these countries.

☆ Fipronil is the main active ingredient of Frontline Top Spot, Fiproguard, and PetArmor (used along with S-methoprene in the 'Plus' versions of these products); these treatments are used in fighting tick and flea infestations in dogs and cats.

☆ In New Zealand fipronil has been used in a trial to control wasps, which are a threat to indigenous biodiversity.

Formulations: 0.3 G, 5 SC, Trade names: Regent, Front line,Tremidor, Zoom, Icon Tempo, Bilgran.

C. Macrocyclic Lactones
Spinosyns–Spinosad

Spinosad is based on a compound found in the bacterial species *Saccharopolyspora spinosa* (*S. spinosa*). The genus of*Saccharopolyspora* was discovered in 1975 by Lacey and Goodfellow, who described isolates from crushed sugar cane.

The spinosyns and spinosoids have a novel mode of action, primarily targeting binding sites on nicotinic acetylcholine receptors (nAChRs) of the insect nervous system kills insects via hyperexcitation of the insect nervous system. The extract of the fermentation broth that contains spinosad is produced by the microorganism, *Saccharopolyspora spinosa*. The primary components are spinosyn A and spinosyn D. Spinosad kills insects by causing rapid excitation by activation of nicotinic acetylcholine receptors of the insect nervous system, leading to involuntary muscle contractions, prostration with tremors, and paralysis. It also effects GABA receptor functioning. Spinosad is a contact and stomach poison with some translaminar movement in leaf tissue. Formulations: 45 SC, 2.5 WSC Trade names: Tracer, Spintor, Precise, Success, Naturalyte, Laser, Credence Caribstar,Boomerang, and Conserve LD_{50} value: 3738 mg/kg.

Avermectins

The avermectins are macrocyclic lactone derivatives with potent anthelmintic and insecticidal properties. These naturally occurring compounds are generated as fermentation products by *Streptomyces avermitilis*, a soil actinomycete. Avermectins activate the GABA gated chloride channel, causing an inhibitory effect, which, when excessive, results in the insect's death. This channel normally blocks reactions in some nerves, preventing excess stimulation of CNS. Emamectin benzoate and abamectin are the two major compounds in this group.contact and stomach poisons. These are used as bait, foliar application against Homoptera, Diptera, Coleoptera, Lepidoptera and mites.

Emamectin Benzoate

Emamectin is the 4"-deoxy-4"-methylamino derivative of abamectin, a 16-membered macrycyclic lactone produced by the fermentation of the soilactinomycete *Streptomyces avermitilis*. It is generally prepared at the salt with benzoic acid, emamectin benzoate, which is a white or faintly yellow powder. Emamectin is derived from avermectin B1, also known as abamectin, a mixture of the natural avermectin B1a and B1b. Emamectin has also shown promising applications in the eradication of fish lice and in fish farming.It is non systemic insecticide which penetrates by translaminar movement and effective against Lepidopterous pests It has low toxicity to non target organisms and environment. Formulations: EC 5, SG 5, Trade names: Proclaim, LD_{50} value: 300 mg/kg.

Abamectin

Abamectin is a mixture of avermectins containing more than 80 per cent avermectin B1a and less than 20 per cent avermectin B1b. These two components, B1a and B1b have very similar biological and toxicological properties. The avermectins are insecticidal and antihelmintic compounds derived from various laboratory broths fermented by the soil bacterium *Streptomyces avermitilis*. Abamectin is a natural fermentation product of this bacterium.It is a broad spectrum insecticide acting on mites of Tetranychidae, Eeriophyidae and Tarsonemidae. It is also effective against tobacco hornworm, diamondback moth, tobacco budworm, serpentine leaf miner and less potent against certain Homoptera (aphids) and Lepidoptera. It is less toxic to beneficial arthropods Formulations: EC 1.8 Trade names: Avid, Agrimec, Vertimec, Argimek, Affirm and Avert.

Milbemectin

Milbemectin occurs naturally in bacteria of the genus Actinomyces which is a white solid and insoluble in water. Milbemectin is as acaricide and insecticide used in orchards. It was developd by the Japanese company Mitsui Chemicals against spider mites in pome fruit, and strawberries as an acaricide and against leafminers used in ornamental plants as an insecticide for use in the greenhouse. On the plant surface Milbemectin is degraded in a short time, which facilitates the applications of the product in ornamental plant in combination with beneficial insects. The effect is based on the interruption of the conduction of the nervous system of spider mites and insects and thus leads to an immediate paralysis and subsequent death of the harmful animals. It is sold under the trade name Mite Knock.

D. Oxadiazines

Indoxacarb

Indoxacarb is an oxadiazine pesticide developed by DuPont that acts against lepidopteran larvae. Its main mode of action is via blocking of nerve sodium channels. The result is impaired nerve function, feeding cessation, paralysis, and death. Indoxacarb is the active ingredient in a number of household insecticides, including cockroach baits, and can remain active after digestion.Indoxacarb is the active ingredient in the new pet product, Activyl from Merck Animal Health. It is marketed to kill fleas on dogs and cats. Formulations: SC 14.5, WDG 30. Trade names: Avaunt, Steward. Torando.

E. Thio-Urea Derivatives

Diafenthiuron

Diafenthiuron is a pro-insecticide, which has first to be converted to its active form. The active compound then acts on a specific part of the energy-producing enzymes in the mitochondria. This results in immediate paralysis of the pest after intake or contact with the product. Excellent whitefly control on cotton and top performance against mites, aphids and jassids through the control of nymphs and adults.controls nymphs and adults resulting in more flexible application timing and longer lasting control. It belongs to a unique chemical group allowing control of insects and mites resistant to major chemical classes such as OPs or Pyrethroids. It is translaminar, allowing control of hidden pests in the plant canopy and on the underside of the leaves.It has vapor action and works well in dense crops and in large fields. It results in quick knockdown through immediate paralysis of the pest. It degrades into a urea derivative resulting in a phytotonic effect. Diafenthiuron is photochemically converted within a few hours in sunlight to its carbodimiide derivative which is much more powerful acaricide/insecticide than diafenthiuron. It is a inhibitor of oxidative phosphorylation, via distruption of ATP formation (inhibitor of ATP synthase). Formulations: 50 WP, Trade names: Polo [Syngenta product] LD_{50} value: 2068 mg/kg.

F. Pyridine Azomethines

Pymetrozine

Pymetrozine is an insecticide derived from a novel type of chemistry which is highly selective against plant sucking insects. It penetrates green leaves and is transported systemically within the plant. Pymetrozine quickly prevents aphid feeding and although there is no immediate knockdown effect, aphids do not feed again, and death occurs through starvation within 1–4 days. Activity against pollen beetle is primarily through direct contact. it blocks the response of stretch receptors (chorodontal mechanoreceptors) in insects.It is a new insecticide highly active and specific against sucking insect pests. Pymetrozine is the only representative of the pyridine azomethine. It has high degree of selectivity, low mammalian toxicity and safety to birds, fish and non-target arthropods. When the insertion of the stylets of sucking insects into the pymetrozine treated plant tissues, stylets are almost immediately blocked. The sucking insects die by starvation a few days later (feeding depressant) Formulations: 50 WDG. Trade names: Full fill, Chess, LD_{50} value: 5693 mg/kg, Maximum Individual Dose is 0.4 kg WG per hectare and Maximum number of treatments is 3 per crop.

G. Pyrroles

Pyrroles are oxidative phosphorylation inhibitors. It works by uncoupling oxidative phosphorylation from electron transport process in mitochondria. (Oxidative phosphorylation is the process through which ATP is synthesized in plants and animals). It interferes with formation of ATP which is essential for muscle contraction.

Chlorfenapyr

Chlorfenapyr was first discovered in 1985 by isolating a toxin from the *Streptomyces fumanus actinomycete bacterium*. In 1995 it was launched as an agricultural insecticide. Chlorfenapyr works by disrupting the production of Adenosine triphosphate, cellular death, and ultimately organism mortality."It is a miticide and insecticide. Chlorfenapyr has broad spectrum of activity against many species of Coleoptera, Lepidoptera, Acarina and Thysanoptera. Belonging to the class of chemistry known as pyrroles, chlorfenapyr has proven remarkably effective against mosquitoes and other vectors, especially those that have developed resistance to other insecticides. It is mainly stomach poison and has contact action also. Formulations: 10 per cent SC, 36 per cent SC. Trade names: Stealth, Phantom, Intrepid, Pirate, Pylon LD_{50} value: 626 mg/kg.

H. Formamidines

The formamidines are a structurally novel group of pesticides of growing importance in the control of mites, cattle ticks and certain orders of insects which have become resistant to conventional acaricides and insecticides. Their mode of action is complex with dose-dependent lethal and sublethal effects. At sublethal levels they cause behavioural changes in the target pest species (for example in feeding and in mating behaviours), changes which are responsible for the protective effects on crops and livestock. Although many suggestions have been made for the underlying biochemical mechanism, including inhibition of monoamine oxidase activity, uncoupling of respiration and blockade of neuromuscular transmission, no direct evidence has been presented. Another possibility is interaction with octopamine receptors in the central nervous system. the formamidine acaricide/insecticide, chlordimeform (CDM), and its demethylated derivative can mimic the actions of octopamine at the locust neuromuscular junction. This gives the clearest evidence to date of the site of action of the formamidines and indicates a novel mode of action for these pesticides.

Chlordimeform

It has marked translaminar and systemic activity. It shows a strong repellent-antifeedant action on both lepidopterous larvae and mites. It has good ovicidal activity. Non toxic to non target organisms except predaceous mites. Formulations: 50 SP, 4 EC Trade names: Galecron, Fundal, Fundal, Spike. LD_{50} value: 340 mg/kg.

Amitraz

Amitraz is a non-systemic acaricide and insecticide. It was first synthesized by the Boots Co. in England in 1969.Amitraz has been found to have an insect repellent effect, works as an insecticide and also as a pesticide synergist. Its effectiveness is traced back on alpha-adrenergic agonist activity, interaction with octopamine receptors of the central nervous system and inhibition of monoamine oxidases and prostaglandin synthesis. Therefore, it leads to overexcitation and consequently paralysis and death in insects. Because amitraz is less harmful to mammals, amitraz is among many other purposes best known as insecticide against mite- or tick-infestation of dogs. Formulations: 50 SP, 20 EC, Trade names: Acarac, Amitraze,Baam LD_{50} value: 523- 800 mg/kg.

I. Ketoenols

In organic chemistry, keto–enol tautomerism refers to a chemical equilibrium between a keto form (a ketone or an aldehyde) and an enol (an alcohol). The enol and keto forms are said to be tautomers of each other. The interconversion of the two forms involves the movement of an alpha hydrogen and the shifting of bonding electrons; hence, the isomerism qualifies as tautomerism.Ketoenols act as insecticide and acaricides against against all developmental stages and is a valuable new tool in the resistance management. They are tetronic acid insecticides with acaricidal action. Their mode of action is to inhibit lipogenesis in treated insects, resulting in decreased lipid contents, growth inhibition of younger insects, and reduced ability of adult insects to reproduce.

Spiromesifen

Spiromesifen is an insecticide and acaricide from the class of tetronic acid derivatives. Spiromesifen acts by inhibiting acetyl CoA carboxylase, an enzyme of the fat metabolism, as a consequence of dry the affected insects die and three to ten days after the treatment. Due to the still short availability are yet no resistance in the target insects known, as there are no cross-resistance with other available insecticides. Spiromesifen is effective against whitefly, spider mites and psyllids. It is is particularly active against juvenile stages. However, it also strongly affects fecundity of mite (and whitefly adults by transovariole effects. Formulations: 2 SC, 4 F 89. Trade names: Oberon, Forbid LD_{50} value: >2000 mg/kg.

Spirodiclofen

Spirodiclofen is a chemical compound from the group of tetronic acid derivatives. Spirodiclofen is a white odorless solid that is insoluble in water. Spirodiclofen is as acaricide fruit and viticulture, as well as insecticide and acaricide (against spider mites and gall mites used). The effect is based on inhibition of the biosynthesis of lipids. Spirodiclofen is a selective, non-systemic foliar insecticide and acaricide. It is effective against mites and sanjose scales. Formulations: 2 SC. Trade names: Envidor LD_{50} value: >2500 mg/kgLD_{50}.

Spirotetramat

Spirotetramat is an insecticide from the group of keto-enols, developed by Bayer CropScience. The pesticidal mechanism of action is disruption of lipogenesis as a result of inhibition of acetyl CoA carboxylase.

Spirotetramat is effective against stinging-sucking insects, including aphids, mites and whiteflies. It is a systemic insecticide.

Formulations: SC 14.5, SC 22.4, Trade names: Bayer brings to the market under the brand names *Movento* and *Ultor* LD_{50} value: >2000 mg/kg.

J. Diamides

The diamides are the most recent addition to the limited number of insecticide classes with specific target site activity that are highly efficacious, control a wide pest spectrum, and have a favorable toxicological profile. Currently available diamide insecticides include chlorantraniliprole and flubendiamide. Flubendiamide, the first diamide insecticidal compound, was discovered by Nihon Nohyaku and co-developed with Bayer. Shortly after, DuPont introduced chlorantraniliprole and cyantraniliprole, which are commercialized by DuPont and Syngenta. Thus, collaboration among companies is critical to prevent or delay the evolution of insect resistance since the four companies competitively sell multiple brands from this novel insecticide class worldwide.

The diamide insecticide molecules share the same target site, the ryanodine receptor. Flubendiamide is a phthalic diamide while chlorantraniliprole and cyantraniliprole are anthranilic diamides. Although structurally distinct, both classes are deemed to bind at the same site in the receptor. Recent research suggested that possibly in housefly, phthalic and anthranilic diamides bind at different allosterically-coupled sites in the ryanodine receptor. Ryanodine receptors are named after ryanodine, a plant secondary compound obtained from the neotropical plant *Ryania speciosa*. The ryanodine receptor is a calcium release channel located on the endoplasmic reticulum. Ryanodine binds to the ryanodine receptor and disturbs calcium flow by locking the channel in a partially opened state. It has been used in the past as an insecticide but high mammalian toxicity of ryanodine precluded its continued field use.

Chlorantraniliprole

Cyantraniliprole is a second-generation anthranilic diamide insecticide discovered by DuPont Crop Protection. This insecticide is currently registered under the active ingredient trade name Cyzapyr™. Anthranilic diamides havea unique mode of action that involves activating ryanodine receptors (RyR), which play a critical role in muscle function.[1-3]Cyantraniliprole binds to the RyR, causing uncontrolled release and depletionof calcium from muscle cells, thus preventing further muscle contractionand ultimately leading to death. Cyantraniliprole is a reduced-risk insecticide, with a very lowtoxicity to vertebrates and non-target organisms. It has root systemic and translaminar activity against a broad spectrum of sucking and chewing.

Chlorantraniliprole is a novel anthranilic diamide insecticide, efficacious for control of lepidopteran insect pests as well as some species in the orders Coleoptera, Diptera and Hemiptera. It is active on chewing pest insects primarily by ingestion and secondarily by contact. It exhibits larvicidal activity as an orally ingested toxicant by targeting and disrupting the Ca^{2+} balance. Chlorantraniliprole activates ryanodine receptors via stimulation of the release of calcium stores from the sarcoplasmic reticulum of muscle cells (i.e for chewing insect pests) causing impaired regulation, paralysis and ultimately death of sensitive species. It has low mammalian toxicity. Chlorantraniliprole can be used as foliar spray on insect pests of fruits and vegetable crops@ 10 to 60 g/ha. Trade names: Coragen 200SC and Altacor 35 WG.

Cyantraniliprole

Cyantraniliprole is another diamide efficacious against a cross spectrum of chewing and sucking pests. It works as toxicant in ingestion orally and has systemic activity. It targets and disrupts Ca^{2+} balance in nervous system. It is formulated as OD (Oil Dispersion) and SC (Suspension Concentrate) and generally applied @ 10-100 g a.i/ha Trade name : Cyazypyr.

Flubendiamide

Flubendiamide, developed by Nihon Nohyaku Co., Ltd. (Tokyo, Japan), is a novel activator of ryanodine-sensitive calcium release channels (ryanodine receptors; RyRs), and is known to stabilize insect RyRs in an open state in a species-specific manner and to desensitize the calcium dependence of channel activity. In this study, using flubendiamide as an experimental tool, we examined an impact of functional modulation of RyR on Ca^{2+} pump.

Formulations: 240 WG, 480 SC Trade names: Fame, Belt, Takumi.

8. Chitin Synthesis Inhibitors

Chitin synthesis inhibitors work by preventing the formation of chitin, a carbohydrate needed to form the insect's exoskeleton. With these inhibitors, an insect grows normally until it molts. The inhibitors prevent the new exoskeleton from forming properly, causing the insect to die. Death may be quick, or take up to several days depending on the insect. Chitin synthesis inhibitors can also kill eggs by disrupting normal embryonic development. Chitin synthesis inhibitors affect insects for longer periods of time than hormonal IGRs. These are also quicker acting but can affect predaceous insects, arthropods and even fish. Chitin synthesis inhibitors disrupt molting by blocking the formation of chitin, the building block of insect exoskeleton. Without the ability to synthesize chitin, molting is incomplete, resulting in malformed insects that soon die. It suppresses egg-laying and causes egg sterility in treated adults through secondary hormonal activity.

Diflubenzuron

Diflubenzuron is a benzoylurea-type insecticide of the benzamide class. It is used in forest management and on field crops to selectively control insect pests, particularly forest tent caterpillar moths, boll weevils, gypsy moths, and other types of moths.The mechanism of action of diflubenzuron involves inhibiting the production of chitin which is used by an insect to build its exoskeleton. One of the metabolites of diflubenzuron, 4-chloroaniline, has been considered as a probable human carcinogen.

It is Stomach and contact poison that acts by inhibiting chitin synthesis so it interferes with formation of cuticle.TN: Dimilin 25 WP.

Flufenoxuron

Broad spectrum Insect and mite growth regulator with contact and stomach action. TN: Cascade 10 WDC. Flufenoxuron was banned in the European Union in 2011 due to its high potential for bioaccumulation in the food chain and high risk to aquatic organisms. Flufenoxuron is marketed as having 'high persistence' in the environment and the product data-sheet states that it does not biodegrade easily.

Chlorfluazuron

Chlorfluazuron is used in termite baiting stations. TN: Atabron5 SC.

Chlorfluazuron is an insect development inhibitor, which is registered for use in termite bait matrix systems. It is generally applied to bait stations and to active termite galleries where termites are encouraged to feed on cellulose material impregnated wit Chlorfluazuron. Chlorfluazuron is an unscheduled termiticide and posses no threat to human health when used in accordance with label directions.

Teflubenzuron

It is effective against Lepidoptera, Coleoptera, Diptera, Hymenoptera Aleyrodidae, and Psyllidae. Nomolt,Dart,Nemolt, 15 SC.

Novaluron

Novaluron, is a benzoylphenyl urea developed by *Makhteshim-Agan Industries Ltd.* the compound is being used on food crops, including apples, potatoes, brassicas, ornamentals and cotton and pose low risk to the environment and non-target organisms. It acts mainly by ingestion, but has shown some contact activity. It does not have ovicidal activity, but a high percentage of mortality of first instars hatching from eggs laid on sprayed foliage occurs. TN: Rimon 10 EC.

Buprofezin

Buprofezin is an Insect Growth Regulator is effective against the nymph stages of whitefly, scales, mealybugs, planthoppers, and leafhoppers by inhibiting chitin biosynthesis, suppressing oviposition of adults, and reducing viability of eggs. Treated susceptible pests may remain alive on the plant for three to seven days, but feeding damage during this time is typically very low.

Applaud® 70 WP Insect Growth Regulator is not disruptive to beneficial insects and mites. Applaud® 70 WP Insect Growth Regulator is labeled for use with almonds, bananas, citrus, cotton, cucumbers, grapes, lettuce, melons, pumpkins, squash, and tomatoes. Contact and stomach, persistent chitin synthesis inhibitor with miticidal action. Effective against specifically on Homopteran pests. TN: Applaud 25 SC, 70 WP.

Flufenoxuron

Contact and stomach, inhibits chitin synthesis in nymphal mites and lepidopteran larvae.Compatible with α-Cypermethrin. TN: Cascade, Casette Tenope, 10 per cent EC, 5 per cent EC.

9. Juvenile Hormone (JH)Mimics

Juvenile Hormone analogues as potential insect control molecules was first recognized by Williams (1965). The compounds showing JH activity, 'Juvenoids'. Four types of JHs (JH= 0, I, II, and III) are known with their structural variations.JH-0 is known from the eggs of *Manduca sexta* only.JH-I and JH-II are from all lepidopterans and are said to be morphogenetic in action *i.e* to retain the larval characters.JH-III present in all insect orders and are said to be gonado tropic *i.e.* for stimulating the ovaries to mature in the female. The so called 'Paper factor' (Karel Slama and C.M. Williams, 1966) against the bug *Pyrrhocoris apterus* (Heteroptera) was subsequently traced to be balsam fir trees, *Abis balsamea* from the wood pulp of which the paper towels were manufactured and scientifically named 'Juvabione'. These findings opened up the flood gates for JH research chemists to get busy in synthesizing JH mimics (Juvenoids).

Mode of Action

i) Antimorphic effect: Do not allow Metamorphosis to take palace there by forcing larva to continue as a larva. There fore if the Juvenoids are provided exogenously the larvae will undergo an extra larval moult (change in to super larva) or moult in to defective intermediate forms which may suffer from a failure to successfully moult, feed or mate. ii) Larvicidal effect, iii) Ovicidal effect, iv) Diapause disrupting effect.,v) Embryogenesis inhibiting effect.

Juvenoids

Juvenoids acts as ovicides when applied directly on eggs and indirectly on ovipositing females. They block embryogenic development of blastokinesis stage. When applied before hatching, they show morphogentic effect at the time of metamorphosis. They inhibit ecdysone synthesis by effecting prothorasic glands. If applied to the last instar larvae, they could prevent pupa from entering in to diapause. They could terminate pupal diapause by activating the inactive PTG of diopausing pupa.

i) Juvabione (Fernesol - extracted from excreta of *Tenebrio* sp.)

ii) **Methoprene** (Altosid) : Methoprene is a juvenile hormone (JH) analog which is an amber-colored liquid with a faint fruity odor which is essentially nontoxic to humans when ingested or inhaled. It is used in drinking water cisterns to control mosquitoes which spread dengue fever and malaria.Methoprene is also used in the production of a number of foods including meat, milk, mushrooms, peanuts, rice and cereals. It also has several uses on domestic animals (pets) for controlling fleas. Methoprene is considered a biochemical pesticide because rather than controlling target pests through direct toxicity, methoprene interferes with an insect's life cycle and prevents it from reaching maturity or reproducing. Methoprene is also used as a food additive in cattle feed. This is done to prevent fly breeding in the dung piles.

iii) **Hydroprene** (Gentrol, Altozar) : Hydroprene is an insect growth regulator used as an insecticide. It is used against cockroaches, beetles, and moths. Products using hydroprene include Gencor, Gentrol, and Raid Max Sterilizer Discs.

iv) Kinoprene (Enstar).

10. Anti JH orJH Agonists or Precocenes

Anti-juvenile hormones found in plants that induce reversible precocious metamorphosis and sterilization in insects by suppressing the function of the corpora allata gland Precocenes are the compounds which would antagonize the JH activity and dearrange the insect development. These compounds induce the precocious metamorphosis of immature insects. Precocenes affect insect diapause, reproduction and behaviour. These compounds first extracted from the plant *Ageratum houstonium.* It contains two simple chromene compounds precocene –I and II. 1) Fenoxycarb (Insegar 50 per cent WP, Award,Comply Logic, Torus, Pictyl and Varikill) @ 0.025 per cent 2) Pyriproxyfen (Tiger 10 per cent EC, Distance, Esteem, Sumilarv and Admiral) @ 0.0125 per cent.

Fenoxycarb is a carbamate insect growth regulator. It has a low toxicity for bees, birds, and humans, but is toxic to fish. The oral LD_{50} for rats is greater than 16,800 mg/kg. Fenoxycarb is non-neurotoxic and does not have the same mode of action as other carbamate insecticides. Instead, it prevents immature insects from reaching maturity by mimicking juvenile hormone.

Pyriproxyfen is a pyridine-based pesticide which is found to be effective against a variety of arthropoda. It was introduced to the US in 1996, to protect cotton crops against whitefly. It has also been found useful for protecting other crops. It is also used as a prevention for fleas on household pets.Pyriproxyfen is a juvenile hormone analog, preventing larvae from developing into adulthood and thus rendering them unable to reproduce.In the US pyriproxyfen is often marketed under the trade name *Nylar.* In Europe pyriproxyfen is known under the brand names Cyclio (Virbac) and Exil Flea Free TwinSpot (Emax).

11. Ecdysone or Moulting Hormone (MH) Agonists

MH contains two hormones, α-Ecdysone and β-Ecdysone. The α-ecdysone is a prohormone produced by PTG which is converted into β-ecdysone in the peripheral tissues of the gland and is also called 20-Hydroxy Ecdysone, which actually brings about molting in insects and is the true MH. Synthetic analogues of ecdysones are called ecdysoids.After absorption into haemolymph it binds the ecdysone receptor proteins which initiates moulting process.The normal moulting process is disrupted. Larvae is prevented from shedding of old cuticle and it will die due to dehydration and starvation.

a) **Tebufenozide** (Mimic, Confirm): Tebufenozide is an insecticide that acts as a molting hormone. It is an agonist of ecdysone that causes premature molting in larvae. It is primarily used against caterpillar pests. Because it has high selectivity for the targeted pests and low toxicity otherwise, the company that discovered tebufenozide, Rohm and Haas, was given a Presidential Green Chemistry Award for its development.

b) **Halofenozide (Mach):** Controls all grubs like White grub larvae, Japanese beetles, European chafers, Oriental beetles, black turfgrass ataenius and northern and southern masked chafers,Lepidoptera larvae, such as cutworms, sod webworms and armyworms. Offers a wide application window (apply as a preventative treatment prior to egg hatch through the second instar).Does not require immediate irrigation after application. Granular and on-fertilizer formulations are labeled for use on golf courses, residential lawns and other established turf. Liquid formulations are registered for use on commercial sites, including golf courses.

c) **Methoxyfenozide (Prodigy).**

Chapter 7

Toxicology of Insecticides

7.1 Definition

Toxicology is a branch of biology, chemistry, and medicine (more specifically pharmacology) concerned with the study of the adverse effects of chemicals on living organisms. A toxicologist is a scientist or medical personal who specializes in the study of symptoms, mechanisms, treatments and detection of venomsand toxins; especially the poisoning of people.

It also studies the harmful effects of chemical, biological and physical agents in biological systems that establishes the extent of damage in living organisms.Toxicology is the branch of medical sciences that deals with the nature, properties, effects and detection of poison.It is therefore, the science of poisons.Whereas, Insect toxicology is th study of economic poisons, their effects, mechanismsof toxicant in insects.

Dioscorides, a Greek physician in the court of the Roman emperor Nero, made the first attempt to classify plants according to their toxic and therapeutic effect. Ibn Wahshiya wrote the *Book on Poisons* in the 9th or 10th century. Mathieu Orfila is considered the modern father of toxicology, having given the subject its first formal treatment in 1813 in his *Traité des poisons*, also called *Toxicologie générale*. In 1850, Jean Stas gave the evidence that the Belgian Count Hippolyte Visart de Bocarmé killed his brother-in-law by poisoning him with nicotine.

Theophrastus Phillipus Auroleus Bombastus von Hohenheim (1493–1541) (also referred to as Paracelsus, from his belief that his studies were above or beyond the work of Celsus – a Roman physician from the first century) is also considered "the father" of toxicology. He is credited with the classic toxicology maxim, *"Alle Dinge sind Gift und nichts ist ohne Gift; allein die Dosis macht, dass ein Ding kein Gift ist."* which translates as, "All things are poison and nothing is without poison; only the dose makes a thing not a poison." This is often condensed to: "The dose makes the poison" or in Latin "Sola dosis facit venenum".

The relationship between dose and its effects on the exposed organism is of high significance in toxicology. The chief criterion regarding the toxicity of a chemical is the dose, *i.e.* the amount of exposure to the substance. All substances are toxic under the right conditions. The term LD_{50} refers to the dose of a toxic substance that kills 50 percent of a test population (typically rats or other surrogates when the test concerns human toxicity).

Factors that influence chemical toxicity: Dosage [Both large single exposures (acute) and continuous small exposures (chronic) are studied], Route of exposure [Ingestion, inhalation or skin absorption], Other factors [Species, Age, Sex, Health, Environment, Individual characteristics.

Toxic interaction of a chemical with a given biological system is dose related.Hence, toxicology can be termed as *Science of doses*.

In toxicology, the median lethal dose, LD_{50} (abbreviation for "lethal dose, 50 per cent "), LC_{50} (lethal concentration, 50 per cent) or LCt_{50} (lethal concentration and time) of a toxin, radiation, or pathogen is thedose required to kill half the members of a tested population after a specified test duration. LD_{50} figures are frequently used as a general indicator of a substance's acute toxicity. The test was created by J.W. Trevan in 1927. The term semilethal dose is occasionally used with the same meaning, in particular in translations from non-English-language texts, but can also refer to a *sub*lethal dose; because of this ambiguity, it is usually avoided. LD_{50} is usually determined by tests on animals such as laboratory mice.

LD_{50} : The median lethal dose of a substance, or the amount required to kill 50 per cent of a given test population. LD_{50} is a measurement used in toxicology studies to determine the potential impact of toxic substances on different types of organisms. It provides an objective measure to compare and rank the toxicity of substances. The LD_{50} measurement is usually expressed as the amount of toxin *per kilogram or pound of body weight*. When comparing LD_{50} values, a lower value is regarded as more toxic, as it means a smaller amount of the toxin is required to cause death. The LD_{50} test involves exposing a population of test animals, typically mice, rabbits, guinea pigs, or even larger animals such as dogs, to the toxin in question. The toxins might be introduced orally, through injection, or inhaled. Because this testing kills a large sample of the animals, it is now being phased out in the United States and some other countries in favor of newer, less lethal methods.Pesticide studies involve LD_{50} testing, usually on rats or mice and on dogs. Insect and spider venoms can also be compared using LD_{50} measurements, to determine which venoms are the most deadly to a given population of organisms. **Examples:**LD_{50} values of insect venom for mice:Honey bee, *Apis mellifera* - LD_{50} = 2.8 mg per kg of body weight, Yellowjacket, *Vespula squamosa* - LD_{50} = 3.5 mg per kg of body weight.

It must be understood that higher the LD_{50} value lesser is the toxic nature of the chemical and vice - versa. Acute toxicity refers to the toxic effect produced by a single dose of a toxicant where as chronic toxicity is the effect produced by the accumulation of small amounts of toxicant over a long period of time. Here the single dose produces no ill-effect.

The LC_{50} (Lethal concentration) is the concentration of the chemical in the external medium required to kill 50 per cent of the test population. This value is used when exact dose given to the insect cannot be determined and is usually determined by potters tower and probit analysis.

ED_{50}/EC_{50} (Effective Dose/Concentration 50): Chemicals that gives desirable effects in 50 per cent of test animals. The "median effective dose" is the dose that produces a quantal effect (all or nothing) in 50 per cent of the population that takes it (median referring to the 50 per cent population base). It is also sometimes abbreviated as the ED_{50}, meaning "effective dose, for 50 per cent of people receiving the drug". The ED_{50} is commonly used as a measure of the reasonable expectancy of a drug effect, but does not necessarily represent the dose that a clinician might use. This depends on the need for the effect, and also the toxicity. The toxicity and even the lethality of a drug can be quantified by the TD_{50} and LD_{50} respectively. Ideally, the effective dose would be substantially less than either the toxic or lethal dose for a drug to be therapeutically relevant.

LT_{50} (Lethal time 50): A calculated period of time within which a specific concentration ofa chemical is expected to cause death in 50 per cent of a defined experimental animal population.

KD_{50}/KT_{50} (Knockdown Dose/Time 50) Dose/): Time required for 50 per cent of population having knockdown effect.

7.2 Why LD_{50} is Taken as Criterion for Comparing the Relative Toxicities of Toxicants?

In an experiment if the percentage of mortality of an insect is plotted against logarithm of the corresponding dose, a sigmoid curve is obtained. The curve indicates that mortality will not increase significantly by increasing the dose of the toxicant of lower and higher levels of kill. A carefull examination of the graph shows that dose giving 100 per cent kill and zero cannot be read out with this graph.

Further the corresponding dose giving 1 to 5 per cent and 95 per cent or more than 95 per cent kill also cannot be worked out with accuracy. The reason being that in both these situations the graph becomes more or less parallel to the abscissa and even the major change in dose will not produce an effect on the mortality. On the other hand the graph is steepest in the region of 50 per cent

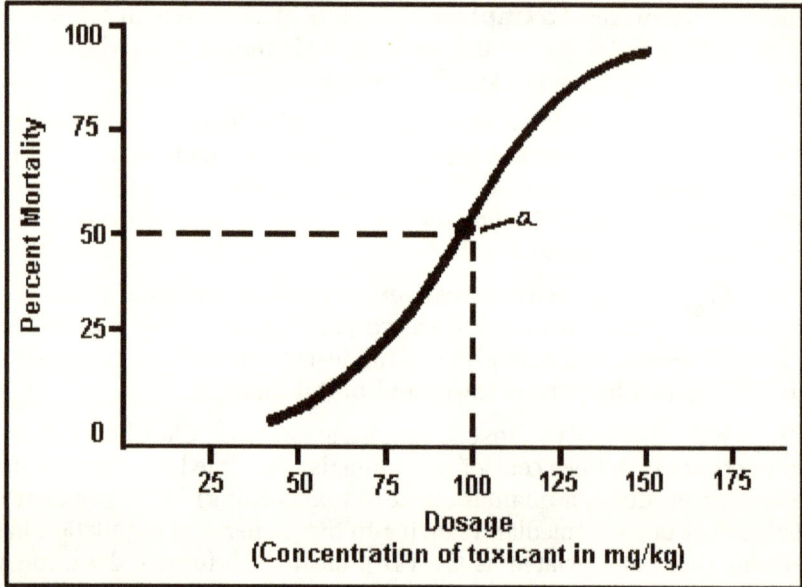

response as a result slightest change in the dose will produce large change in the mortality. Thus the dose LD_{50} is the most sensitive and reliable index of toxicity. It is due to this reason that LD_{50} is adopted as the standard for comparing the relative toxicity of toxicants.

☆ **Hazard**: It is the probability of being harmed due to the use/exposure/ handling of toxic substances.

☆ **Risk**: It is the degree of physicall, biochemical and histochemical changes acceptable in terms of usefulness of a chemical and its possible effects on the public health.

7.3 Bioassay of Insecticides

Study of response of individual or group of organisms exposed to the toxicant is called 'Bioassay' or Any quantitative procedure used to determine the relationship between the amount (dose or concentration) of an insecticide administered and the magnitude of response in a living organism. Suitable for studying the biological effects of chemicals, both when applied as direct spray on organisms or as a residual film. Bioassays are used for screening of potential insecticides, for determination of values LD_{50} and LC_{50}. Estimation of residues, and quality testing of formulated insecticides.

1. Topical Application

A most commonly employed method where the insecticide is dissolved in a relative non toxic solvents and measured droplets are applied at a chosen location on the body surface by Potter spraying tower apparatus. This air operated spraying apparatus applies an even deposit of spray over a circular area of 9 cm diameter.

2. Injection Method

When knowledge of the exact amount of insecticide inside the body of the insect is needed, the injection method is used. A very fine stainless steel needles of 27 or 30 gauge [0.41 or 0.30 mm dia] are used. Insecticides are generally dissolved in propylene glycol or peanut oil and injection is made intraperitoneally [into the body cavity]. Care must b taken to avoid bleeding by the insect.

3. Dipping Method

This method is employed when topical application or injection are impractical, for example, with small plant feeding insects, stored product insects, house fly larvae, insect eggs etc. The insects are dipped in aqueous solutions, emulsions, or suspensions of the chemical for short period of time. In this case LC_{50} is used to express the results.

4. Contact or Residual Method

The insecticide in a volatile solvent is applied to a glass container such as vial or a jar. The solvent is allowed to evaporate by rotating th container so that the insecticide is spread evenly over the entire surface leaving a residual film. The deposits are expressed as milligrams or grams of a.i/m^2.

5. Fedding or Drinking Method

These methods are used to evaluate the toxicity of ingested chemicals.

Feeding: coated leaves [discs, sandwiches, squares, strips]

Drinking: Sugar syrups and drinking through membranes such as plant juices or blood.

7.4 Formulations of Insecticides

A formulation is a mixture of the active ingredient in a pesticide with other inert (inactive) substances. Different formulations may be used differently. Some are to be used direct from the package, while others need to be diluted with water, oil, or other carriers. The reason for mixing the active ingredient with other substances is to make handling and application safer, easier, and more accurate. Some active ingredients do not dissolve in water or oil. Others can only be manufactured as solids. Still others are liquids or gases in their original forms. By mixing the active ingredient with other materials such as solvents, wetting agents, stickers, powders, or granules, manufacturers produce formulations that can be handled accurately and safely by application machinery. A few pesticides are now formulated for controlled release. These pesticides allow the active ingredient to be slowly released after application. This provides better control for certain pests at possibly lower rates and over a longer period of time.

7.4.1 What Makes up a Formulation?

A. The pesticide formulation is a mixture of active and other ingredients (previously called inert ingredients). An active ingredient is a substance that prevents, kills, or repels a pest or acts as a plant regulator, desiccant, defoliant, synergist, or nitrogen stabilizer. Pesticides come in many different formulations due to variations in the active ingredient's solubility, ability to control the pest, and ease of handling and transport.

B. Synergists are a type of active ingredient that are sometimes added to formulations. They enhance another active ingredient's ability to kill the pest while using the minimum amount of active ingredient, but do not themselves possess pesticidal properties. For example, insecticides containing the active ingredient pyrethrins often contain piperonyl butoxide or n-octyl bicycloheptane dicarboximide as a synergist.

C. Other (or inert) ingredients may aid in the application of the active ingredient. Other ingredients can be solvents, carriers, adjuvants, or any other compound, besides the active ingredient, which is intentionally added. There are many types of other ingredients: solvents are liquids that dissolve the active ingredient, carriers are liquids or solid chemicals that are added to a pesticide product to aid in the delivery of the active ingredient, and adjuvants often help make the pesticide stick to or spread out on the application surface *(i.e.,* leaves). 5 Other adjuvants aid in the mixing of some formulations when they are diluted for application.

7.4.2 How to Choose a Formulation

A single pesticide is often sold in different formulations. Different formulations of the same active ingredient often behave differently. For example, some types of formulation may mix in water better, while others may increase the chance of crop injury. Choose the formulation that is suitable for the job. Things to consider include:

1. Per cent of active ingredient
2. Ease in handling and mixing
3. Personal safety risk.
4. Type of environment (agriculture, forest, urban, etc.)
5. Effectiveness against the pest.
6. Habits of the pest.
7. The crop to be protected.
8. Type of application machinery.
9. Danger of drift or runoff.
10. Possible injury to crop
11. Cost.

7.4.3 Different Types of Formulations

1. Dusts (D)

These are ready to use insecticides in powder form. In a dust formulation the toxicant is diluted either by mixing with or by impregnation on a suitable finely divided carrier which may be an organic flour or pulverized mineral like lime, gypsum, talc etc., or clay like attapulgite bentonite etc. The toxicant in a dust formulation ranges from 0.15 to 25 per cent and the particle size in dust formulations is less than 100 microns and with the decrease in particle size the toxicity of the formulation increases. Dusts are easy to apply, less labour is required and water is not necessary. However if wind is there, loss of chemical occurs due to drift hence dusting should be done in calm weather and also in the early morning hours when the plant is wet with dew. *e.g.* HCH 10 per cent dust; Endosulfan 4 per cent D.

2. Granules or Pelleted Insecticides (G)

These are also ready to use granular or pelleted forms of insecticides. In this formulation the particle is composed of a base such as an inert material impregnated or fused with the toxicant which released from the formulation in its intact form or as it disintegrates giving controlled release. The particle size ranges from 0.25 to 2.38 mm, or 250 to 1250 microns and contains 1 to 10 per cent concentration of the toxicant. The granules are applied in water or whorls of plants or in soil. Action may be by vapour or systemic. In application of granules there is very little drift and no undue lose of chemical. Undesirable contamination is prevented. Residue problem is less since granules do not adhere to plant surface. Release of toxicant is achieved over a long period. Easy for application as water is not required for application. Less harmful for natural enemies. *e.g.,* Carbofuran 3G, Phorate 10 G, Cartap hydrochloride 4G.

3. Wettable Powders (WP)

It is a powder formulation which is to be diluted with water and applied. It yields a stable suspension with water. The active ingredient (toxicant) ranges from 15 to 95 per cent. It is formulated by blending the toxicant with a diluent such as attapulgite, a surface active agent and an auxiliary material. Sometimes stickers are added to improve retention on plant surface. Loss of chemical due to run off may be there and water is required for application. *e.g.,* Carbaryl 50 per cent WP, Thiodicarb 75 per cent WP.

4. Emulsifiable Concentrates (EC)

Here the formulation contains the toxicant, a solvent for the toxicant and an emulsifying agent. It is a clear solution and it yields an emulsion of oil-in water type, when diluted with water. The active ingredient (toxicant) ranges from 2.5 to 100 per cent. When sprayed the solvent evaporates quickly leaving a deposit of toxicant from which water also evaporates. The emulsifying agents are alkaline soaps, organic amines alginates, Carbohydrates, gums, lipids, proteins etc. *e.g.,* Endosulfan 35EC, Profenophos 50EC.

5. Soluble Powder or Water Soluble Powder (SP or WSP)

It is a powder formulation readily soluble in water. Addition of surfactants improves the wetting power of the spray fluid. Sometimes an anti-caking agent is added which prevents formation of lumps in storage. This formulation usually contains a high concentration of toxicant and therefore convenient to store and transport. *e.g.,* Acephate 75 SP.

6. Suspension Concentrate (SC)

Active ingredient is absorbed on to a filler which is then suspended in a liquid matrix (water). It is not dusty and easier to disperse in water. *e.g.,* Imidacloprid 50 SC.

7. Flowables (F)

When an active ingredient is insoluble in either water or organic solvents, a flowable formulation is developed. The toxicant is milled with a solid carrier such as inert clay and subsequently dispensed in a small quantity of water. Prior to application it has to be diluted with water. Flowables do not usually clog nozzles and require only moderate agitation. *e.g.* Methoxyfenozide (Intrepid 2F).

8. Water Dispersible Granules (WDG)

This formulation appears as small pellets or granules. It is easier and safer to handle and mix than wettable powders. When the granules are mixed with spray water, they break apart and, with agitation, the active ingredient becomes distributed throughout the spray mixture. *e.g.* Thiamethoxam 25 WDG.

9. Solutions

Many of the synthetic organic insecticides are water insoluble but soluble in organic solvents like amyl acetate, kerosene, xylene, pine oil, ethylene dichloride etc., which themselves possesses some insecticidal properties of their own. Some toxicants are dissolved in organic solvents and used directly for the control of household pests. *e.g.* Baygon.

10. Concentrated Insecticide Liquids

The technical grade of the toxicant at highly concentrated level is dissolved in non-volatile solvents. Emulsifier is not added. Generally applied from high altitudes in extremely fine droplets without being diluted with water at ultra volume rates. There is greater residual toxicity and less loss through evaporation. Active ingredient ranges from 80-100 per cent *e.g.,* Malathion, Bifenthrin, Fenitrothion.

11. Insecticide Aerosels

The toxicant is suspended as minute particles 0.1 to 30 microns in air as fog or mist. The toxicant is dissolved in a liquified gas and if released through a small hole causes the toxicant particles to float in air with rapid evaporation of the released gas. *e.g.,* Allethrin.

12. Fumigants

A chemical compound which is volatile at ordinary temperature and sufficiently toxic is known as fumigant. Most fumigants are liquids held in cans or tanks and quite often they are mixtures of two or more gases. Advantage of using fumigant is that the places not easily accessible to other chemicals can be easily reached due to penetration and dispersal effect of the gas. *e.g.* Aluminium phosphide.

13. Microencapsulation

Microencapsulated formulations consist of dry and liquid pesticide particles enclosed in tiny plastic capsules which are mixed in water and sprayed. After spraying, the capsule slowly releases the pesticide. The encapsulation process can prolong the active life of the pesticide by providing timed release of the active ingredient. *e.g.* Lambda-cyhalothrin

14. Insecticide Mixtures

Insecticide mixtures involve combinations of two or more insecticides in the right concentration into a single spray solution. Insecticide mixtures are widely used to deal with the array of arthropod pests encountered in greenhouse and field production systems due to the savings in labor costs. Furthermore, the use of pesticide mixtures may result in synergism or potentiation (enhanced efficacy) and the mitigation of resistance. However, antagonism (reduction in efficacy) may also occur due to mixing two (or more) pesticides together.

15. Baits

In baits a.i is mixed with edible substance. These are always stomach poisons and are used for poison baiting which is chiefly made up of 3 components, Poison (Insecticide carbaryl), Carrier or base (Rice bran), and Attractant (Jaggery) at ratio of 1: 10: 1. Poison should be strong and easily soluble. Base is the filler like rice bran with just enough water.

7.5 Synergists

A synergist is a chemical formulated in pesticide prod-ucts, in addition to the active and inert ingredients, that increases the potency of the active ingredient. While the increased potency makes the pesticide more deadly to their targets, synergists may also compromise the de-toxifying mechanisms of non-target species, including humans.

What is Piperonyl Butoxide [PBO] ?

1. Piperonyl butoxide is a synergist used in a wide variety of pesticides. Synergists are chemicals that lack pesticidal effects of their own but enhance the pesticidal properties of other chemicals. Piperonyl butoxide is used in pesticides containing chemicals such as pyrethrins, pyrethroids, rotenone, and carbamates.

2. Researchers developed piperonyl butoxide in 1947 using naturally-occurring safrole as a key raw material.

3. Piperonyl butoxide is a colorless to pale yellow liquid. It does not dissolve in water and is stable to breakdown by water and ultraviolet light. Researchers consider piperonyl butoxide to be noncorrosive.

Synergists in Resistance Management

The mode of action of the majority of synergists is to block the metabolic systems that would otherwise break down insecticide molecules. They interfere with the detoxication of insecticides through their action on polysubstrate monooxygenases (PSMOs) and other enzyme systems. The role of synergists in resistance management is related directly to an enzyme-inhibiting action, restoring the susceptibility of insects to the chemical, which would otherwise require higher levels of the toxicant for their control. For this reason synergists are considered straightforward tools for overcoming metabolic resistance, and can also delay the manifestation of resistance.

7.6 Adjuvants

An adjuvant is *any* compound that facilitates the action of pesticides or modifies characteristics of pesticide formulations or spray solutions. The terminology for pesticide additives is confusing. It is often assumed that any material that lowers the surface tension of water (*i.e.*, a surfactant) in the spray mixture or increases the wettability of the spray solution on surfaces is an adjuvant. Adjuvants are used in pesticide spray solutions as: Wetting agents, Co-solvents, Penetrants, Stickers, Spreaders, Stabilizing agents.

It is obvious that the term *adjuvant* encompasses a wider meaning than wetting agent or surfactant. There are many adjuvants that have little, if any, effect on pesticidal activity. These types of adjuvants include: Anti-foam agents, Buffering agents, Compatibility agents, Drift control/deposition aids.

Adjuvants can be classified according to their type of action. There are three basic kinds:

1. Activator adjuvants include surfactants, wetting agents, penetrants, and oils. Activator agents are the best known class of adjuvants because they are normally purchased separately by the user and added to the pesticidal solution in the spray tank.

2. Spray modifier agents include stickers, film formers, spreaders, spreaders/stickers, deposit builders, thickening agents, and foams.

3. Utility modifiers include emulsifiers, dispersants, stabilizing agents, coupling agents, co-solvents, compatibility agents, and anti-foam agents.

Spray modifier agents and utility modifier agents usually are found as part of the pesticide formulation and thus are added to the pesticide product by the manufacturer.

Combining Different Formulations

Sometimes various pesticides are combined. Some pesticides are registered for use in combination with a liquid fertilizer. If pesticides may be combined safely and effectively, they are called *compatible*. If not, they are called *incompatible*. Incompatibility can be physical or chemical.

Physical Incompatibility

It means that the chemicals cannot physically be mixed together. Solid materials may become deposited at the bottom of the spray tank or the ingredients may become separated into two or more layers following agitation.

Chemical Incompatibility

Even if some chemicals can be mixed together physically, there may be other kinds of incompatibility that may reduce effectiveness or cause injury to the plant. Pesticide manufacturers try to anticipate combinations that farmers want to use and provide warnings on the label for incompatible mixtures.

7.7 Mode of Action of Insecticides

7.7.1 Target site: Nervous System

The nervous system functions as a fast acting means of transmitting important information throughout the body. This system has two components: 1.) the *peripheral nervous system* to receive and transmit incoming signals (taste, smell, sight, sound, and touch) and to transmit outgoing signals to the muscles and other organs, effectively telling them how to respond, and 2.) the *central nervous system (CNS)* to interpret the signals and coordinate the body's responses and movements. A *neuron* is a single nerve cell. It connects with other neurons and with muscle fibers (the basic units of muscles) through gaps at the end of each neuron. The gap between neurons, or between a neuron and a muscle fiber, is called a *synapse*. Incoming signals (the pain from a sharp object, the sight of a predator, or the odor of food, etc.) are transformed by the neuron into an *electrical charge* which then travels down the length of the neuron. These charged particles (called ions) move through *channels* in the membrane of the neurons.

There are four main types of channels to allow different ions to move along the neuron: *sodium channels, potassium channels, calcium channels*, and *chloride channels*. Many of the channels have gates that open or close in response to a certain stimulus, which is an important mechanism through which some pesticides work. When an electrical charge reaches the end of the neuron, it stimulates a chemical messenger, called a *neurotransmitter*, to be released from the end of the neuron. This neurotransmitter crosses the synapse and binds to a *receptor* on the receiving end of the next neuron. Binding to the receptor causes the signal to be converted back into an electrical charge in the second neuron, and the signal is transmitted along the length of that neuron. After transmitting its message across the synapse, the neurotransmitter is resorbed back into its originating neuron, and the nerve cell is then in a resting stage until the next signal is received. This process repeats over and over until the signal has reached the CNS (the brain and spinal cord in humans and a series of ganglia, or nerve bundles, in the insect) to be interpreted.

Impulses from the CNS to the peripheral nervous system continue in the same way until the signal reaches the appropriate muscles or organs. Both humans and insects have many different neurotransmitters that work at different sites throughout the nervous system. Some neurotransmitters are *excitatory* (they result in the signal being sent on through the synapse to a connecting neuron), and some are *inhibitory* (they result in the reaction being blocked from travelling to a connecting neuron). In this way, the body ensures that the signal has the desired effect in each muscle or organ, since many different reactions are involved in even a simple movement. Of the many neurotransmitters that both insects and humans have, *acetylcholine (ACh)* and *gamma-aminobutyric acid (GABA)* are important targets of some insecticides. ACh can either excite or inhibit its target neurons – depending on the particular neuron and the specific receptors at the site, ACh can cause particular neurons to "fire," ontinuing the nerve impulse transmission, or it can cause the nerve impulse to stop at that particular site. In contrast, GABA is an inhibitory neurotransmitter – when GABA is the neurotransmitter activated at a synapse, the nerve impulse stops. Some insecticides interfere with the normal action of these neurotransmitters. Other insecticides attacking the nervous system work by other means. The most common mechanisms are explained below.

7.7.2 Cholinesterase Inhibition

Organophosphate and *carbamate insecticides* are known as *cholinesterase inhibitors*. They bind to the enzyme that is normally responsible for breaking down ACh after it has carried its message across the synapse. When an insect has been poisoned by a cholinesterase inhibitor, the cholinesterase is not available to help break down the ACh, and the neurotransmitter continues to cause the neuron to "fire," or send its electrical charge. This causes overstimulation of the nervous system, and the insect dies. Like insects, humans also use ACh as a neurotransmitter and cholinesterase to break it down, and cholinesterase poisoning in humans can be very severe. Upon each exposure to an organophosphate or carbamate insecticide, more cholinesterase becomes bound and is unavailable to do its job. Although cholinesterase inhibition by carbamates is somewhat reversible, organophosphate poisoning is not reversible. This means the insecticide does not release the bound cholinesterase. Fortunately, the body continually produces cholinesterase, although it may take several weeks to again reach the desirable circulating level. Applicators using cholinesterase inhibiting pesticides regularly should consider having their cholinesterase level monitored.

What are the Basic Reactions Involved in the Hydrolysis of Acetylcholine?

In synaptic transmission the impulses are transmitted by acetylcholine. As soon as as the impulses are transmitted the acetylcholine is hydrolysed by cholinesterase enzyme in three stages in which acetylcholine is hydrolysed and enzyme is recovered.

☆ **I Stage:** When acetylcholine reacts with cholinesterase an enzyme complex known as "Michaelis Complex" is formed

Acetylcholine + cholinesterase = enzyme complex [Michaelis Complex]

☆ **II Stage**: The enzyme complex then yields choline and acetylated enzyme

Enzyme complex = Choline + Acetylated enzyme

☆ **III Stage**: Deacetalyzation takes place in which acetylated enzyme is hydrolysed to give free enzyme and acetic acid

Acetylated enzyme = Acetic acid + enzyme

The Choline which is proposed in II Stage is converted into Acetylcholine by the action of of acetyl co-enzyme A under the influence of cholinacetylase

Choline + acetyl co-enzyme A = Acetylcholine + co-enzyme A

What is the difference in Mode of Action of Carbamates and OP Insecticides?

Organophospahte Insecticides	*Carbamate Insecticides*
Organophospahte inhibit the enzyme choline esterases which is essential for the hydrolysis of acetylcholine. All OP compounds mimic the gross molecular shape of acetylcholine. Therefore, the reaction between the cholinesterase enzyme and OP compounds is essentially analogous. By the action of enzyme on paraoxon a reversible complex is formed and in this process the hydroxyl group of enzyme attacks the phosphorus atom of paraoxon. The hydrogen atom of acidic group of the enzyme is transferred to part of the paraoxn to give *p*-nitrophenol. The remaining product is phoshorylated enzyme. This reaction is called phosphorylation. In this reaction because of high value of K [dissociation constant] and phosphorylation constant k+ 2, the reversible complex is immediately converted into phosphorylated enzyme and *p*-nitrophenol. Hydrolysis of this phosphorylated enzymeis called as dephosphorylation. In normal reaction Acetylcholine + cholinesterase the k+3 is very fast so that the enzyme is recovered very soon [295000 molecules/active centre per minute] but in case of reaction between OP + cholinesterase the k+3 is very SLOW [3 moleculs/active centre per minute] With the result th almost no recovery or negligible recovery of enzyme takes place. Consequently insect dies	The mechanism for carbamate reaction with enzyme choline esterases is analogous to the reaction of OP compounds. The only difference is that in the carbamates the value of K [dissociation constant] is low and the reaction by which reversible complex is carbamylated is slower [k+2 is slow] than the OP compounds as a result, a smaller amount of enzyme is inhibited

7.7.3 Acetylcholine Receptor Stimulation

a. **Neonicotinoid** insecticides act as agonists of the acetylcholine receptor. That is, they mimic the action of the neurotransmitter, acetylcholine (ACh). Although cholinesterase is not affected by these insecticides, the nerve is continually stimulated by the neonicotinoid itself, and the end result is similar to that caused by cholinesterase inhibitors – overstimulation of the nervous system leads to poisoning and death. Fortunately, the neonicotinoids are a closer mimic for the insect's ACh than for human ACh, giving this class of insecticides more specificity for insects and less ability to poison humans.

b. **Spinosad** is also an acetylcholine receptor agonist. The exact mechanism of spinosad is somewhat different than that of the neonicotinoid class, but the end result is the same.

7.7.4 Chloride Channel Regulation

a. **Avermectins** are derived from a soil microorganism and belong to a group called the acrolactones. Avermectins bind to the chloride channel. This channel normally blocks reactions in some nerves, preventing excessive stimulation of the central nervous system (CNS). Avermectins activate the chloride channel, causing an inhibitory effect, which, when excessive, results in the insect's death.

b. **Organochlorine insecticides of the cyclodiene type** affect the chloride channel by inhibiting the GABA receptor. As explained above, the GABA receptor has an inhibitory function at its site. When a cyclodiene insecticide binds to the GABA molecule, the neurotransmitter can no longer close the chloride channel for which it acts as a gate. Thus there is nothing to stop the electrical charge from continuing down the neuron. The end result is overstimulation of the nervous system.

c. **Bifenazate** affects the GABA-gated chloride channel as an agonist. That is, it causes the gate to have the same action as GABA would cause, which closes the gate. Nerve impulses are then unable to travel down the chloride channel.

7.7.5 Sodium Channel Modulators

a. **Pyrethrins** are naturally-occurring compounds derived from members of the chrysanthemum family. While they have a quick knock-down effect against insects, they are unstable in the environment, so may not last long enough to kill the pest.

b. **Pyrethroids** are synthetic versions of pyrethrins, specifically designed to be more stable in the environment (although still lasting only days or weeks), and thus provide longer-lasting control. Pyrethrins and pyrethroids act on tiny channels through which sodium is pumped to cause excitation of neurons. They prevent the sodium channels from closing, resulting in continual nerve impulse transmission, tremors, and eventually, death. Pyrethrins and pyrethroids are wellknown irritants of humans' respiratory systems as well as of the skin and eyes. Applicators who have an allergic reaction to these insecticides must either increase the amount of personal protective equipment worn during handling, or stop working with this class of insecticides.

7.7.6 Growth and Development

Unlike humans, insects must shed their skin in order to grow and to develop into their next life stage. Insects' skin is a hard exoskeleton, also called the *cuticle*, which provides both protection and structure. Molting is necessary not only for the insect to grow, but also for the insect to reach the adult stage so that it can reproduce. Hormones play various roles in molting. Disruption of, or interference with, any of these hormones inactivates the molting process. Some insecticides target the insect's growth and development processes through interfering with hormones, and others through blocking the production of a structural component of the exoskeleton.

a. Chitin Synthesis Inhibitors (CSIs)

Chitin is an important component of the insect's cuticle. Some insecticides, called **chitin synthesis inhibitors**, block the production of chitin. An insect poisoned with a CSI cannot make chitin and so cannot molt. Because molting must take place for the insect to reach the adult stage, a CSI poisoned insect also cannot reproduce. Eventually, the insect dies. Because humans do not make chitin, CSIs are not considered toxic to humans. However, CSIs are very toxic to any organism that has an exoskeleton, such as crustaceans (shellfish), and should be used with great care, if at all, in areas where they could contaminate the environment.

b. Insect Growth Regulators (IGRs)

Insect Growth Regulators, or **IGRs**, attack the insect's endocrine system, which produces the hormones needed for growth and for development into an adult form. Insects poisoned with IGRs cannot molt or reproduce, and eventually they die. Many of the currently available IGRs mimic a special protein called *juvenile hormone*. In a normal insect, juvenile hormone is circulated throughout the insect's body and "tells" the insect to stay in its current stage. After a certain amount of time, the insect stops producing juvenile hormone, and the insect *metamorphoses*, or changes, into its next life stage. When an insect is poisoned by an IGR that mimics juvenile hormone, the insect doesn't receive the signal to metamorphose because, even though the insect may have stopped producing juvenile hormone, the IGR is still circulating throughout its body and sending the signal to stay in the current stage. Another hormone important in metamorphosis is *ecdysone*. The insecticide tebufenozide interferes with the production of ecdysone, causing the insect to be unable to molt.

c. Prothoracicotropic Hormone (PTTH)

It is another insect development hormone. The insecticide azadirachtin, which is derived from neem oil, interferes with synthesis of PTTH. Besides its ability to kill through interfering with growth and development, azadirachtin also acts as a feeding deterrent, as discussed later in this leaflet. Humans do not make or use the hormones insects use in molting. Because of this, IGRs are considered to have little human toxicity. Nonspecific Growth Regulators The exact mode of action of the mite growth regulator hexythiazox is not well understood. Hexythiazox kills the eggs before the mites hatch and also some immature mites. Adult mites are not killed, although adults exposed to residues may lay eggs that are not viable.

7.7.7 Energy Production

All organisms must generate energy from the food they take in. As organisms digest the nutrients in the food they consume, they store the energy from those nutrients in molecules known as *adenosine triphosphate (ATP)*. Some insecticides inhibit or disrupt energy production. Initially, the insect can mobilize enough stored energy to continue its basic functions. While it can eat and digest food in the initial stages after being poisoned, it cannot produce more energy from the food. Eventually, the insect "runs out of steam," stops eating and even moving, and dies. Two main processes in energy production, *electron transport* and *oxidative phosphorylation*, which are normally linked together, are described below. Electron Transport Inhibition Electron transport is an important process in the production of energy in plants and animals. When

Table 14: Mode of Action of various Insecticide Groups.

Sl.No.	Main Group and Primary Site of Action	Chemical Subgroup	Active Ingredients (With trade names)
1.	Acetylcholine esterase inhibitors	Carbamates Organophosphates	Carbaryl, Carbofuvars, Carbosulfars, Profenophos, Quinalphos, Triazophos
2.	GABA – gated chloride channel antagonists	Cyclodiene organochlorines Phenylpyrazoles (Fiproles)	Endosulfan Fipronil
3.	Sodium channel modulators	Pyrethroids DDT	Fipronil, Bifenthrin, Beta-cyfluthrin, Lambda – cyhalothrin, Cypermethrin, Alpha – methrin, Deltamethrin, Fenvalerate
4.	Nicotinic acetylcholine receptor agonists	Neonicotinoids	Acetamiprid, Imidacloprid
5.	Nicotinic acetylcholine receptor allosteric activators	Spinosyns	Spinosad
6.	Chloride channel activators	Avermectins	Abamectin, Emamectin benzoate
7.	Ecdysone agonists/moulting	Azadirachtin	Neem commercial formulations
8.	Voltage – dependent sodium channel blockers	Indoxacarb	Indoxacarb

this process is disrupted, oxidative phosphorylation is inhibited, and energy (ATP) cannot be stored for later use. Organochlorine insecticides of the aliphatic type interfere with electron transport, effectively shutting down the target organism's ability to produce energy from its food.

a. **Oxidative Phosphorylation Disruption:** Oxidative phosphorylation is the process through which ATP is synthesized in plants and animals. Organotin miticides inhibit oxidative phosphorylation directly, while pyrroles work by uncoupling oxidative phosphorylation from electron transport. The end result for both groups is that the cell is unable to produce ATP for energy.

b. **Metabolism:** Some insecticides block feeding. Different classes of insecticides work through different mechanisms, as described below.

7.8 Safe Use of Insecticides

The importance of taking safety precautions while handling and applying pesticides is often underestimated. An effort must be made to give a comprehensive account of the various aspects of the safe use of pesticides, especially for staff operating spraying equipment and handling chemicals.

7.8.1 Pesticide Selection

The most important step in pesticide safety is its proper selection. First of all, the pest problem must be correctly identified. Control measures need not be taken if the pest is not of economic importance. Once economic damage due to a pest has been established, the appropriate pesticide and method of treatment can be chosen. Buying an excess of pesticide should be avoided.

7.8.2 Handling and Mixing

The following safety guidelines should be followed while handling pesticides:

1. Read label carrying out the necessary calculations for the required dilution of the insecticide.
2. Obtain proper equipment, including protective clothing.
3. Mix insecticides outside or in a well ventilated area.
4. Always stand upwind when mixing or loading the insecticides.
5. Clean up spilled insecticide immediately from skin, clothing etc.
6. Persons handling, applying insecticides should not smoke, eat or drink while working.
7. Do not use mouth to siphon an insecticide from the container.
8. Avoid drift.
9. Guard against drift of insecticides on to near by crops, field, fish pond, stream.
10. Do not spray when it is windy.
11. Do not spray or dust when it is likely to rain.

12. Do not use poor quality or leaky equipment.

13. Do not blow out the clogged nozzles with the mouth.

14. Never allow the children to apply insecticides.

15. Take the most needed parts/tools to the field (site of application).

16. Cleanliness and maintenance of insecticide application equipments.

17. Never leave insecticides and equipments unattended in the field.

18. The insecticides should always be stored in their original containers.

19. These should be kept away from food or feed stuffs and medicines.

20. Instructions found on the labels should be carefully read and strictly followed.

21. The empty containers, after the use of the insecticide, should be destroyed.

22. Persons engaged in handling insecticides should undergo regular medicinal check-up.

23. In case poisoning due to insecticides, the nearest physician should be called immediately.

7.8.3 Recognizing Pesticide Poisoning

It is easier to prevent poisoning than to treat it. Different pesticides act differently on the human body, and the mechanism and mode of action varies for different insecticides. Some general symptoms however apply. They are listed below.

1. Symptoms of Mild Poisoning
1. Headache
2. A feeling Of Sickness (nausea)
3. Dizziness
4. Fatigue
5. Irritation of the Skin, Eyes, Nose and Throat,
6. Perspiration
7. Loss of Appetite

2. Symptoms of Moderate Poisoning
1. Vomiting
2. Blurred vision
3. Stomach cramps
4. Rapid pulse
5. Difficulty in breathing, constricted pupils of the eyes
6. Excessive precipitation
7. Trembling and twitching of muscles, fatigue and nervous distress headache

3. Symptoms of Severe Poisoning
1. Convulsions
2. Respiratory failure
3. Loss of consciousness
4. Loss of pulse.

4. Symptoms Due to Chlorinated Hydrocarbons Poisoning
1. Uneasiness
2. Headache
3. Nausea
4. Vomiting
5. Dizziness and tremors
6. Convulsions
7. Respiratory arrest followed by coma
8. Leucocytosis and rise in blood pressure.

5. Symptoms Due to Organophosphate and Carbamate Insecticides Poisoning
1. Headache, giddiness, vertigo, weakness, excessive mucous discharge from nose and sense of tightness are symptoms of inhaled exposures.
2. Nausea followed by vomiting, abdominal contraction, diarrhea and salivations are symptoms of ingestion.
3. Loss of muscle coordination, speech defects; twitching of muscles; difficulty in breathing; hypertension; jerky movements; convulsions and coma indicate seriousness of poisoning.
4. Death may occur due to depressions of respiratory centre
5. Headache, giddiness, vertigo, weakness, excessive mucous discharge from nose and sense of tightness are symptoms of inhaled exposures.
6. Nausea followed by vomiting, abdominal contraction, diarrhea and salivations are symptoms of ingestion.
7. Loss of muscle coordination, speech defects; twitching of muscles; difficulty in breathing; hypertension; jerky movements; convulsions and coma indicate seriousness of poisoning.
8. Death may occur due to depressions of respiratory centre

6. Zinc Phosphide
1. Nausea
2. Vomiting
3. Diarrhea
4. Severe abdominal pain followed by symptom free period of eight hours or longer

7. Aluminium Phosphide

1. Headache
2. Giddiness
3. Nausea
4. Diarrhea and mental confusion
5. If treatment is delayed, coma, loss of reflexes may develop and death may occur from respiratory or circulatory collapse

7.8.4 First Aid Operations

Many accidental pesticide deaths are caused by eating or drinking the chemical. Some applicators die or are injured when they breathe pesticide vapors or get pesticides on their skin. Repeated exposure to small amounts of some pesticides can cause sudden, severe illness. All pesticide handlers should know and thoroughly understand first aid treatment for pesticide poisoning.

1. Emergency

1. Call local emergency response provider and local emergency medical facility immediately and Remove patient to fresh air.
2. Loosen all knots of clothes and change overalls.
3. Flush eyes with copious cold water till irritation subsides.
4. Wash the patient thoroughly with plenty of soap and water.
5. Keep the patient calm, comfortable and warm.
6. In case of accidental ingestion, induce vomiting by administering a glass of warm water mixed with two spoons of common salt or putting the forefinger at the base of plate.
7. Show label leaflet of pesticide for identification.
8. If breathing is stopped provide artificial breathing.

2. Swallowed Poisoning

1. Remove poison from the patient's stomach immediately by inducing vomiting.
2. Give common salt 15 g in a glass of warm water as anemetic and repeat until vomit fluids is clear.
3. Gently stroking or touching the throat with the finger or the blunt end of a spoon will aid in inducing vomiting when the stomach is full of fluid.
4. If the patient is already vomiting, do not give emetic butgive large amounts of warm water and then follow the specific directions suggested.

3. Inhaled Poisons

1. Carry the patient to fresh air immediately.
2. Open all doors and windows.
3. Loosen all tight clothing.

4. Apply artificial respiration if breathing has stopped or is irregular and avoid vigorous application of pressure to the chest.

5. Prevent chilling and wrap the patient in a blanket.

6. Keep the patient as quiet as possible.

7. If the patient is convulsing, keep him in bed in some dark room.

8. Do not give alcohol in any form.

4. Skin Contamination

1. Drench the skin with water.

2. Apply a stream of water on the skin while removing clothing.

3. Rapid washing is most important for reducing the extent of injury

5. Eye Contamination

1. Hold eye lids open

2. Wash the eyes gently with a stream of running water immediately

3. Delay of even a few second greatly increase the extent of injury

4. Continue washing until physician reaches

5. Do not use chemicals as they may increase the extent of injury.

7.8.5 Antidotes

1. General antidotes

1. Remove poison by inducing vomiting

2. **Universal Antidote:** It is a mixture of 7 g of activated charcoal, 3.5 g of magnesium oxide and 3.5 g of tannic acid in half a glass of warm water may be used to absorb or neutralize poisons. Except in cases of poisoning by corrosive substances, it should be fallowed by gastric lavage.

3. Removal of stomach contents (Gastric lavage.)

4. Demulcents: After removal of stomach contents as completely as possible, give one of the following:

   Raw egg white mixed with water, Gelatine 9 g to 18 g dissolved in 570 ml of warm water, Butter, Cream, Milk or Mashed potato.

2. Specific Antidotes

1. **Atropine** is the usual antidote for organophosphate and carbamate poisoning. It can be given orally and in severe cases, injections are given. Repeated injections may be required.

2. **2 PAM** is injected intravenously as an antidote in organophosphate poisoning. It should not be used in case of carbamate poisoning

3. **Calcium gluconate** is recommended as an antidote for some organochlorine insecticides.

4. **Vitamin K** is the preferred antidote for anticoagulant poisoning such as warfarin.

5. **Dimercaprol (BAL)** is recommended for arsenic poison.

Table 15: The Status of Number of Pesticides in India Registered.

1.	Number of total pesticides registered under insecticide act, 1968 as on 17/06/2011	241
2.	Number of pesticides refused registration	18
3.	Number of pesticides restricted for use in India	14
4.	Number of pesticides banned for manufacture, import and use	28
5.	Number of pesticides formulations banned and their manufacture is allowed to export	2
6.	Pesticides formulation banned for import, manufacture and use	4
7.	Number of pesticides withdrawn	7
8.	Number of pesticides Registered	13
9.	Number of combination is insecticides approved	17

Table 16: Trade Names of Insecticide Mixtures.

Sl.No.	Insecticide Combination	Trade Name
1.	Acephate 50 per cent + Imidacloprid 1.8 per cent	Lancer gold
2.	Buprofezin 5.65 per cent + Deltamethrin 0.72 per cent	Dadeci
3.	Chlorpyriphos + BPMC	Brodan 31.5 EC
4.	Chlorpyriphos 16 per cent + Alphamethrin 1 per cent EC	Duet 17 EC
5..	Chlorpyriphos 50 per cent + Cypermethrin 5 per cent EC	Nurelle D 505
6.	Deltamethrin 0.72 per cent + Buprofezin 5.65 per cent	Dadeci, Tusker
7.	Deltamethrin 1 per cent + Triazophos 35 per cent EC	Spark 36 EC
8.	Ethion 40 per cent + Cypermethrin 5 per cent EC	Nagata 45 EC
9.	Fenvalerate + Fenthrothion	Sumicomb 20 EC
10.	Fenvalerate 3 per cent + Acephate 25 per cent EC	Koranda 28 EC
11.	Indoxacarb 14.5 per cent + Acetamiprid 7.7 per cent	Almighty
12.	Profenophos 40 per cent + Cypermethrin 3 per cent	Polytrin C 44 EC
13.	Quinalphos 20 per cent + Cypermethrin 3 per cent	Virat, Cannon plus

List of Pesticides which are Banned, Refused Registration and Restricted in Use [As on 31th Dec, 2012]

I. Pesticides/Formulations Banned in India

A.   Pesticides Banned for manufacture, import and use.

1. Aldicarb
2. Aldrin
3. Benzene Hexachloride
4. Calcium Cyanide
5. Chlorbenzilate
6. Chlordane
7. Chlorofenvinphos
8. Copper Acetoarsenite

9. Dibromochloropropane

10. Dieldrin

11. Endrin

12. Ethyl Mercury Chloride

13. Ethyl Parathion

14. Ethylene Dibromide

15. Heptachlor

16. Lindane (Gamma-HCH) (Banned vide Gazette Notification No S.O. 637 (E) Dated 25/03/2011)-Banned for Manufacture, Import or Formulate w.e.f. 25th March,2011 and banned for use w.e.f. 25th March,2013.

17. Maleic Hydrazide

18. Menazon

19. Metoxuron

20. Nitrofen

21. Paraquat Dimethyl Sulphate

22. Pentachloro Nitrobenzene

23. Pentachlorophenol

24. Phenyl Mercury Acetate

25. Sodium Methane Arsonate

26. TCA (Trichloro acetic acid)

27. Tetradifon

28. Toxaphene (Camphechlor)

B. Pesticide formulations banned for import, manufacture and use

1. Carbofuron 50 per cent SP

2. Methomyl 12.5 per cent L

3. Methomyl 24 per cent formulation

4. Phosphamidon 85 per cent SL

C. Pesticide/Pesticide formulations banned for use but continued to manufacture for export

1. Captafol 80 per cent Powder

2. Nicotin Sulfate

D. Pesticides Withdrawn (Withdrawal may become inoperative as soon as required complete data as per the guidelines is generated and submitted by the Pesticides Industry to the Government and accepted by the Registration Committee. (S.O 915 (E) dated 15th Jun, 2006).

1. Dalapon

2. Ferbam

3. Formothion

4. Nickel Chloride

5. Paradichlorobenzene (PDCB)

6. Simazine

7. Warfarin

II. Pesticides Refused Registration

Sl.No.	Name of Pesticides	Sl.No.	Name of Pesticides
1.	Ammonium Sulphamate	2.	Azinphos Ethyl
3.	Azinphos Methyl	4.	Binapacryl
5.	Calcium Arsenate	6.	Carbophenothion
7.	Chinomethionate (Morestan)	8.	Dicrotophos
9.	EPN	10.	Fentin Acetate
11.	Fentin Hydroxide	12.	Lead Arsenate
13.	Leptophos (Phosvel)	14.	Mephosfolan
15.	Mevinphos (Phosdrin)	16.	2,4, 5-T
17.	Thiodemeton/Disulfoton	18.	Vamidothion

III. Pesticides Restricted for Use in the Country

Sl.No.	Name of Pesticides	Details of Restrictions
1.	Aluminium Phosphide	The Pest Control Operations with Aluminium Phosphide may be undertaken only by Govt./Govt. undertakings/Govt. Organizations/ pest control operators under the strict supervision of Govt. Experts or experts whose expertise is approved by the Plant Protection Advisor to Govt. of India except [1]Aluminium Phosphide 15 per cent 12 g tablet and [2]Aluminum Phosphide 6 per cent tablet. *[RC decision circular F No. 14-11 (2)-CIR-II (Vol. II) dated 21-09-1984 and G.S.R. 371 (E) dated 20th may 1999]*. [1]*Decision of 282nd RC held on 02-11-2007 and*, [2]*Decision of 326th RC held on 15-02-2012*. The production, marketing and use of Aluminium Phosphide tube packs with a capacity of 10 and 20 tablets of 3 g each of Aluminium Phosphide are banned completely. (S.O.677 (E) dated 17thJuly, 2001).
2.	Captafol	The use of Captafol as foliar spray is banned. Captafol shall be used only as seed dresser. (S.O.569 (E) dated 25thJuly, 1989). The manufacture of Captafol 80 per cent powder for dry seed treatment (DS) is banned for use in the country except manufacture for export. (S.O.679 (E) dated 17thJuly, 2001).
3.	Cypermethrin	Cypermethrin 3 per cent Smoke Generator, is to be used only through Pest Control Operators and not allowed to be used by the General Public. [Order of Hon, ble High Court of Delhi in WP (C) 10052 of 2009 dated 14-07-2009 and LPA-429/2009 dated 08-09-2009]
4.	Dazomet	The use of Dazomet is not permitted on Tea. (S.O.3006 (E) dated 31st Dec, 2008)
5.	Diazinon	Diazinon is banned for use in agriculture except for household use. (S.O.45 (E) dated 08th Jan, 2008)
6.	Dichloro Diphenyl Trichloroethane (DDT)	The use of DDT for the domestic Public Health Programme is restricted up to 10,000 Metric Tonnes per annum, except in case of any major outbreak of epidemic. M/s Hindustan Insecticides Ltd., the sole manufacturer of DDT in the country may manufacture

Contd...

Sl.No.	Name of Pesticides	Details of Restrictions
		DDT for export to other countries for use in vector control for public health purpose. The export of DDT to Parties and State non-Parties shall be strictly in accordance with the paragraph 2 (b) article 3 of the Stockholm Convention on Persistent Organic Pollutants (POPs). (S.O.295 (E) dated 8th March, 2006)
		Use of DDT in Agriculture is withdrawn. In very special circumstances warranting the use of DDT for plant protection work, the state or central Govt. may purchase it directly from M/s Hindustan Insecticides Ltd. to be used under expert Governmental supervision. (S.O.378 (E) dated 26thMay, 1989);
7.	Fenitrothion	The use of Fenitrothion is banned in Agriculture except for locust control in scheduled desert area and public health. (S.O.706 (E) dated 03rdMay, 2007).
8.	Fenthion	The use of Fenthion is banned in Agriculture except for locust control, household and public health. (S.O.46 (E) dated 08th Jan, 2008).
9.	Lindane (Gamma-HCH)	Lindane is banned for manufacture, import or formulate. However it is allowed for use up to 24th march, 2013 for termite control in Building including wood, and termite control in Agriculture as per approved label claims by the Registration Committee and for exports. [S.O.637 (E) dated 25thMarch, 2011 AND S.O.1472 (E) dated 29th Aug., 2007].
10.	Methoxy Ethyl Mercuric Chloride (MEMC)	The use of MEMC is banned completely except for seed treatment of potato and sugarcane. (S.O.681 (E) dated 17thJuly, 2001).
11.	Methyl Bromide	Methyl Bromide may be used only by Govt./Govt. undertakings/ Govt. Organizations/Pest control operators under the strict supervision of Govt. Experts or Experts whose expertise is approved by the Plant Protection Advisor to Govt. of India. [G.S.R.371 (E) dated 20thMay, 1999 and earlier RC decision].
12.	Methyl Parathion	Methyl Parathion 50 per cent EC and 2 per cent DP formulations are banned for use on fruits and vegetables. (S.O.680 (E) dated 17thJuly, 2001). The use of Methyl Parathion is permitted only on those crops approved by the Registration Committee where honeybees are not acting as a pollinators. (S.O.658 (E) dated 04th Sep., 1992.)
13.	Monocrotophos	Monocrotophos is banned for use on vegetables. (S.O.1482 (E) dated 10thOct, 2005).
14.	Sodium Cyanide	The use of Sodium Cyanide shall be restricted for Fumigation of Cotton bales under expert supervision approved by the Plant Protection Advisor to Govt. of India. (S.O.569 (E) dated 25thJuly, 198.]

Chapter 8

Rodenticides and Acaricides

8.1 Rodenticides

Compounds, which kill the rodents by their chemical action, are known as rodenticides. Rodents belong to order Rodentia, class Mammalia and phylum Vertebrata. Rodents such as rats, mice, gophers and ground squirrels spread diseases like plague, rat bite fever and leptospiral jaundice in human beings. They damage the standing crops and cause substantial loss during storage of the produce. Rodenticides are classified into two groups, anticoagulants and non-anticoagulants.

A. Anticoagulant Rodenticides

Anticoagulant rodenticides were first discovered in the 1940' s and have since become the most widely used toxicants for rodent control. Rodents poisoned with anticoagulants die from internal bleeding, the result of loss of the blood s clotting ability and damage to the capillaries. Prior to death, the animal exhibits increasing weakness due to blood loss, though appetite and body weight are not specifically affected. Because anticoagulant baits are slow in action (several days following the ingestion of a lethal dose), the target animal is unable to associate its illness with the bait eaten. Therefore, bait shyness does not occur. This delayed action also has a safety advantage because it provides time to administer the antidote (vitamin K1) to save pets, livestock, and people who may have accidentally ingested the bait.

1. First-generation Anticoagulants or Multiple-Feed Rodenticides

The first anticoagulants include, warfarin, pindone, diphacinone and clorophacinone. These compounds are chronic in their action, requiring multiple feedings over several days to a week or more to produce death. In order to achieve this multiple feeding, the bait must be made available on a continuous basis until the desired control is reached. Where anticoagulants have been used over long periods of time at a particular location, there is an increased potential for a population to become somewhat resistant to the lethal effects of the baits. Resistance of rats to warfarin was first noted in Scotland in 1958, some years following its repeated use. Shortly thereafter, anticoagulant resistance was identified in both rats and house mice in other European countries. Rats and mice that are resistant to warfarin also show some resistance to all first generation anticoagulants, rendering control with these compounds less effective. Although relatively uncommon, a few instances of resistance have been reported in the United States.

a. **Warfarin** (Final®andothers)3- (alpha-acetonylbenzyl)-4-hydroxy-coumarin): Warfarin was the first marketed anticoagulant and therefore became the best known and most widely used. It has relatively limited sales today, due to the availability of more potent anticoagulants.

b. **Pindone** (Pival®, Pivalyn®)2-pivalyl-1,3-indandione: Pindone is also one of the early anticoagulants which is still available for use in commensal rodent control. Although regarded as slightly less effective than warfarin, it has some properties that resist insects and growth of mold. For optimal control using warfarin or pindone, bait must be available to rodents over a period of several days, so that there is no longer than 48 hours between feedings. Ideally, daily feedings should occur.

c. **Chlorophacinone** (RoZol®) 2-[(p-chlorophenyl)phenylacetyl]-1,3-indandione.

d. **Diphacinone** (Ramik®, Ditrac®) 2-diphenylacetyl-1,3-indandione:

e. Chlorophacinone and diphacinone are similar in potency and are significantly more toxic than the anticoagulant compounds developed earlier. Consequently, they are formulated at lower concentrations. Chlorophacinone and diphacinone may kill some rodents in a single feeding, but multiple feedings are needed to give adequate control of an entire population. With these compounds, feeding does not always have to be on consecutive days. When anticoagulants are eaten daily, however, death may occur as early as the third or fourth day. For optimal lethal effects, several feedings should occur within a 10-day period with no longer than 48 hours between feedings.

2. Second-Generation or Single-Feed Anticoagulants

Warfarin resistance led to the development of the second-generation anticoagulants, bromadiolone and brodifacoum. These compounds are much more potent than the first-generation anticoagulants, making them effective for the control of warfarin-resistant rats and mice. As one feeding can produce death if a sufficient amount of bait is consumed, they are often referred to as single-feed anticoagulants. In commensal situations where rodents are often marginal or reluctant feeders, these compounds can be extremely effective. The effects of these compounds are also cumulative and will result in death after several feedings of even small amounts. As in the case of all anticoagulants, death is delayed for several days following the ingestion of a lethal dose.

Where anticoagulant resistance is known or suspected, the use of first-generation anticoagulants should be avoided in favor of the second-generation anticoagulants or one of the non-anticoagulant rodenticides like bromethalin or cholecalciferol. Because of their similarity in mode of action, all anticoagulant baits are used in a similar fashion. Label directions commonly instruct the user to "maintain a continuous supply of bait for 15 days or until feeding ceases", thus ensuring that the entire rodent population has ample opportunity to ingest a lethal dose of the bait. Anticoagulants have the same effect on nearly all warm-blooded animals, but the sensitivity to these toxicants varies among species. If misused, anticoagulant rodenticides can be lethal to nontarget animals such as dogs and cats. Additionally, residues of anticoagulants which are present in the bodies of dead or dying rodents can cause toxic effects to scavengers and predators. In general, however, the secondary poisoning hazard from anticoagulants is relatively low.

Second-Generation Anticoagulants

    a. Brodifacoum: Brodifacoum is the most potent rodenticide currently available for rodents. It is available in 0.005 per cent pellet formulations and in wax blocks. Because of its acute toxicity, a lethal dose can be obtained in a single feeding, although death is delayed for 4 or 5 days.

    b. Bromadiolone: Bromadiolone is available in 0.005 per cent pellet formulations, powder and in wax blocks. It act by depressing the hepatic vitamin K dependent synthesis of substances essential to blood clotting. A single dose anticoagulant rodenticide from Coumarin group The technical material (97 per cent pure) is an odourless, yellow-white powder. Bromadiolone is vitamin K antagonist. The main site of its action is the liver, where several of the blood coagulation precursors under vitamin K dependent post translation processing take place before they are converted into the respective procoagulant zymogens. The point of action appears to be the inhibition of K1 epoxide reductase. Formulation: Solids Trade names: Roban, Moosh moosh, Bromard; Bromatrol; BromoneR; Bromorat, Deadline; Hurex LD_{50}: 1-3 mg/kg.

Anticoagulant Bait Formulations

Most of the anticoagulant baits used today are commercial ready-to-use baits in grain, pelleted or wax form. Grain and pelleted anticoagulant baits are used extensively in tamper-resistant bait boxes or stations for a permanent baiting program. Paraffin-type bait blocks provide an alternative to bait stations containing pelleted or loose cereal bait. If permitted by the label, bait blocks can be placed or fastened in locations where bait boxes with loose grain or pelleted bait would be difficult to place, and where they are readily accessible to rats.

Reasons for Anticoagulant Bait Failure

Resistance is one of the reason for failure in the control of rodents with anticoagulant baits. Control with baits that are highly accepted may fail for one or more of the following reasons:

a. Too short a period of bait exposure.

b. Insufficient bait and insufficient replenishment of bait (none remains from one baiting to the next).

c. Too few bait stations and/or too far apart. In some situations, stations may have to be within 20 to 30 feet (7 to 10 m) of one another.

d. Too small a control area, permitting rodents to move in from untreated adjacent areas.

e. Genetic resistance to the anticoagulant. Although this is unlikely, it should be suspected if about the same amount of bait is taken for a number of weeks. Control with anticoagulant baits that are poorly accepted may fail for one or more of the following reasons:

f. Poor bait choice, or bait is formulated improperly. Other more attractive foods are available to rodents.

g. Improperly placed bait stations. Other foods are more convenient to rodents.

h. Abundance of other food choices.

i. Tainted bait: the bait has become moldy, rancid, insect-infested, or contaminated with other material that reduces acceptance. Discard old bait periodically, and replace it with fresh bait.

Occasionally, rodents accept bait well and an initial population reduction is successful. Then bait acceptance appears to stop although some rodents remain. In such instances it is likely that the remaining rodents never accepted the bait either because of its formulation or placement. The best strategy is then to switch to a different bait formulation, place baits at different locations, and/ or use other control methods such as traps.

B. Non-anticoagulant Rodenticides

The older rodenticides, formally referred to as the acute toxicants *(e.g.,* arsenic, red squill and phosphorus) are either no longer registered or of little importance in rodent control. Newer rodenticides are much more effective and have resulted in the phasing out of these older materials over the last 20 years. At present there are three non-anticoagulant rodenticides–zinc phosphide, cholecalciferol (Vitamin D3) and bromethalin. Since none of these are anticoagulants, all can be used to control anticoagulant resistant rodent populations. Of these active ingredients, bromethalin and cholecalciferol are formulated to serve as chronic rodenticides, applied so that rodents will have the opportunity to feed on the baits one or more times over the period of one to several days. Because they are slow-acting in comparison to zinc phosphide, bait shyness is not usually a problem, nor is prebaiting necessary to get good control in most situations. Zinc phosphide differs in that prebaiting (offering rodents similar but non-toxic bait prior to applying the toxicant-treated bait) is recommended to increase bait acceptance. Zinc phosphide is not designed to be left available to rodents for more than a few days, as continued exposure is likely to result in bait shyness within the population. Non-anticoagulant rodenticides, particularly zinc phosphide, remain useful tools to achieve rapid reductions in rodent populations. When population levels are high, the cost of baiting with these materials may be lower than for the anticoagulants.

a. Bromethalin

Bromethalin is a single-dose rodenticide that causes central nervous system depression and paralysis, leading to death in 2 to 4 days. Bait should be renewed at intervals of several days. Continuous bait availability (as with anticoagulants) is not required, but bait needs to be present long enough to allow all animals in the area to feed. The amount of bait needed is usually about one-third that used with anticoagulants, since an animal ingesting a lethal dose does not feed again. This effect is unlike that of anticoagulants, in which rodents continue to consume bait after they have ingested a lethal dose. Bait shyness has not been reported.

b. Cholecalciferol

Cholecalciferol is a single-dose or multiple-dose rodenticide that causes mobilization of calcium from the bone matrix to plasma and death from hypercalcemia. Time to death is 3 to 4 days after ingestion of a lethal dose. As the toxicant is slow-acting, bait shyness apparently does not occur. As with Bromethalin, once a rodent consumes a lethal dose, all food intake ceases.

c. Zinc Phosphide

Zinc phosphide is a dark gray powder, insoluble in water, that has been used extensively in the control of rodents. When zinc phosphide comes into contact with dilute acids in the stomach, phosphine (PH_3) is released. It is this substance that probably causes death. Rats and mice that ingest lethal amounts of bait usually succumb overnight with terminal symptoms of convulsions, paralysis, coma, and death from asphyxia. They typically die in a prone position with their legs and tails outstretched. Because zinc phosphide is not stored in muscle or other tissues of poisoned animals, there is no secondary poisoning with this rodenticide. The bait, however, remains toxic up to several days in the gut of a dead rodent. Other animals can be poisoned if they eat enough of the gut content of rodents recently killed with zinc phosphide. Zinc phosphide, acute stomach poison is used as a bait at 2 per cent strength, mixed with popcorn, rice, dry fish, onions etc., Pre baiting is essential as rats exhibit bait shyness to this. About 500 g of poison is needed per hectare. Formulation: 80 per cent powder Trade name: Zintox LD_{50}: 46 mg/kg. Its strong garlic-like odor appears to be attractive to rodents that are not bait-shy and apparently makes the bait unattractive to some other animals. Bait shyness can be a problem; so prebaiting is recommended or necessary for achieving good bait acceptance.

C. Fumigants

Fumigants are used to kill rodents in their burrows. As a result, homeowners are much less likely to encounter the use of these chemicals but they are worthy of mention. The two most commonly used gasses to kill rodents are phosphine gas and methyl bromide.

a. Phosphine Gas

Available in a variety of forms including aluminum phosphide and magnesium phosphide, phosphine gas is extremely toxic. When aluminum phosphide is dropped into a rodent burrow it reacts with moisture to form phosphine gas. The signs and symptoms of exposure to phosphine gas are described above under zinc phosphide.

Aluminium Phosphide

Aluminum phosphide is an inorganic phosphide used to control insects and rodents in a variety of settings. Aluminum Phosphide is available in pellet and tablet form. Under optimum moisture conditions, it liberates 'Phosphine' gas, which is highly toxic. $1/4$ to $1/2$ of a 3 g tablet is put in a liveburrow; a little water is added if necessary and the burrow closed with mud. Also used to fumigate the godowns @ 1tab/ton/5 days and also@ 1 or $1/2$ tab/tree against red palm weevil. Formulation: Tablets (3g) Trade name: Celphos, Fumitoxin, Phostoxin, and Quic k Phos. LD_{50}: 11.5 mg/kg.

b. Methyl Bromide

Methyl bromide's potential to destroy ozone, As a result, methyl bromide is scheduled to be phased out by 2005, although there is political pressure to extend or reopen the phase out. Long-term exposure studies have found that methyl bromide is a mutagen, and neurotoxin that causes liver and kidney damage.

8.2 Acaricides

The substances exercising toxic effects on mites are specifically called as miticides or may be generalized as 'Acaricides'. Most of the organophosphatic insecticides are also effective acaricides, whereas, most of the organochlorines (except dicofol) are ineffective against mites. Thus, use of organochlorines in situations where mites are present, may aid in the increase of mite population substantially by killing their natural enemies. The following are some of the specific acaricides:

Phytophagous mites became important pests of cultivated plants in the mid-20th century, during the golden age of insecticide discovery that was marked by intensive use of *organochlorines, organophosphates* and *carbamates*, broad spectrum insecticides which, as it was later discovered, included many acaricidal compounds. Those were the neuroactive compounds which disrupt the transmission of impulses between nerve cells of an insect by blocking the action of the enzyme acetylcholinesterase (organophosphates, carbamates) or interferring with ion channels in the nerve membrane (organochlorines) chlorpyrifos, probably the most common commercialized organophosphate today, carbaryl, the first synthesized carbamate, and endosulfan, one of the rare organochlorine compounds still in use.

The first serious and widespread spider mite outbreaks following applications of neuroactive insecticides, observed at the end of 1940s and beginning of 1950s, initiated the research and development of specific acaricides. These compounds, exclusively or primarily effective against mites, were gradually taking over the organochlorines, organophosphates and carbamates.

First specific acaricides like bromopropylate, chloropropylate, chlorobenzilate, chlorfenethol, dicofol, tetradifon are Bridged diphenyls which, established themselves on the market in the 1950s. Sulphur is a fungicide and acaricide. It is formulated as a fine dust (80–90 per cent) to which about 10 per cent inert material is added to prevent 'balling'. Flowability of the dust is increased by adding 3 per cent tricalcium phosphate. The finer the dust the more effective it will be. Also available as a Wettable Powder (50 per cent). Dicofol is a hydroxylated product of DDT. It is non-systemic acaricide effective against all stages of mites. It has long residual action. and low mammalian toxicity. Formulation: 18 EC. Trade names: Kelthane, Hifol LD_{50}: 850 mg/kg. Tetradifon is a persistent acaricide, kills all stages of mites except adults. Formulation: 20 EC Trade names: Tedion LD_{50}: >5000 mg/kg.

During the 1960s and early 1970s, the second generation of structurally rather different specific acaricides emerged, the most important of which were propargite, organotins (cyhexatin, fenbutatin-oxide) and formamidines (amitraz, chlordimeform). Most of first and second generation acaricides are not used any longer. Specific acaricides of the third generation are represented by mite growth inhibitors (clofentezine, hexythiazox), commercialized in the first half of the 1980s. In addition to specific acaricides, several structurally diverse synthetic acaro-fungicides (dinocap, dinobuton, chinomethionate, dichlofluanid) were introduced; on the other hand, the use of sulphur products (that had been exploited as acaro-fungicides since 19th century) was largely displaced by novel synthetic compounds.

Introduction of specific acaricides reduced the adverse impact on beneficial insects (predators of insect and mite pests, polinators) to the minimum; at the same time, many specific acaricides proved to be selective, *i.e.* less toxic to predaceous mites than phytophagous mites. These acaricides effectively control populations of phytophagous mites resistant to neuroactive compounds, since they are compounds having different biochemical modes of action, with targets mostly being outside the nervous system (Krämer and Schirmer, 2007). Moreover, specific acaricides are far more safer for humans, non-target organisms and the environment in comparison to neuroactive compounds, in particular organochlorines that were almost all severely restricted or banned in developed countries in the 1970s. Organophosphates and carbamates, however, remain to be the predominant group of insecticides by accounting for 35 per cent of the global market (van Leeuwen *et al.*, 2009).

In addition to specific acaricides, two new groups of synthetic insecto-acaricides were placed on the market in the 1970s and 1980s: pyrethroids (neuroactive compounds, sodium channel modulators) and benzoylureas (compounds acting on growth and development by inhibition of biosynthesis of chitin, a biopolymer present in the cuticle of arthropods). Another new commercial product was abamectin, neuroactive insecto-acaricide (chloride channel activator), a mixture of macrocyclic lactones avermectin B1a and avermectin B1b, natural products isolated from the fermentation of *Streptomyces avermitilis*, a soil Actinomycete. These compounds increased the biochemical diversity of acaricides and insecto-acaricides, but beside the partly expected resistance, some other problems emerged, such as the pyrethroid-induced spider mite outbreaks.

New Synthetic Acaricides Acting on Respiration Targets

Similar to nervous system of insects, nervous system of mites has also long been the target for most chemicals used for their control (Casida and Quistad, 1998). The situation has somewhat changed during the last two decades due to commercialization of large Number of acaricidal compounds acting on mitochondrial respiration process, that produces most of the energy in cells. This process includes two coupled parts: mitochondrial electron transport (MET) and oxidative phosphorylation. Although some of the older acaricides were known to inhibit respiration, the real exploitation of this target started no sooner than after the 1990s, with the prospects for expanding and developing new, more effective and safer products. Throughout the mitochondrial electron transport chain there are various potential sites for inhibition, but only three have been used so far as target sites of acaricidal activity, at transmembrane enzyme complexes. In the period 1991-93, four compounds from different chemical classes were successively commercialized: fenpyroximate (pyrazole), pyridaben (pyridazinone), fenazaquin (quinazoline) and tebufenpyrad (pyrazolecarboxamide), whose mode of action was inhibition of MET at complex I. These compounds, also known as METI acaricides, quickly gained the popularity worldwide owing to the high efficacy against both tetranychid and eriophyoid mites, quick knockdown effect and long-lasting impact. In addition, these substances have low to moderate mammalian toxicity and short to moderate environmental persistence (van Leewen *et al.*, 2010). Complex I inhibitors also include pyrimidifen (pyrimidinamine), commercialized in 1995, as well as insecto-acaricide tolfenpyrad, another pyrazolecarboxamide, commercialized in 2002, and flufenerim, more recent derivative of pyrimidifen. The only known complex II inhibitor is the recently introduced insecto-acaricide cyenopyrafen, a compound from the acrylonitrile class of chemistry. Complex III inhibition is mode of action of acequinocyl, fluacrypyrim and bifenazate. Acequinocyl, a naphthoquinone compound commercialized in 1999, is a proacaricide which is bioactivated via deacetylation. It is effective against all stages of spider mites, with low mammalian toxicity and short environmental persistence.

Bifenazate, a carbazate compound is highly effective against immatures and adults of spider mites, with rapid knockdown effect (Ochiai *et al.,* 2007). Although it was first considered to be a neurotoxin, more recent experimental results indicate complex III as target site. Bifenazate is a pro-acaricide which is bioactivated via hydrolysis of ester bonds, so the organophosphorous compounds, as inhibitors of esterase hydrolitic activity, can antagonize the toxicity of this acaricide. Fluacrypyrim, introduced in 2002, shows acaricidal effect against all stages of tetranychids. This is the first strobilurin not commercialized as a fungicide and more compounds with acaricidal effect from this group are anticipated (Li *et al.,* 2010).

Insecto-acaricide diafenthiuron, a novel thiourea compound launched in 1991, is the only modern representative of compounds that disrupt oxidative phosphorylation by inhibition of the mitochondrial ATP synthase, an enzyme with essential role in cellular bioenergetics (this mode of action has been recognized in propargites, tetradifons and organotin compounds). Diafenthiuron is a pro-acaricide, its carbodiimide metabolite inhibits the enzyme. It is effective against motile stages of spider mites and also provides good eriophyoid control. Diafenthiuron has low mammalian toxicity and short environmental persistence (van Leewen *et al.,* 2010).

Another insecto-acaricide, chlorfenapyr, a pyrrole compound commercialized in 1995, at biochemical level acts as uncoupler of oxidative phosphorylation via disruption of the proton gradient. Chlorfenapyr is effective against all stages of spider mites and eriophyoid mites. This compound is a pro-acaricide activated by N-dealkylation. Chlorfenapyr is a compound of moderate mammalian toxicity, but long environmental persistence.

Acaricides Acting on Growth and Development Targets

Another direction in research and development of synthetic acaricides is directed towards compounds affecting developmental processes. Etoxazole, a oxazoline compound is acaricide highly effective against eggs and immatures of spider mites, non-toxic to adults, but it considerably reduces fertility of treated females (Kim and Yoo, 2002; Dekeyser, 2005). This acaricide, launched in 1998, is usually classified among mite growth inhibitors, together with clofentezine and hexythiazox, older acaricides that cause similar symptoms, but whose exact mode of action is unknown. On the other hand, Nauen and Smagghe (2006) provided experimental evidence that etoxazole acts as a chitin synthesis inhibitor similar to benzoylureas.

Discovery of spirodiclofen and spiromesifen, tetronic acid derivatives launched in 2002-2004, broadened the biochemical diversity of acaricides by introducing a completely new mode of action. These compounds act as inhibitors of acetyl-CoA-carboxylase, a key enzyme in fatty acid biosynthesis. Spirodiclofen and spiromesifen are highly toxic to eggs and immatures of spider mites, while their effects on adult females are slower with fecundity and fertility reduction; their acaricidal effect is long-lasting and stable (Marèiæ *et al.,* 2010). These two acaricides are the only new compounds used for control of eriophyoid mites as well (van Leeuwen *et al.,* 2010). In addition to acaricidal effect, spirodiclofen has also shown considerable insecticidal activity against eggs and larvae of pear psylla and scales (Krämer and Schirmer, 2007; Marèiæ *et al.,* 2007), while spiromesifen provides effective control of whiteflies (Krämer and Schirmer, 2007; Kontsedalov *et al.,* 2008). Both compounds have low mammalian toxicity and short environmental persistence. Spirotetramat, a tetramic acid derivate recently introduced, belongs to inhibitors of acetyl-CoA-carboxylase. Although initially developed for control of whiteflies and aphids (Brück *et al.,* 2009), the studies of its effects on *T. urticae* indicate that spirotetramat could potentially be an effective acaricide as well.

Natural Acaricides and other Alternative Solutions

The use of natural products for plant and crop protection dates back to times long before the introduction of synthetic pesticides which imposed themselves as the main means for suppression of harmful organisms. In recent times, the significance of natural pesticides is constantly growing, primarily in organic agriculture, but also in the framework of biorational pest control programs which insist on use of environmentally-friendly pesticides and exploatation of novel biochemical modes of action (Horowitz *et al.,* 2009). Some of the natural products are substances that have significant acaricidal effect. Probably the most studied botanical insecticide in the last twenty years is a triterpenoid. Azadirachtin is the major active ingredient of extracts, oils and other products derived from the seeds of the Indian neem tree *(Azadirachta indica).* Neem-products are registered in over 40 countries as products for suppression of arthropod pests important in growing of fruit, vegetables and ornamental plants. The effects of azadirachtin on treated insects manifest slowly and they include complete or partial antifeedant response, delayed and/or disrupted moulting, inhibited reproduction The studies on spider mites (Martinez-Villar *et al.,* 2005) indicate that azadirachtin, in addition to being toxic to various development stages, acts as antifeedant, reduces fecundity and fertility and shortens the life span of adult insects. Beside on spider mites, azadirachtin also exhibits acaricidal effect on some acarid and tarsonemid mites (Venzon *et al.,* 2008). Azadirachtin is considered to be non-toxic to mammals and is not expected to have any adverse effects on the environment (Copping and Duke, 2007).

Products isolated from soil actinomycetes are an important source for deriving natural insecticides and acaricides. In early 1990s, several years after introduction of abamectin, another fermentation product, milbemectin, was commercialized. Milbemectin is a mixture of milbemycin A3 and milbemycin A4, natural products isolated from the fermentations of *Streptomyces hygroscopicus* subsp. *aureolacrimosus.* Milbemectin is a neuroactive acaricide (chloride channel activator), effective against tetranychid and eriophyoid mites, relatively safe compound owing to the rapid uptake into treated plants combined with fast degradation of surface residues (Krämer and Schirmer, 2007). The more recent example is spinosad, a mixture of spinosyn A and spinosyn D, secondary metabolites of *Saccharopolyspora spinosa,* introduced in 1997 as neuroactive insecticide, nicotinic acetylcholine receptor agonist. This insecticide exerts significant systemic acaricidal effect.

Essential oils, secondary metabolites abundant is some aromatic plants from families Lamiaceae, Apiaceae, Rutaceae, Myrtaceae and others, have been suggested as alternative sources for pest control products. Predominant bioactive ingredients of essential oils are monoterpenes and sesquiterpenes. Besides exerting acute toxicity to insects and mites, essential oils show sublethal effect as repellents, antifeedants and reproduction inhibitors. Petroleum oils have been used for more than a century to control a wide range of crop pests, including spider mites. Because of their high phytotoxicity, the use of petroleum oils was limited to dormant or delayed dormant application against overwintering pest stages, to avoid injury to green plant tissue. Advances in petroleum chemistry considerably reduced phytotoxicity in newer, highly-refined petroleum-derived spray oils (PDSO), which are recognized today as an important alternative to synthetic pesticides. PDSO are highly effective against spider mites and eriophyoid mites in various field and greenhouse crops (Chueca *et al.,* 2010). Beside mineral, plant oils proved to be effective acaricides as well, such as cottonseed oil, soybean oil, and rapeseed oil.

Chapter 9

The Sterile Insect Technique

The basic principle in genetic control of insects is to utilize factors which will lead to reproductive failure. Genetic control of insects is not limited to the use of insects sterilized by radiation or chemicals but also include cytoplasmic incompatibility, induced sterility, hybrid sterility etc.

The Sterile Isect Technique [SIT]

The sterile insect technique is a method of biological control, whereby overwhelming numbers of sterile insects are released. The released insects are normally male as it is the female that causes the damage, usually by laying eggs in the crop, or, in the case of mosquitoes, taking blood from humans. The sterile males compete with the wild males for female insects. If a female mates with a sterile male then it will have no offspring, thus reducing the next generation's population. Repeated release of insects can diminish small populations, though it could be impossible to eradicate it and is not efficient against dense insect populations.

The technique has successfully been used to eradicate the Screw-worm fly (*Cochliomyia hominivorax*) in areas of North America. There have also been many successes in controlling species of fruit flies, most particularly the Medfly (*Ceratitis capitata*), and the Mexican fruit fly (*Anastrepha ludens*). Insects are mostly sterilized with radiation, which might weaken the newly sterilized insects, if doses are not correctly applied, making them less able to compete with wild males. However, other sterilization techniques are under development which would not affect the insects' ability to compete for a mate.The technique was pioneered in the 1950s by American entomologists Dr. Raymond C. Bushland

and Dr. Edward F. Knipling. For their achievement, they jointly received the 1992 World Food Prize.

Development of SIT

Raymond Bushland and Edward Knipling first developed the technique to eliminate screwworms preying on warm-blooded animals, especially cattle herds. With larvae that invade open wounds and eat into animal flesh, the flies were capable of killing cattle within 10 days of infection. In the 1950s, screwworms caused annual losses to American meat and dairy supplies that were projected at above $200 million. Screwworm maggots are also known to parasitize human flesh. Since a female screwworm mates only once in her lifetime, this physiological phenomenon has been exploited by biologists in breaking its life cycle. After mating with a sterile male, the screwworm female will not mate again or lay any eggs.

Their work in this area was interrupted by World War II, but Drs. Bushland and Knipling resumed their efforts in the early 1950s with their successful tests on the screwworm population of Sanibel Island, Florida. The sterile insect technique worked; near eradication was achieved using X-ray sterilized flies.In 1954, the technique was used to completely eradicate screwworms from the 176-square-mile (460 km²) island of Curaçao, off the coast of Venezuela. Screwworms were eliminated in a span of only seven weeks, saving the domestic goat herds that were a source of meat and milk for the island people.

During the 1960s and 1970s, SIT was used to control the screwworm population in the United States. The 1980s saw Mexico and Belize eliminate their screwworm problems through the use of SIT, and eradication programs have progressed through all of Central America, with a biological barrier having been established in Panama to prevent reinfestation from the south. In 1991, Knipling and Bushland's technique halted a serious outbreak in northern Africa. Similar programs against the Mediterranean fruit fly in Mexico and California use the same principles. In addition, the technique was used to eradicate the melon fly from Okinawa and has been used in the fight against the tsetse fly in Africa.

The technique has been able to suppress insects threatening livestock, fruit, vegetable, and fiber crops. The technique has also been lauded for its many environmentally sound attributes: it uses no chemicals, leaves no residues, and has no negative effect on non-target species.Proven effective in controlling outbreaks of a wide range of insect pests throughout the world, the technique has been a boon in protecting the agricultural products to feed the world's human population. Both Bushland and Knipling received worldwide recognition for their leadership and scientific achievements, including the World Food Prize. Their research and the resulting Sterile Insect Technique were hailed by former U.S. Secretary of Agriculture Orville Freeman as "the greatest entomological achievement of (the 20th) century."

Table 17: List of Sterile Insect Technique Trials Worldwide.

Target	Yr	Location	Method	Outcome
Tsetse fly	1944–1946	Tanzania	Release of *Glossina morsitans centralis* into a *Glossina swynnertoni* population	Hybrid males were sterile and the female hybrids partially sterile. 99 per cent suppression in 26 km.
Cochliomyia hominivorax	1951	United States: Sanibel Island (47 km^2), Florida	Release 39 sterile male flies per km^2 per week for several weeks	Field evaluation pilot test. Resulted in up to 100 per cent sterility of the egg masses, greatly reduced the wild population, incomplete eradication because of the wild fertile flies flying from the mainland.
Cochliomyia hominivorax	1954	Netherlands Antilles: Curaçao (435 km^2)	Released 155 sterile males per km^2 per week	100 per cent egg sterility after 2 generations. Evident eradication was accomplished within 14 weeks. Releases were stopped after 22 weeks.
Cochliomyia hominivorax	1958–1959	United States: Florida	Release 155–1160 sterile flies per km^2 per week	Eradication. Total cost was $11M, about 50 per cent of the annual losses.
Cochliomyia hominivorax	1962–1966	United States: Texas and western states	Release 200–1000 sterile flies per km^2 per week	Declared eradication in Texas and New Mexico in 1964 and in the entire USA in 1966. Thereafter, the program goal changed to population containment from the initial eradication.
Cochliomyia hominivorax	1984–2001	Central America	Sterile flies release	Declared eradication in Mexico, 1991, Guatemala, 1994, El Salvador 1995, Honduras 1996, Nicaragua 1999, Costa Rica 2000, Panama, 2001.
Cochliomyia hominivorax	1990–1992	Libya	Release 40 million sterile flies per week	Operated by a joint FAO/IAEA Division. Only 6 instances of wounds infested with screwworm larvae were found in 1991, compared with more than 12000 cases in 1990. Eradication was declared in June 1992.

Contd...

Table 17–*Contd...*

Target	Yr	Location	Method	Outcome
Mexican fruit fly	1964–current	United States: Southern California and Texas	For eradication, release 96,000 and 61,500 sterile flies per km^2 per week in CA and TX, respectively	Started to eradicate in CA in 1964 and to exclude in TX a decade later. Continued as containment program.
Bactrocera tryoni	1962–	Australia	Released 1600 million sterile flies in 1990. For containment method, release 60,000 sterile flies per km^2 for 12 weeks after catching the last wild fly.	Field trials began in 1962. Population was suppressed strongly, but not eradicated because of long-range immigrants. Eradication was achieved in Western Australia in 1990. Since the mid-1990s, it has been used as containment method.
Ceratitis capitata	1978–	Mexico and Guatemala	Produced 500 million and 3,500 million sterile flies per week in Mexico and Guatemala, respectively	First large-scale fruit fly AW-IPM program using SIT. Eradication in 1982. For over 25 years, this program kept Mexico, the USA, and half of Guatemala free of the pest. Genetic sexing strains were later introduced.
Melon fly	1972–1993	Japan	Released up to 4 million sterile fly pupae per week, total 264 million during the pilot test. Total 50,000 million sterile flies were released.	A pilot experiment began in 1972 and eradication was declared in 1978. An operation program started in 1984. Complete eradication achieved in 1993.
Ceratitis capitata	1980s–	Israel	Released males	Genetic sexing strain.
Ceratitis capitata	1994–	United States: California and Florida	Release sterile males of the tsl sexing strain VIENNA 7	Started as eradication program. It was so successful and cost-effective and thereafter (1996) applied as a permanent preventative program in CA, FL, and Guatemala.
Ceratitis capitata	1997–	Jordan-Israel-Palestine	Released genetic sexing strain VIENNA 7	As population suppression rather than eradication[19]
Onion maggot	1981–	Netherlands	Sterile insects are provided from a private source	The programme has not been able to expand beyond 16 per cent of the onion production area due to free-riders. Ongoing long-term suppression program over 20 years.

Contd...

Table 17–*Contd...*

Target	Yr	Location	Method	Outcome
Tsetse fly	1970–1990s	Tanzania (1,650 km²)	Combination method with attractant traps and insecticides	Eradication.
Anopheles quadrimaculatus	1959–1960	United States: Florida	Release adult males after sterilizing in pupal stage. 430,000 males over 48 wks at 2 locations	Poor competitiveness. No population reduction.
Culex quinque-fasciatus	1967	Myanmar: Okpo	Release 5000 daily for 9 wks. Sterility from cytoplasmic incompatibility	Population eliminated.
Culex quinque-fasciatus	1969	United States: Florida	Release 930,000 males over 12 wks after chemosterilization with thiotepa	Population suppressed and eliminated partially due to the sterile males released.
Culex pipiens	1970	France	Release hundreds of thousands over 8 wks after sterilizing with chromosome translocation	Population reduced due to the persistent translocation.
Culex quinque-fasciatus	1973	India: Delhi	Release 300,000 sterile males daily over 14 wks, total 23 million. Sterilization with cytoplasmic incompatibility, and chromosome translocation.	Population reduced due to the established sterility from cytoplasmic incompatibility and translocation.
Culex quinque-fasciatus	1973	India: Delhi	Release total 38 million sterile males over 25 wks. Chemoterilization with thiotepa.	Up to 90 per cent sterile eggs, but no clear population suppression due to immigration.
Aedes aegypti	1974	Kenya: Mombasa	Release 57,000 genetically modified males over 10 wks. Sterilization with chromo-some translocation	Partial sterility, but no long-term persistent translocation.
Culex tarsalis	1981	United States: California	Released 85,000 males over 8 wks after sterilizing with adult irradiation	Assortative mating was observed, but no population reduction.
Cockchafers	1959, 1962	Switzerland	Released 3,109 and 8,594 males after radiation sterilization.	Field trials. The population was reduced by 80 per cent and 100 per cent.

Contd...

Table 17–*Contd...*

Target	Yr	Location	Method	Outcome
Boll weevil	1971–1973	United States: Mississippi	Combined methods of insecticide and SIT	Large pilot field experiment. Population was suppressed below detection levels in 203 of 236 fields. The remainder were close to uncontrolled area (less than 40 km).
Sweetpotato weevil	1994–1999	Japan	Released sterile weevils after insecticide application.	Complete eradication.
Lepidoptera	1994	Canada: British Columbia	Released irradiated codling moths	As a population suppression method.
Aedes albopictus	2012	Reunion Island	Semi field condition test using the sterilizing dose of 40 Gy with cesium-137 irradiator	Two-fold reduction of the wild population's fertility.

Sterile Fly for African Trypanosomiasis

Sleeping sickness or the African trypanosomiasis is a parasitic disease in humans. Caused by protozoa of genus *Trypanosoma* and transmitted by the Tsetse fly, the disease is endemic in certain regions of Sub-Saharan Africa, covering about 36 countries and 60 million people. It is estimated that 50,000 - 70,000 people are infected, and about 40,000 die every year. Three majorepidemics have occurred in the past hundred years, in 1896 - 1906, 1920, and 1970.Studies of the tsetse fly show that females generally only mate once in their lifetimes and very rarely mate a second time. Once a female fly has mated, she can then produce continual offspring throughout her short life.

The sterile fly is an innovative solution to the problem of the African trypanosomiasis. Specially bred male Tsetse flies are sterilized through irradiation process. These sterilized male flies are then released into areas where sleeping sickness is prevalent, and then mate with the females. Because the male is sterile, and the females mate only once, the population of Tsetse flies in the affected area will drop. Studies have shown that this process has been very effective in preventing sleeping sickness in people who live in the area.Since sleeping sickness is fatal without treatment and infected people can be without symptoms for months, the release of sterile flies into affected areas leads to greater levels of health and economic activity.

Drawbacks

1. As with insecticide treatment, repeated treatment is sometimes required to suppress the population before the use of sterile insects.

2. Sex separation could be difficult for some species, though this can be easily performed on Medfly and screwworm, for example.

3. Radiation treatment in some cases affects the health of males, so sterilized insects in such cases are at a disadvantage when competing for females.

4. Standard operating procedures of mass rearing and irradiation do not leave room for mistakes. Since the fifties, when SIT was first used as a means for pest control, several failures have occurred in different places around the world where non-sterilized artificial produced insects were released before the problem was spotted.

5. Application to large areas should be long lasting, otherwise migration of wild insects from outside the control area could repopulate.

6. The major drawback to this technique is that the cost of producing such a large number of sterile insects is often prohibitive in poorer countries.

Chapter 10

Insecticide Resistance: Mechanisms and Management

Resistance to insecticides by insects is one of the most well documented cases of evolutionary adapatations to environmental changes and especially notable for rapid selection, in evolutionary terms. This phenomenon is largely supported by the enormous reproductivecapacity and genetic flexibility of insects. The first documented case of arthropod resistance to insecticides dates back to 1914 following the observations made on limesulphur resistance of San Jose scale in Washington, USA in 1908 (Melander, 1914). This was followed by a gradual increase in the number of reported cases until the introduction of the first effective synthetic insecticide DDT in 1946, which resulted in the sudden appearance of new cases of resistance over the next five years. The subsequent wide scale use of new insecticides further increased the number of reported cases and by 1980 resistance had been detected in populations of at least 428 species representing 14 orders and 83 families of insects and acarines (Forgash. 1984). The rate of discovery and development of new insecticides has declined, so that alternatives are unlikely to be available for the population that are resistant to the existing insecticides, Therefore an understanding of the mechanisms which underlie these resistances is of great importance for the development of rational strategies for the management of resistant populations. This understanding would also be helpful in designing new insecticides to use against resistant strains and to develop strategies to revert resistant strains back to their initial sensitivity levels.

10.1 Definition

a. Resistance (technical) – a genetic change in an organism in response to selection by pesticides, which may impair control in the field.

b. Resistance (practical) – a heritable change in the sensitivity of a pest population that is reflected in the repeated failure (more than one instance) of a product to achieve the expected level of control when used according to the label recommendation for that pest species and where problems of product storage, application and unusual climatic or environmental conditions can be eliminated as causes of the failure.

c. Resistance mechanism – biological processes used by the pest to avoid the lethal action of the pesticide. Resistant organisms may have more than one resistance mechanism.

d. Resistance selection – the survival of resistant individuals in a population while susceptible individuals are killed by the pesticide treatment. The resistant individuals are "selected" to survive and produce resistant offspring. The net result is that continued use of the pesticide "selects" a pest population that becomes less and less susceptible to the pesticide. The selection process can be rapid, one or two seasons, or develop slowly over a number of years, depending on the pest, its exposure to the pesticide, and the genetics of resistance to a particular pesticide.

e. Pest resurgence: The rapid numerical rebound of a pest population after use of a broad-spectrum pesticide, brought about usually by the destruction of natural enemies that were otherwise holding the pest in check.

f. Cross-resistance: When resistance to one pesticide confers resistance to another pesticide, even where the pest has not been exposed to the latter product. Cross-resistance occurs because two or more compounds are acting on the same target site and/or are affected by the same resistance mechanism. Cross-resistance develops most commonly with compounds having the same mode of action and that are usually, but not always, chemically related from the same chemical group. It may be complete or partial (if more than one mechanism is responsible for the resistance).

g. Multiple resistance – the simultaneous presence of several different resistance mechanisms in the same organism. The different resistance mechanisms may combine to provide resistance to multiple classes of pesticides. In the field, multiple resistance and cross-resistance may appear, but the former developed from separate selection events, while the latter is the result of shared resistance mechanisms.

10.2 Resistance Mechanisms

Agricultural pests use a variety of mechanisms to survive exposure to toxicants. Resistance can develop more easily when two or more of these mechanisms are used at the same time. The resistance mechanisms fall into the following general categories:

a. **Metabolic detoxification (enzymatic):** Resistance through metabolic detoxification is most often found in insects and is less common in weeds and pathogens. It is based on enzyme systems that insects have developed to detoxify naturally occurring toxins found in their host plants and in the blood ingested by blood feeding insects. These systems include esterases, cytochrome P450 mono-oxygenases, and glutathione S-transferases. Resistant insects may have elevated levels of a particular enzyme or altered forms of the enzyme that metabolize the pesticide at a much faster rate than the non-altered form. In either case the resistant insect can detoxify the pesticide before the pesticide kills it. Metabolic resistance can range from compound specific resistance to very general resistance to a broad range of compounds. Similarly, the level of resistance provided to the insect can range from very low to very high, and can vary from compound to compound. This mechanism often cleaves the pesticide molecule or adds molecules to the pesticide, *e.g.* glutathione transferase, which detoxifies the compound. Enhanced metabolism is also a common resistance mechanism in weeds. For example, enhanced rates of metabolism of acetyl-CoA carboxylase (ACCase), acetolactate synthase (ALS), and photosystem 2 (PS2) herbicides have been reported.

b. **Reduced sensitivity at the target site:** With this mechanism the binding site of the pesticide is changed so that it cannot effectively bind to the target site, thus eliminating or significantly reducing the pesticide's effectiveness. This is the most common mechanism in fungi and weeds, and is also very common in insects. There are four general categories of target site resistance in insects:

1. *kdr* (knock-down resistance) interferes with the sodium channel in nerve cells. This is a common mechanism used for resistance to DDT and pyrethroids, *e.g.* in *Anopheles gambiae*, *Blattella germanica*. There are several mutations that produce *kdr* and super *kdr*.

2. *MACE* (modified acetylcholinesterase) modifies the structure of acetylcholinesterase so that it is no longer affected by the insecticide. This is, for example, the mechanism for pirimicarb resistance in *Phorodon humuli* and is responsible for resistance in *Tetranychus urticae*.

3. *Rdl* (resistance to dieldrin) is a point mutation that reduces dieldrin binding at the GABA receptor. It is responsible for dieldrin resistance in *Anopheles quadrimaculatus* mosquitoes and in *Lucilia cuprina*, the sheep blowfly.

4. *Bt* resistance occurs through loss of cadherin, which has important roles in cell adhesion, ensuring that cells within tissues are bound together. This mechanism is found, for instance, in *Bt*-resistant diamondback moth *(Plutella xylostella)*.

c. **Altered target-site resistance.** Chitin is a major component of an insect's exoskeleton. In this example, enzyme H is necessary for chitin production (i). To prevent molting, the insecticide binds with the target site (ii). In a resistant insect, the target site is altered (iii) and prevents the insecticide from binding with the enzyme.

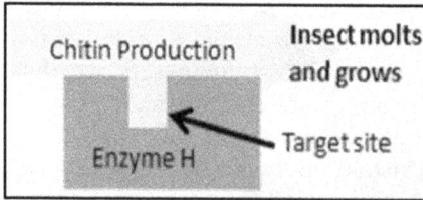

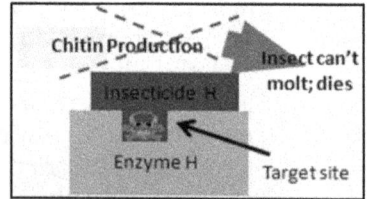

Chitin Production | Insect molts and grows | Enzyme H | Target site

Chitin Production | Insecticide H | Insect can't molt; dies | Enzyme H | Target site

d. **Reduced penetration:** This mechanism slows the penetration of the pesticide through the cuticle of resistant insects. Alone, this mechanism produces only low levels of resistance. However, by slowing

Insecticide H | Chitin Production | Insect molts and grows | Enzyme H | Altered target site

the penetration of the toxicant through the cuticle it can greatly enhance the impact of other resistance mechanisms. For example, an insect without any penetration resistance might be 25-fold resistant, whereas if penetration of the pesticide were reduced two-fold then the overall resistance could be nearly 50-fold.

e. **Sequestration:** In plants, the pesticide is removed from sensitive parts of the organism to a tolerant site, such as a vacuole, where it is effectively harmless to the target organism. This type of resistance has been demonstrated for the herbicides glyphosate, paraquat and 2,4-D. In insects (aphids, *Culex* mosquitoes, etc.) metabolic enzymes are significantly amplified (up to 15 per cent of the total body protein) and bind to the insecticide but the insecticide is not metabolised, *i.e.* the insecticide is sequestered.

f. **Behavioural resistance:** Behavioural resistance is limited to insects, mites and rodents. It refers to any modification in the organism's behaviour that helps to avoid the lethal effects of pesticides. This mechanism of resistance has been reported for several classes of insecticides, including organochlorines, organophosphates, carbamates and pyrethroids. Insects may simply stop feeding if they come across certain insecticides, or leave the area where spraying occurred (for instance, they may move to the underside of a sprayed

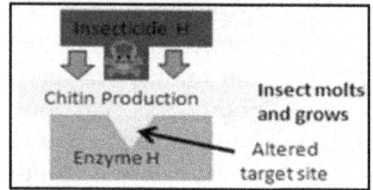

leaf, move deeper into the crop canopy, or fly away from the target area). Behavioural resistance has also been reported in mice. Behavioural resistance does not have the same importance as the physiological resistance mechanisms mentioned above but can be considered to be a contributing factor, leading to the avoidance of lethal doses of a pesticide.

10.3 Insecticide Resistance Management [IRM]

Pest management is practical and works in concert with pesticide-use strategies to lessen resistance selection by facilitating prudent, as-needed pesticide use. Pesticide-use strategies work best when implemented as a new pesticide comes into commerce. Pesticide manufacturers, IPM scientists, and growers have come to recognize that using resistance management from the beginning works best. Collecting baseline susceptibilities, defining probable resistance problems beforehand, and proposing pesticide-use strategies to forestall resistance development are the province of manufacturers and IPM scientists. Biologically and economically sound resistance management plans offered pre-sale give growers the best hope for managing resistance. IRM strategies are often grouped as follows: (1) management by moderation, (2) rotation and mixtures, and (3) saturation (Georghiou, 1983; Denholm and Rowland, 1992; Metcalf, 1994).

A. Moderation

Moderation means limiting the use of a pesticide. Moderation is employed in concert with IPM practices, such as using treatment thresholds, spraying only specific pest generations or growth stages, maintaining unsprayed wild host reservoirs to act as refuges for genetically susceptible individuals, and using pesticides with shorter residualor lower toxicity to important beneficial populations, etc. Moderation should be used to the fullest extent that will provide commercially acceptable control.

B. Rotation

An individual pest is less likely to be resistant to two or more differing classes of toxins. In theory, most individual pests resistant to one pesticide will be killed when exposed to a different class of toxin. Rotations depend on having effective, labeled materials with different modes of action. Material cost is a key practical consideration that favors rotation. Mixtures of fungicides have been used successfully to combat disease resistance, although cost lessens the attractiveness of this approach. Mixtures of insecticides and miticides have typically performed poorly. Rotation is seen as the desired approach for insecticides, miticides, and some fungicides.

C. Saturation

The use of higher pesticide rates to control resistant individuals, is the least attractive resistance management approach, although it has been used to manage resistance to DMI fungicides. Saturation is generally a last resort, when there are no other effective, labeled alternatives. In this scenario, higher rates will often provide control for a time, although at greater cost. Synergists, chemicals that increase the toxicity of pesticides, have sometimes been effective in boosting the efficacy of resistance-prone pesticides. As with simple rate increases, saturation with synergists typically provides only short-term benefits.

The best strategy to avoid insecticide resistance is *prevention*. More and more pest management specialists recommend insecticide resistance management programs as one part of a larger integrated pest management (IPM) approach.

Monitor Pests

Scouting is one of the key activities in the implementation of an insecticide resistance management strategy. Monitor insect population development in fields (with the assistance of a crop consultant or advisor if necessary) to determine if and when control measures are warranted. Monitor and consider natural enemies when making control decisions. After treatment, continue monitoring to assess pest populations and their control.

Focus on Economic Thresholds

Insecticides should be used only if insects are numerous enough to cause economic losses that exceed the cost of the insecticide plus application. An exception would be in-furrow, at-planting treatments for early season pests that usually reach damaging levels each year. Consult local crop advisors about economic thresholds for target pests in your area.

Take an Integrated Approach to Managing Pests

Use as many different control measures as possible. Effective IPMbased programs will include the use of synthetic insecticides, biological insecticides, beneficial arthropods (predators and parasites), cultural practices, transgenic plant varieties, crop rotation, pest-resistant crop varieties and chemical attractants or deterrents. Select insecticides with care and consider the impact on future pest populations and the environment. Avoid broad-spectrum insecticides when a narrow-spectrum or more specific insecticide will work.

Time Applications Correctly

Apply insecticides when the pests are most vulnerable. For many insects this may be when they have just emerged. Use application rates and intervals recommended by the manufacturer or a local pest management expert *(i.e.,* university insect management specialist, county Extension agent, or crop consultant).

Mix and Apply Carefully

As the potential for resistance increases, the accuracy of insecticide applications in terms of dose, timing, coverage, etc. assumes greater importance. The pH of water used to dilute some insecticides in tank mixes may need to be adjusted to the product manufacturer's specifications. In aerial application, the swath widths should be marked, preferably by permanent markers. Sprayer nozzles should be checked for blockage and wear, and should be able to handle pressure adequate for good coverage. Spray equipment should be properly calibrated and checked on a regular basis. In tree fruits, proper and intense pruning will allow better canopy penetration and tree coverage. Use application volumes and techniques recommended by the manufacturers and local crop advisors.

Alternate Different Insecticide Classes

Avoid the repeated use of the same insecticide or insecticides in the same chemical class, which can lead to resistance and/or cross-resistance (1). Rotate insecticides across all available classes to slow resistance development. In addition, do not tank-mix products from the same insecticide class. Rotate insecticide classes and modes of action (see Insert (s)), consider the impact of pesticides on beneficial insects, and use products at labeled rates and spray intervals.

Protect Beneficial Arthropods

Select insecticides in a manner that is the least damaging to populations of beneficial arthropods. For example, applying insecticides in-furrow at planting or in a band over the row rather than broadcasting will help maintain certain natural enemies.

Preserve Susceptible Genes

Preserve susceptible individuals within the target population by providing a haven for susceptible insects, such as unsprayed areas within treated fields, adjacent "refuge" fields, or habitat attractions within a treated field that facilitate immigration. These susceptible individuals may outcompete and interbreed with resistant individuals, diluting the resistant genes and therefore the impact of resistance.

Consider Crop Residue Options

Destroying crop residue can deprive insects of food and overwintering sites. This cultural practice will kill insecticide-resistant pests (as well as susceptible ones) and prevent them from producing resistant offspring for the next season. However, review your soil conservation requirements before removing crop residue.

Chapter 11

Pesticide Residues in Food Commodities

11.1 Pesticide Residues: Impact on Indian Trade

Chemical pesticides have become the most important form of pest control Since the post-World War II era. There are two categories of pesticides, first-generation pesticides and second-generation pesticide. The first-generation pesticides, which were used prior to 1940, consisted of compounds such as arsenic, mercury, and lead. These were soon abandoned because they were highly toxic and ineffective. The second-generation pesticides were composed of synthetic organic compounds. The growth in these pesticides accelerated in late 1940s after Paul Müller discovered DDT in 1939. The effects of pesticides such as aldrin, dieldrin, endrin, chlordane, parathion, captan and 2,4-D were also found at this time. Those pesticides were widely used due to its effective pest control. However, in 1946, people started to resist to the widespread use of pesticides, especially DDT since it harms non-target plants and animals. People became aware of problems with residues and its potential health risks. In the 1960s, Rachel Carson wrote *Silent Spring* to illustrate a risk of DDT and how it is threatening biodiversity.

In 2010, the European Union rejected three okra consignments from India due to high levels of Monocrotophos, Acephate and Triazaphos. All three of these pesticides can cause headaches, vomiting, nausea, abdominal cramps and cardiac problems. EU Maximum Residue Limit is 0.03mg/kg, but tests revealed levels of 0.13mg/kg. India's MRL for Monocrotophos is considerably higher at 0.2mg/kg, but it is recommended only for use on cotton crops, as it is toxic to birds and humans. Nevertheless, levels detected in food for sale on the domestic market are far higher than for exports. The Indian Ministry of Commerce and Industry's drive to increase grape exports from 37,000 to 44,000 tonnes is being hampered by differing MRLs in exporting countries.Last year, exports to the EU were threatened by a deadlock caused by Chlormequat, just one of 98 pesticides for which grape consignments to the EU are tested. The UK and Sweden allowed import of Indian grapes by introducing their own MRL.A child weighing 16.15kg needed to eat just 211.5g of grapes to be at risk, No warning was issued in the UK.

Aldrin was detected in brinjal, cauliflower, tomato, okra, banana, apple, wheat and milk. Chlordane, which is banned in 47 countries was found in apples, bananas and cabbage. Chlorfenvinfos_was detected in bitter gourd, cabbage, cauliflower, tomatoes, rice and wheat. Heptachlor was detected in brinjal, okra, tomatoes, rice, milk and butter. These four substances are among the persistent organic pollutants (POPs) identified by the Stockholm Convention as the 'dirty dozen.' DDT is not supposed to be used on vegetable crops, was found in tomatoes in Uttar Pradesh at over 100 times the MRL. Fenpropathrin is not recommended for use on tea plants, was detected in Assam tea at more than twice the CODEX MRL of 2ppm.

11.2 Definitions

☆ **Pesticide residue**: Pesticide residue refers to the pesticides that may remain on or in food after they are applied to food crops. The levels of these residues in foods are often stipulated by regulatory bodies in many countries. Exposure of the general population to these residues most commonly occurs through consumption of treated food sources, or being in close contact to areas treated with pesticides such as farms or lawns around houses.

☆ **Maximum Residue Levels**: MRL is defined as the maximum concentration of pesticide residue (expressed as milligrams of residue per kilogram of food/animal feeding stuff) likely to occur in or on food and feeding stuffs after the use of pesticides according to Good Agricultural Practice (GAP), *i.e.* when the pesticide has been applied in line with the product label recommendations and in keeping with local environmental and other conditions). MRLs are primarily trading standards, but they also help ensure that residue levels do not pose unacceptable risks for consumers.

☆ **Acceptable daily intake:** ADI is a measure of the amount of a specific substance (usually a food additive, or a residue of a veterinary drug or pesticide) in food or drinking water that can be ingested (orally) on a daily basis over a lifetime without an appreciable health risk. ADIs are expressed by body mass, usually in milligrams (of the substance) per kilograms of body mass per day.

☆ **Pre-harvest interval:** The pre-harvest interval (PHI) plays very important role in reducing the pesticide residues in food. As this is based on scientific studies for detection of pesticide residue compared with Maximum Residue Level (MRL) before the harvest. This gives guideline to the extension workers, farmers and pesticide users to stop the pesticide use in production well before the harvest date. Safe PHI is depend on the MRL, residue detection facilities, The dose of pesticide, use of combination pesticide.

11.3 Strategies for Reducing Pesticide Residues

A. Pesticides Mainly Enter into Food Products due to Following Reasons

1. Indiscriminate use of chemical pesticides
2. Non-observance of prescribed waiting periods
3. Use of sub-standard pesticides
4. Wrong advice and supply of pesticides to the farmers by pesticide dealers
5. Continuance of DDT and other uses of pesticides in Public Health Programmes
6. Effluents from pesticides manufacturing units
7. Wrong disposal of left over pesticides and cleaning of PP equipments
8. Pre-marketing pesticides
9. Treatment of fruits and vegetables.

B. Ways of Controlling Pesticide Residue

1. Modification of Treatment or level of exposure to Pesticide.
2. Studying Rate of dissipation of pesticide residues.
3. Organic farming

Modification of Treatment or Level of Exposure to Pesticide

a. Use of minimum addition of chemicals in Crop cultivation
b. Diagnosis of pest and diseases to apply correct plant protection measures.
c. Effective use of pesticide
d. Right time of pesticide application.
e. Right type of pesticide application.
f. Right mode of pesticide application.
g. Use of bio-pesticides.
h. The nozzle which has to be used should have capacity to formulate fine droplets.

Studying Rate of Dissipation of Pesticide Residues

Dissipation rates are the function of pesticide's physical and Chemical properties. The dissipation of residues occur through number of processing Includes following

1) Volatilazation to the atmosphere
2) Washing of by rainfall or overhead irrigation
3) Chemical Degradation
4) Growth dilution
5) Metabolism

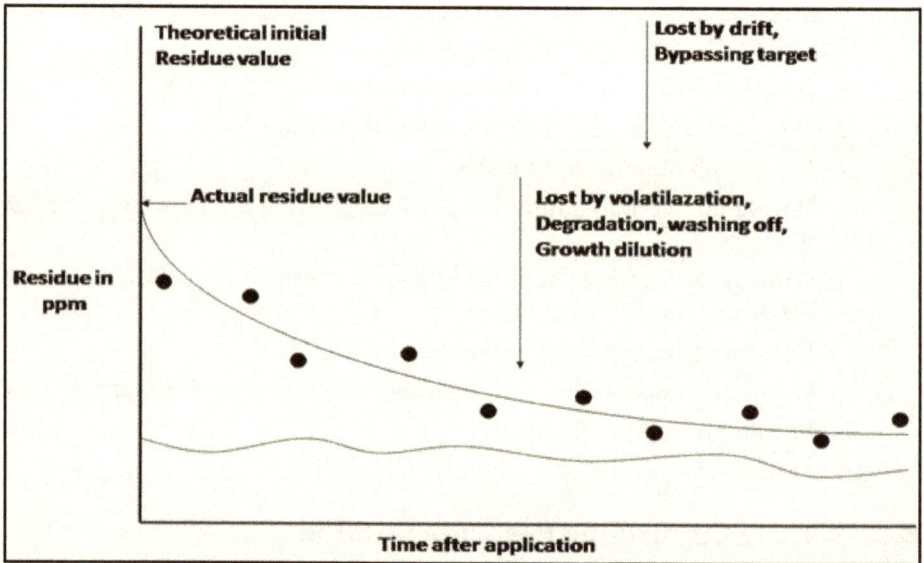

Hypothetical Pesticide Residue Dissipation Curve.

Other Strategies for Reducing Pesticide Residues

1. Training to Dealer and the person working to sale the product.
2. Training to farmers regarding uses, safe handling, and disposal of pesticide.
3. Label to be followed for name of crops on which the pesticide can be used, Correct doses with stages of crops and PHI, Name of the pest and diseases against which the pesticides can be used, Name of the crops on which it should not be used (if any specific).
4. Approved use of pesticides are very important, because scientific studies revealed that particular pesticide if used on specific crop then only it is safe.
5. The Pre-harvest Interval (PHI) was also studied to fix the spray schedule.

6. The doses are also determined on these issues, care has to be taken for such field trials after which the doses of particular pesticides are recommended.

7. Feed back to farmers regarding residue detected in their produce.

11.4 Pesticide Regulations in India

Each country adopts their own agricultural policies and Maximum Residue Limits (MRL) and Acceptable Daily Intake (ADI). The level of food additive usage varies by country because forms of agriculture are different in regions according to their geographical or climatical factors.

Pesticides Regulations are Governed in India under Acts/Rules

1. The Insecticides Act, 1968 and Rules, 1971

2. The Environment (Protection) Act, 1986

3. Hazardous Waste (Management and Handling) Rules, 1989

4. Water (Prevention and Control of Pollution) Act, 1974

5. Air (Prevention and Control of Pollution) Act, 1981

6. Prevention of Food Adulteration Act, 1954

7. The Factories Act, 1948

8. Bureau of Indian Standards Act

1. The Codex Alimentarius Commission (CAC)

CAC was created in 1961/62 by Food and Agriculture Organization of the United Nations (FAO) and the World Health Organization (WHO), to develop food standards, guidelines and related texts such as codes of practice under the Joint FAO/WHO Food Standards Programme. The main purpose of this Programme is to protect the health of consumers, ensure fair practices in the food trade, and promote coordination of all food standards work undertaken by international governmental and non-governmental organizations. "**Codex India**" the National Codex Contact Point (NCCP) for India, is located at Food Safety and Standards Authority of India (Ministry of Health and Family Welfare), FDA Bhawan, Kotla Road, New Delhi -110002, India. It coordinates and promotes Codex activities in India in association with the National Codex Committee and facilitates India's input to the work of Codex through an established consultation process.

2. The Food Safety and Standards Authority of India (FSSAI)

It has been established under Food Safety and Standards Act, 2006 which consolidates various acts and orders that have hitherto handled food related issues in various Ministries and Departments. FSSAI has been created for laying down science based standards for articles of food and to regulate their manufacture, storage, distribution, sale and import to ensure availability of safe and wholesome food for human consumption.

Highlights of the Food Safety and Standard Act, 2006

Various central Acts like Prevention of Food Adulteration Act, 1954, Fruit Products Order, 1955, Meat Food Products Order, 1973,Vegetable Oil Products (Control) Order, 1947,Edible Oils Packaging (Regulation)Order 1988, Solvent Extracted Oil, De- Oiled Meal and Edible Flour (Control) Order, 1967, Milk and Milk Products Order, 1992 etc will be repealed after commencement of FSS Act, 2006. The Act also aims to establish a single reference point for all matters relating to food safety and standards, by moving from multi- level, multi-departmental control to a single line of command.

11.5 Tolerence Limits of Insecticides

The food safety and standards regulations (contaminants, toxins and residues) of ministry of health and family welfare (food safety and standards authority of india) new delhi, dated the 1st august, 2011 confers restrictions on the use of insecticides in the following manner " Subject to the Provisions of regulation, no insecticides shall be used directly on articles of food Provided that nothing in this regulation shall apply to the fumigants which are registered and recommended for use as such on articles of food by the Registration Committee, constituted under section 5 of the Insecticides Act, 1968 ". The amount of insecticide mentioned in Column 2 on the foods mentioned in column 3, shall not exceed the tolerance limit prescribed in column 4 of the Table No. 18.

**Table 18: Tolerence Limits of Insecticides given by
Food Safety and Standards Authority of India.**

Sl.No.	Name of Insecticides	Food	Tolerance Limit (mg/kg. ppm)
1.	Aldrin, dieldrin	Foodgrains	0.01
		Milled Foodgrains	Nil
		Milk and Milk products	0.15
		Fruits and Vegetables	0.1
		Meat	0.2
		Eggs	0.1
2.	Carbaryl	Fish	0.2
		Foodgrains	1.5
		Milled food grains	Nil
		Okra and leafy vegetables	10.0
		Potatoes	0.2
		Other vegetables	5.0
		Cottonseed (whole)	1.0
		Maize cob (kernels)	1.0
		Rice	2.50

Contd...

Table 18–*Contd...*

Sl.No.	Name of Insecticides	Food	Tolerance Limit (mg/kg. ppm)
		Maize	0.50
		Chillies	5.00
3.	Chlordane	Food grains	0.02
		Milled food grains	Nil
		Milk and milk products	0.05
		Vegetables	0.2
		Fruits	0.1
	Sugar beet	0.3	
4.	D.D.T.	Milk and milk products	1.25
		Fruits and vegetables	3.5
		Meat, poultry and fish	7.0
		Eggs	0.5
5.	D.D.T. (singly)	Carbonated Water	0.001
6.	D.D.D. (singly)	Carbonated Water	0.001
7.	D.D.E. (singly)	Carbonated Water	0.001
8.	Diazinon	Foodgrains	0.05
		Milled foodgrains	Nil
		Vegetables	0.5
9.	Dichlorvos	Foodgrains	1.0
		Milled foodgrains	0.25
		Vegetables	0.15
		Fruits	0.1
10.	Dicofol	Fruits and Vegetables	5.0
		Tea (dry manufactured)	5.0
		Chillies	1.0
11.	Dimethoate	Fruits and Vegetables	2.0
		Chillies	0.5
12.	Endosulfan	Fruits and Vegetables	2.0
		Cottonseed	0.5
		Cottonseed oil (crude)	0.2
		Bengal gram	0.20
		Pigeon Pea	0.10
		Fish	0.20
		Chillies	1.0
		Cardamom	1.0
13.	Endosulfan A	Carbonated Water	0.001
14.	Endosulfan B	Carbonated Water	0.001

Contd...

Table 18–_Contd..._

Sl.No.	Name of Insecticides	Food	Tolerance Limit (mg/kg. ppm)
15.	Endosulfan-Sulphate	Carbonated Water	0.001
16.	Fenitrothion	Foodgrains	0.02
		Milled foodgrains	0.005
		Milk and Milk Products	0.05
		Fruits	0.5
		Vegetables	0.3
		Meat	0.03
17.	HCH and its Isomers	Food grains except rice	0.10
	Gamma Isomer	Milled foodgrains	Nil
	(Known as Lindane)	Rice grain Unpolished	0.10
		Rice grain polished	0.05
		Milk	0.01
		Milk products	0.20
		Fruits and vegetable	1.00
		Fish	0.25
		Eggs	0.10
		Meat and poultry	2.00
		Carbonated Water	0.001
18.	Malathion	Foodgrains	4.0
		Milled foodgrains	1.0
		Fruits	4.0
		Vegetables	3.0
		Dried fruits	8.0
		Carbonated Water	0.001
19.	Parathion	Fruits and Vegetables	0.5
20.	Parathion methyl	Fruits	0.2
		Vegetables	1.0
21.	Phosphamidon residues	Foodgrains	0.05
		Milled foodgrains	Nil
		Fruits and Vegetables	0.2
22.	Chlorpyrifos	Foodgrains	0.05
		Milled foodgrains	0.01
		Fruits	0.5
		Potatoes and Onions	0.01
		Cauli Flower and Cabbage	0.01
		Other vegetables	0.2
		Meat and Poultry	0.1
		Milk and Milk Products	0.01

Contd...

Table 18–*Contd...*

Sl.No.	Name of Insecticides	Food	Tolerance Limit (mg/kg. ppm)
		Carbonated Water	0.001
23.	Ethion	Tea (dry manufactured)	5.0
		Cucumber and Squash	0.5
		Other Vegetables	1.0
24.	Monocrotophos	Food grains	0.025
		Milled Food grains	0.006
		Citrus fruits	0.2
		Other fruits	1.0
		Carrot, Potatoesa, Sugar beet	0.05
		Onion and Peas	0.1
		Other Vegetables	0.2
		Cottonseed	0.1
		Cottonseed oil (raw)	0.05
		*Meat and Poultry	0.02
		*Milk and Milk Products	0.02
		Eggs	0.02
		Coffee (Raw beans)	0.1
		Chillies	0.2
		Cardamom	0.5
25.	Acephate	Safflower seed	2.0
		Cotton Seed	2.0
26.	Aldicarb	Potato	0.5
		Chewing Tobacco	0.1
27.	Carbofuran	Food grains	0.10
		Milled food grains	0.03
		Fruit and Vegetables	0.10
		Oil seeds	0.10
		Sugarcane	0.10
		Milk and Milk Products	0.05
28.	Cypermethrin (sum (sum of isomers)	Wheat grains	0.05
		Milled wheat grains	0.01
		Brinjal	0.20
		Cabbage	2.00
		Bhindi	0.20
		Oil seeds except groundnut	0.20
		Meat and Poultry	0.20
		Milk and Milk Products	0.01

Contd...

Table 18–*Contd...*

Sl.No.	Name of Insecticides	Food	Tolerance Limit (mg/kg. ppm)
29.	Decamethrin/Deltamethrin	Cotton Seed	0.10
		Food grains	0.50
		Milled Foodgrains	0.20
		Rice	0.05
30.	Fenvalerate	Cauliflower	2.00
	(fat soluble residue)	Brinjal	2.00
		Okra	2.00
		Cotton Seed	0.20
		Cotton seed oil	0.10
		Meat and Poultry	1.00
		Milk and Milk Product	0.01
31.	Phorate	Foodgrains	0.05
		Milled foodgrains	0.01
		Tomatoes	0.10
		Other vegetables	0.05
		Fruits	0.05
		Oil seeds	0.05
		Edible oils	0.03
		Sugarcane	0.05
		Eggs	0.05
		Meat and Poultry	0.05
		Milk and Milk Products	0.05
32.	Cartaphydrochloride	Rice	0.50
33.	Chlormequatchloride	Grape	1.00
		Cotton Seed	1.00
34.	Chlorothalonil	Groundnut	0.10
		Potato	0.10
35.	Diflubenzuron	Cotton Seed	0.20
36.	Quinolphos	Rice	0.01
		Pigeon pea	0.01
		Cardamom	0.01
		Tea	0.01
		Fish	0.01
		Chillies	0.2

Contd...

Table 18–*Contd...*

Sl.No.	Name of Insecticides	Food	Tolerance Limit (mg/kg. ppm)
37.	Triazophos	Chillies	0.2
		Rice	0.05
		Cotton seed oil	0.1
		Soyabean oil	0.05
38.	Profenofos	Cotton seed oil	0.05
39.	Fenpropathrin	Cotton seed oil	0.05
40.	Carbosulfan	Rice	0.2
41.	Imidacloprid	Cotton seed Oil	0.05
		Rice	0.05
42.	Lambdacyhalothrin	Cotton seed Oil	0.05
43.	Spinosad	Cotton seed oil	0.02
		Cabbage	0.02
		Cauliflower	0.02
44.	Thiamethoxam	Rice	0.02
45.	Acetamiprid	Cotton seed oil	0.1
46.	Ethofenprox	Rice	0.01
47.	Bifenthrin	Cotton seed	0.05
48.	Buprofezin	Rice	0.05
49.	Chlorfenopyr	Cabbage	0.05
50.	Indoxacarb	Cotton seed	0.1
		Cottonseed oil	0.1
		Cabbage	0.1
51.	Lufenuron	Cabbage	0.3
52.	Novaluron	Cottonseed	0.01
		Cottonseed oil	0.01
		Tomato	0.01
		Cabbage	0.01
53.	Thiochlorprid	Cotton seed	0.05
		Cotton seed oil	0.05
		Rice	0.01

11.6 Basic Procedures in Pesticide Residue Analysis

    A. Sample Preperation

    B. Extraction

    C. Clean up

    D. Analysis

11.6.1 Sample Preperation

1. Sampling, Transport, Processing and Storage of Samples

Sampling

Laboratory samples should be taken in accordance with superseding legislation. Where it is impractical to take primary samples randomly within a lot, the method of sampling must be recorded.

Laboratory Sample Transportation

Samples must be transported to the laboratory in clean containers and robust packaging. Polythene bags, ventilated if appropriate, are acceptable for most samples but low-permeability bags (*e.g.* nylon film) must be used for samples to be analysed for residues of fumigants. Very fragile or perishable products (*e.g.* ripe raspberries) may have to be frozen to avoid spoilage and then transported in "dry ice" or similar, to avoid thawing in transit. The use of marker pens containing organic solvents should be avoided for labelling bags containing samples to be analysed for fumigant residues, especially if an electron capture detector is to be used.

2. Sample Preparation and Processing Prior to Analysis

On receipt, each laboratory sample must be allocated a unique reference code by the laboratory. Sample preparation, sample processing and sub-sampling to obtain analyti-cal portions must take place before visible deterioration occurs. Analyses for residues of very labile or volatile pesticides should be started, and the procedures in-volved in potential loss of analyte completed, on the day of sample receipt. If a single analytical portion is unlikely to be representative of the ana-lytical sample, replicate portions must be analysed, to provide a better estimate of the true value.

3. Pesticide Standards, Calibration Solutions, etc.

Identity, Purity, and Storage of Standards

Pure standards of analytes and internal standards should be of known purity and each must be uniquely identified and the date of receipt recorded. They should be stored at low temperature, preferably in a freezer, with light and moisture excluded, *i.e.* under conditions that minimise the rate of degradation.

Preparation and Storage of Stock Standards

When preparing stock standards (solutions, dispersions or gaseous dilutions) of "pure" standards of analytes and internal standards, the identity and mass (or volume, for highly volatile compounds) of the "pure" standard and the identity and amount of the solvent (or other diluents) must be recorded. Stock standards must be labelled indelibly, allocated an expiry date and stored at low temperature in the dark in containers that prevent any loss of solvent and entry of water.

Preparation, Use and Storage of Working Standards

When preparing working standards, a record must be kept of the identity and amount of all solutions and solvents employed. The solvent(s) must be appropriate to the analyte (solubility, no reaction) and method of analysis.

11.6.2 Extraction Methods in Pesticide Residue Analysis

Extraction Techniques

Extraction means separation of pesticide residues from the matrix by using solvent. The extraction procedure should be such that it quantitatively removes pesticides form matrix (high efficiency), does not cause chemical change in pesticide and use inexpensive and easily cleaned apparatus. The extraction method and solvent type determine the extraction efficiency from substrates.

Choice of Extraction Method

The main objective behind employing a particular method for a specific substrate is to bring the solvent to close proximity of the pesticide residues for sufficient period so that pesticide residues gets solubilised in the solvent. The choice of method depends on the type of substrate and ageing of residues. The substrates in pesticide residue analysis could be liquids like water, fruit juices, body fluids (urine, blood etc.) and solids like soil, flesh, green plant materials (leaves, fruit etc.), dry fodder, grains etc.

Liquid Substrates

Partitioning

Samples like water, body fluids, juices are extracted by partitioning with water immiscible solvent. The addition of sodiumchloride in aqueous samples improves the extraction efficiency by reducing the solubility of pesticide in water. It also prevents the emulsion formation, which is frequently encountered during partitioning.

Use of Absorbent

The pesticide residues from aqueous samples can be extracted by passing the sample through solid adsorbents packed in glass column. The adsorbents have high affinity for pesticide molecules, therefore, they are held up on the absorbent whereas water passes out. The solid adsorbents are then extracted with organic solvent. The solid adsorbents normally used for removal of pesticide from aqueous samples *e.g.* Activated charcoal.

Solid Substrates

Fresh Residues

Dipping, tumbling, shaking: This method is usually employed for solid substrate when pesticide residues are present on the surface as in case of freshly applied pesticide.

Weathered Residues

When sufficient time has elapsed after the application (weathering), the residues are not present on the surface but they penetrate the substrate matrix and are in adsorbed form. The substrate matrix needs to be broken down in fine particles before extraction with solvent. The methods that employ these techniques are macerating/blending, macerating/blending followed by column extraction, soxhlet extraction, etc.

Choice of Solvent

The choice of solvent for extraction depends on the a) nature of the substrate and b) the type of pesticide to be extracted. However, the solvent should satisfy the following conditions.

☆ Should have high solubility for the pesticide and least solubility for coextractives.

☆ Should not change the pesticide chemically or react with it.

☆ Economical

☆ Low boiling.

☆ Easily separated from the substrate.

☆ Compatible to the method of final determination.

Choice of Solvent Depending on Type of Substrate

The solvent for extraction of pesticide in different substrate is chosen as follows.

Aqueous Substrate

Water immiscible solvent like hexane, petroleum ether, benzene, dichloromethane, chloroform, ethyl acetate, etc.

Solid Substrates (Soil, fodder etc.)

Different solvents like acetonitrile, hexane and mixed solvents are used for dry sample with low moisture like grains, samples with high moisture content (green plant samples and substrates with high fat content (grains, oilseeds, egg, meat, fish etc.).

Choice of Solvent Depending on Nature of Pesticides

The pesticide molecules can be broadly divided into two groups namely nonionic and ionic type. The non-ionic pesticides also differ in their polarity. For nonionic type of pesticides, organic solvents with varying polarity depending on the polarity of pesticide molecules are employed.

Recent Techniques of Extraction

a. Solid Phase Extraction (SPE)

Solid phase extraction technique is based on the concept of selective retention by the device for the analyte, in this case the pesticide. SPE can be made to work on either the batch or column mode. This method has not become popular as there is loss of the pesticide as it tends to adhere to the surface of the beaker and tubes. The method is modified by the use of adsorbents contained in cartridges of various sizes usually made of plastic such as polyethylene or polypropylene of extremely high purity and is termed as column-liquid solid extraction (CLSE), however for simpliciaty it is referred to a SPE cartridges.

b. Solid Phase Micro-Extraction (SPME)

In this technique a droplet of extractant, or a fiber coated with the extractant is suspended in the solution to be extracted and then transferred to an analytical device.

c. Accelerated Solvent Extraction (ASE)

The extracting solvent is passed under amabient temperature or pressure through the matrix, removing the analyte using a smaller volume of the solvent.

d. Microwave-Assisted Solvent Extraction (MASE)

The technique employs the use of microwave energy and a suitable solvent to extract the analyte from the matrix, water is commonly preferred solvent in this procedure.

e. Supercriticial Fluid Extraction (SFE)

In the supercritical fluid extraction (SFE) method carbon dioxide gas is passed under supercritical temperature and pressure (liquefied carbon dioxide) through the matrix to extract the pesticide and then transferred to the analytical device for quantification.

11.6.3 Clean Up Methods in Pesticide Residue Analysis

Cleanup refers to a step or series of steps in the analytical procledure in which the bulk of the potentially interfering coextractives are removed by physical or chemical methods. During extraction, the solvent comes in contact with the substrate matrix, to enable extraction of the pesticide along with some of the constituents of the substrate matrix also get solubilized. The extract not only contains pesticide residues but also other constituents, which are called co-extractives. The removal of interfering coextractives from extract is called clean up. The co-extractive generally extracted along with pesticide from various substrates are moisture, coloured pigment like chlorophyll, xanthophylls and anthocyanins, colourless compounds like oil, fat and waxes etc. When dry substrate is extracted with water immiscible solvent, it contains traces of moisture, which can be removed by passing the extract through anhydrous sodium sulfate. High moisture containing substrate are extracted with water miscible solvent,the extract contains lot of water and water soluble compounds, the extract is concentrated to remove organic solvent, the aqueous phase is diluted with saturated sodium chloride solution and then extracted with water immiscible solvent just like water samples. After removal of moisture, the other coextractives are removed by using various separation techniques.

a. Liquid-liquid Partitioning

In this technique, co-extractives from the extract are removed by partitioning the residues between two immiscible solvents.

b. Acetonitrile-Hexane Partitioning

Acetonitrile-hexane partitioning is used for the removal of oil and fat from the extract. This technique is used for the cleanup of extracts of oil seeds, milk, butter etc.

c. Partitioning with Acid/Base Treatment

This technique can be used for the pesticides, which are either acidic or basic in nature. This technique can not be used for neutral type of pesticides.

d. Chemical Treatment

In these techniques, the co-extractives are either precipitated and separated by filtration or made water-soluble so that pesticide can be partitioned into water miscible organic solvent.

e. Saponification

This technique is employed to remove fats and oils from the extract. The fat and oil is saponified or hydrolysed by treatment with alkaline aqueous solution. This method can be employed for the pesticides that are stable to alkali treatment or the pesticides, which give definite product that can be analysed easily.

f. Precipitation

This technique can be used only for the pesticides having some water solubility. In this technique, the co-extractives are precipitated with a coagulating agent like ammonium chloride.

g. Oxidation

In this technique, the co-extractives are oxidized with concentrated sulphuric acid. This technique can be used for pesticides, which are stable to acid. For example this techniques has been used for the clean up of milk extracts containing HCH, DDT, aldrin and dieldrin.

h. Chromatographic Techniques

Chromatography is a technique used for the separation of constituents from the mixture. In Chromatography, two phases are involved in separation. The extract from the sample contains mixture of pesticide and co-extractives; various Chromatographic techniques can be employed for separating them or removal of the co-extractives from the pesticides.

i. Thin Layer Chromatography (TLC)

For clean up, preparative TLC plates (20 x 20 cm) with thick layer of adsorbent (~ 2 mm) are used. Silica gel plates are normally used but other adsorbents like alumina can also be used.

j. Ion Exchange Chromatography

Ion exchange resins can also be used for clean up of ionic pesticide. For cationic pesticides like paraquat and diquat, cation exchange resins (H+) while for anionic pesticide like 2,4-D, anion exchange resins (Ch) are used. The matrix contains fixed charged groups are the counter ions of opposite charge. These counter ions can be exchanged from other ions of similar charge in the mobile phase. The aqueous extract containing pesticide is passed through a column of ion exchange resin. The exchange resin holds up the pesticide being ionic, whereas non-ionic coextractives pass out of the column. The held up pesticide is eluted out using suitable electrolyte solution.

k. Gel Permeation Chromatography and Molecular Sieves

The separation in gel permeation Chromatography and molecular sieves is based on the principle of size exclusion. Both gel and molecular sieves have tubular structures with inner diameter (id) similar to the molecular sizes. The molecules having size greater than the tube id do not pass through it. The molecules having size less than the tube id pass through it. Molecules having greater size moves faster than the smaller ones, enabling separation of molecules occur depending on their sizes. The co-extractives like chlorophyll, other pigments, etc. have molecular sizes greater than most of the pesticide, therefore, they are easily separated. Also the co-extractives having molecular size less than pesticide molecule will elute later than pesticide.

l. Adsorption Column Chromatography

Adsorption column Chromatography is the most common and widely used technique for clean up. different type of adsorbent or mixture of adsorbent have been used for clean up. The adsorbent generally used for clean up are Silica gel (80-100 mesh), alumina (acidic, basic, neutral), polyethylene coated alumina, Florisil, Charcoal and mixtures of charcoal + Celite + MgO.

m. Solid Phase Extraction Cartridges

Serves the dual purpose of extraction and clean up. Advantages of SPE device over other conventional solvent extraction and clean up of pesticides includes better reproducibility, reduction in solvent use, high speed, versatility, freedom from interferences and field applications.

11.6.4 Analysis and Reporting Analytical Results

It is extremely important in pesticide residue studies that analytical results be reported in a consistent and unambiguous manner. Often, national or international organisations that summarize the analytical results and calculate dietary intakes, must evaluate and interpret data obtained from different laboratories, each reporting results in a different format. To facilitate these evaluations, laboratories should report enough details about the detection and quantification limits of the analytical method to enable correct interpretations of the data to be made. Laboratories should ensure consistent detection and quantification limits throughout a study. Results of recovery tests for the different contaminants should also be reported. However, analytical findings should be reported as measured, without the use of correction factors that take recovery into account.

a. Detection and Quantification Limits of the Analytical Method

The analytical methods used in the analysis of food samples must be as sensitive as possible. The sensitivity of the overall analytical procedure is usually defined in terms of limit of detection (Ld) and limit of quantification (Lq) or determination.

b. Limit of Detection

The limit of detection (Ld) as it applies to food contaminants may be defined as the minimum concentration of contaminant in a food sample that can just be qualitatively detected, but not quantitatively determined, under a pre-established set of analysis conditions. This is necessarily a very broad definition, since it encompasses all classes of contaminants and all detection techniques. Ld defines the minimum amount of contaminant in a food sample below which no finite value can be reported, *e.g.* not detected at limit of detection of 1 g/kg. In many instances particularly for the determination of pesticides, it is appropriate to place a special restriction on the term detected. Because of the possibility that a small detectable signal at a particular GLC retention time could result from an interference, it is desirable to specify that the identity of a detection be confirmed before it is reported. A detection whose identify cannot be confirmed would not be reported as a positive finding.

c. Limit of Quantification

The limit of quantification (Lq) is the minimum concentration of a contaminant in a food sample that can be determined quantitatively with an acceptable accuracy and consistency. The Codex Committee on Pesticide Residues employs the term limit of determination which is defined as the lowest practical concentration of a pesticide residue on contaminant that can be quantitatively measured and identified in the specified food commodity or animal feed stuff with an acceptable degree of certainty by current regulatory methods of analysis.

d. Reporting Results

Report the exact portion of food taken for analysis and report results in parts per million (ppm).

    ☆ *Raw Agricultural Commodities:* Report the results as ppm on portion examined, except report raw milk on fat basis.

    ☆ *Processed foods:* Report the results as ppm on portion examined, except for dairy products and concentrates.

    ☆ *Dairy products:* Report results as ppm on a fat basis, except report residues in low fat dairy products *(e.g.* skim milk, buttermilk, nonfat dried milk, uncreamed cottage cheese) on a whole as is basis.

    ☆ *Concentrated and Dehydrated Products:* Report results as ppm on as is basis. Where significant residues are encountered on concentrates which must be reconstituted to the whole product basis before consumption, it is useful to calculate to the whole basis and record both results.

e. Reporting Not Detected

The term not detected should be used to indicate that food sample (s) was analysed or a particular contaminant or class of contaminants, *e.g.* organochlorine pesticides, and analytical response was observed. A not detected reporting should be accompanied by a limit of detection for the analytical method used, and a short summary of which compounds or types of compounds were amenable to the method. For example, if a leafy-vegetable composite were analysed using a method capable of recovering several organchlorine pesticides, the report might state organochlorine pesticides; not detected at detection limit of 1 g/kg or heptachlor, 25 g7kg; other organochlorine residues, not detected at detection limit of 1 g/kg. Values of zero should be avoided, or at least qualified by defining the limit of detection.

f. Reporting Trace Values

Occassionally, although not a good practice, laboratorieswill use the termtrace to report the detection of a contaminant. This usually refers to an analytical response that is just above the limit of detection but below the limit of quantification, *i.e.* the compound is detectable (with confirmed identity), but cannot bequantified (detection limit < trace < quantification limit). Laboratories should report such low-level detections, but also should clearly state the meaning of trace, and the analytical uncertainties associated with it.

g. Rounding of Numbers

To report correct significant digits we must round off numbers. To obtain this approximate number you must do off certain of the lower digits in the number, the process known as rounding off. An error is of course introduced in rounding off, and it is desirable to make this error be as small as possible. By using the proper method of rounding, it is always possible to make the absolute error no greater than half a unit in the last place retained. This is done by increasing by one the last digit kept if the discarded part is greater than half a unit in that digit position. If the discarded part is exactly one-half a unit the last place kept, it is best to increase the last kept digit by one sometimes, and not other times, so that if several round offs are made during the problem, all will not introduce errors in the same direction. A simple way to do this is always to round the last kept digit to an even number when the discarded part is exactly half unit in the last kept place.

11.7 Extraction of Pesticide Residue from Fruits and Vegetables [Conventional Method]

Principle

Most non-ionic residues are extracted with acetone and the residues are partitioned from aqueous acetone to DCM/Hexane phase. After removing traces of DCM, the final extract is made up with acetone. The extract is cleaned up using florisil column. For the carbamate pesticide determination, the extract is subjected to C-18 solid phase cartridge clean up. The present method is applicable for low fat [< 2 per cent] commodities.

Reagents/Solvents Required

1. Anhydrous sodium sulphate
2. Sodium Chloride
3. n-Hexane
4. Acetone
5. Dichloromethane
6. Florisil

Apparatus Required

1. Separating funnel [1000 ml]
2. Beaker [100ml]
3. Measuring cylinder [10,100,250 ml]
4. Conical flask
5. Cotton washed with acetone
6. Funnels
7. Analytical weighing balance
8. Volumetric flask [10 ml]
9. Mechanical Shaker
10. Pipettle [1ml, 0.2 ml, 0.1 ml]
11. Dropper
12. Blender
13. GC-ECD

Procedure

A. Extraction

1. Blend 400 g of bhendi and weigh 50 g into 250 ml conical flask.
2. For fortification, spike pesticide mixture [Aldrin, lindane, Cpp] @ 0.3 and 0.6 ppm level and keep it for 30 min.
3. Add 100 ml acetone, shake it for 20 min on mechanical shaker.
4. Filter the extract through the buchner funnel using vacuum.
5. Take an aliquate of 40 ml of the extract to the separating funnel.
6. Add 100 ml mixture of hexane: DCM [1: 1] to the extract and shake well for 1 min.
7. Collect the aqueous [lower layer] into the conical flask.
8. Dry the organic layer through sodium sulphate [20 g approx] supported on prewashed cotton in funnel.
9. To the aqueous layer, add saturated sodium chloride solution [or 50 ml of 15 per cent sodium chloride solution]. Add 50 ml DCM to the aqueous layer. shake vigorously and dry the organic layer through sodium sulphate.
10. Repeat the extraction of aqueous layer with 50 ml DCM and dry the organic layer through sodium sulphate.
11. Pool up all the extracts and concentrate in RVE. The concentration is repeated 3-4 times in presence of hexane to remove all the traces of DCM.
12. Dilute 1ml extract to 10 ml with 10 per cent acetone in hexane.

B. Clean up

1. Plug cotton at the bottom of the sintered chromatography column [22 mm id] and wash with hexane

2. Add approximately 50 ml of n – hexane into the column

3. Pour 2 g sodium sulphate followed by 3 g activated florisil then followed by 2 g sodium sulphate

4. Elute the solution in florisil column and rinse with 3 ml hexane for 2 – 3 times

5. Add 1 ml of the extract which is diluted to 10 ml of 10 per cent acetone in hexane

6. Elute column at 5 ml/min flow rate with 50 ml elutant [50 per cent DCM + 1.5 per cent Acetonitrile +48.5 per cent Hexane]

7. Collect the elute in conical flask and concentrate on RVE upto 1ml

8. Make up the volume to 10 ml for GC-ECD injection.

11.8 QuEChERS Method [Quick, Easy, Cheap, Effective, Rugged and Safe] for Determining Pesticide Residues in Vegetables [*e.g.* Tomato]

Principle

The QuEChERS method uses a single-step buffered acetonitrile (MeCN) extraction while salting out water from the sample by using anh. $MgSO_4$ to induce liquid-liquid partitioning. For cleanup, a simple, inexpensive, and rapid technique called dispersive solid-phase extraction (SPE) is conducted using a combination of primary secondary amine (PSA) sorbent to remove fatty acids among other components and anh. $MgSO_4$ to reduce the remaining water in the extract. Then, the extracts are concurrently analyzed by liquid and gas chromatography (LC and GC) combined with mass spectrometry (MS) to determine a wide range of pesticide residues. The final extract concentration of the method in MeCN is 1 g/mL. To achieve <10 ng/g limit of quantitation (LOQ) in modern GC/MS, large volume injection (LVI) of 8 µL is typically needed, or the final extract can be concentrated and solvent exchanged to toluene (4 g/mL) in which case 2 µL splitless injection provides the anticipated degree of sensitivity.

Reagents/Solvents required

1. Ethyl acetate

2. Sodium sulphate

3. Primary secondary amine [PSA]

4. Magnesium sulphate

5. Methanol

Apparatus Required

1. Food Chopper

2. Blender

3. Container jars.

4. Blank sample–verified to contain no detectable analytes.

5. Samples to be analyzed

6. Freezer

7. Centrifuge tubes

8. Homogenizers

9. Appendoff tubes

10. Analytical weighing balance

11. Homogenizer

12. High speed centrifuge

13. Low speed centrifuge

14. Concentrator 15)GC-MS/LC-MS.

Procedure

1. Take 10 g of blended tomato in 50 ml centrifuge tubes

2. For fortification, spike pesticide mixture @ 0.1 ppm level [0.2 ml of 5ppm for 10g sample]

3. Keep it for 30 min

4. Add 10 ml Ethyl acetate [20 or 30 ml if necessary for easy homogenizing]

5. Add 10 g Sodium sulphate

6. Homogenize the contents in centrifuge tube

7. Centrifuge @ 15000 rpm for 2 min

8. Take 2ml appendoff tube and to it add 25 mg PSA + 150 mg Magnesium Sulphateer + 1ml supernatant of extract after centrifugation

9. Centrifuge @ 10000 rpm for 5 min

10. 1ml of the supernatant after centrifugation is taken for GC-MS injection

11. For LC-MS Injection, supernatant is concentrated and is dissolved in 1 ml methanol

11.9 Instrumentation in PRA

A. Gas Chromatography–Mass Spectrometry (GC-MS)

GCMS is an analytical method that combines the features of gas-liquid chromatographyand mass spectrometry to identify different substances within a test sample. Applications of GC-MS include drug detection, Pesticide residues, fireinvestigation, environmental analysis, explosives investigation, and identification of unknown samples. GC-MS can also be used in airport security to detect substances in luggage or on human beings. Additionally, it can identify trace elements in materials that were previously thought to have disintegrated beyond identification.

The use of a mass spectrometer as the detector in gas chromatography was developed during the 1950s after being originated by James and Martin in 1952. These sensitive devices were originally limited to laboratory settings. The development of affordable and miniaturized computers has helped in the simplification of the use of this instrument, as well as allowed great improvements in the amount of time it takes to analyze a sample. In 1964, Electronic Associates, Inc. (EAI), a leading U.S. supplier of analog computers, began development of a computer controlled quadrupole mass spectrometer under the direction of Robert E. Finnigan. By 1966 Finnigan and collaborator Mike Uthe's EAI division had sold over 500 quadrupole residual gas-analyzer instruments. In 1967, the Finnigan Instrument Corporation was formed and in early 1968, delivered the first prototype quadrupole GC/MS instruments to Stanford and Purdue University. FIC was eventually renamed Finnigan Corporation and went on to establish itself as the worldwide leader in GC/MS systems. In 1996 the top-of-the-line high-speed GC-MS units completed analysis of fire accelerants in less than 90 seconds, whereas first-generation GC-MS would have required at least 16 minutes. By the 2000s computerized GC/MS instruments using quadrupole technology had become both essential to chemical research and one of the foremost instruments used for organic analysis. Today computerized GC/MS instruments are widely used in environmental monitoring of water, air, and soil; in the regulation of agriculture and food safety; and in the discovery and production of medicine.

The GC-MS is composed of two major building blocks: the gas chromatograph and the mass spectrometer. The gas chromatograph utilizes a capillary column which depends on the column's dimensions (length, diameter, film thickness) as well as the phase properties (*e.g.* 5 per cent phenyl polysiloxane). The difference in the chemical properties between different molecules in a mixture and their relative affinity for the stationary phase of the column will promote separation of the molecules as the sample travels the length of the column. The molecules are retained by the column and then elute (come off) from the column at different times (called the retention time), and this allows the mass spectrometer downstream to capture, ionize, accelerate, deflect, and detect the ionized molecules separately. The mass spectrometer does this by breaking each molecule into ionized fragments and detecting these fragments using their mass-to-charge ratio.

These two components, used together, allow a much finer degree of substance identification than either unit used separately. It is not possible to make an accurate identification of a particular molecule by gas chromatography or mass spectrometry alone. The mass spectrometry process normally requires a very pure sample while gas chromatography using a traditional detector (*e.g.* Flame ionization detector) cannot differentiate between multiple molecules that happen to take the same amount of time to travel through the column (*i.e.* have the same retention time), which results in two or more molecules that co-elute.

Sometimes two different molecules can also have a similar pattern of ionized fragments in a mass spectrometer (mass spectrum). Combining the two processes reduces the possibility of error, as it is extremely unlikely that two different molecules will behave in the same way in both a gas chromatograph and a mass spectrometer. Therefore, when an identifying mass spectrum appears at a characteristic retention time in a GC-MS analysis, it typically increases certainty that the analyte of interest is in the sample.

B. Liquid Chromatography–Mass Spectrometry (LC-MS, or alternatively HPLC-MS)

It is an analytical chemistry technique that combines the physical separation capabilities of liquid chromatography (or HPLC) with the mass analysis capabilities of mass spectrometry (MS). LC-MS is a powerful technique that has very high sensitivity and selectivity and so is useful in many applications. Its application is oriented towards the separation, general detection and potential identification of chemicals of particular masses in the presence of other chemicals (*i.e.*, in complex mixtures), *e.g.*, natural products from natural-products extracts, and pure substances from mixtures of chemical intermediates. Preparative LC-MS systems can be used for rapid mass-directed purification of specific substances from such mixtures that are important in basic research, and pharmaceutical, agrochemical, food, and other industries.

Part II

Pest Management in Horticultural Crops

Chapter 12

Pests of Solanacious Crops and Bhendi

12.1 Pests of Brinjal

1. Brinjal Fruit and Shoot Borer, *Leucinodes orbonalis* (*Pyralidae, Lepidoptera*)

Identification

The adult moth is small, white with a pink and bluish tinge,with a few brown spots on its wings. It lives for 6-10 days. Moths do not feed on eggplant, rather they survive on plant exudate or dew drops. They hide under eggplant leaves during the day and are not easily seen. During the night, the moths come out in the open and mate. The female moth tends to curl its abdomen upwards. The adult life span is about a week; the females live longer than males.

Biology

The adult females lay eggs singly or in groups of two to five on the under surfaces of leaves, tender shoots, flower buds, or the base of developing fruits. Each female lays about 250 eggs, which are creamy white soon after laying, but turn red before hatching. The egg period is three to five days. The grown-up larva is pink with sparse hairs on the warts on the body and a dark brown or blackish head. Full-grown larva measures about 16-23 mm in length. Larva after hatching bores into the tender shoots and fruits and completes 6 larval instars in 16-22 days. Mature larva comes out and pupates outside the bore hole or in the stem. Pupa is present inside a dark brown tough silken cocoon.

Pupal period varies from 9-14 days. The total life cycle is completed in 36-38 days on an average.

Damage

Larva is a internal feeder it immediately bore into the nearest tender shoot or flower or fruit just after hatching,. Soon after boring into shoots or fruits, they plug the entrance hole with excreta. As a result, the affected twigs, flower and fruits dries up and may drop off. Larval feeding, inside shoots, result in wilting of the young shoot. Presence of wilted shoots in an eggplant field is the surest sign of damage by this pest. The damaged shoots ultimately wither and drop off. This reduces plant growth, which in turn, reduces fruit number and size. New shoots can arise but this delays crop maturity and the newly formed shoots are also subject to larval damage. Larval feeding in flowers results in failure to form fruit from damaged flowers. Larval feeding inside the fruit results in destruction of fruit tissue. The feeding tunnels are often clogged with frass. This makes even slightly damaged fruit unfit for marketing. The yield loss varies from season to season and from location to location.

Management

1. Avoid monoculture and follow crop rotation.

2. Avoid growing eggplant seedlings near fields with standing crops, in or near fields where the crop was grown previously, or near dried eggplant heaps. If seedlings must be grown in those areas, cover the beds with 30-mesh nylon net to prevent the entry of FSB moths.

3. Choose resistant or moderately resistant cultivars available in the region. For instance, accessions or varieties such as EG058, Pusa Purple Long, Pusa Purple Cluster, Pusa Purple Round, H- 128, H-129, Aushey, Thorn Pendy, Black Pendy, H-165, H-407, Dorley, PPC-17-4, PVR-195, Shyamla Dhepa, Banaras Long Purple, Arka Kesav, Arka Kusmakar, Punjab Barsati, Punjab Chamkila, Kalyanpur-2 and Gote-2 have been reported to be tolerant or resistant (Parker *et al.,* 1995; Alam *et al.,* 2003; Shivalingaswamy and Satpathy 2007).

4. Remove and destroy infested shoots and fruit at regular intervals until final harvest.

5. weekly releases of egg parasitoid, *Trichogramma chilonis* Ishii @ 1g parasitized eggs/ha/week and larval parasitoid, *Bracon habetor* Say @ 800-1000 adults/ha/week could be followed.

6. Install FSB sex pheromone lures in traps at the rate of 100 traps per hectare.

7. Spray Azadirachtin 1.0 per cent EC (10000 ppm) @2.0 ml/lit., Chlorantraniliprole 18.5 SC @ 60 ml/ha, Emamectin benzoate 5 per cent SG @4 g/10 lit, Flubendiamide 20 WG@5 g/10 lit., Novaluron 10 per cent EC @7.5 ml/10 lit., Phosalone 35 per cent EC @13 ml/10 lit.

2. Brinjal Stem Borer, *Euzophera perticella (Puralidae, Lepidoptera)*

Identification

The medium-sized moth is pale in color. The forewing is pale yellow or grayish-brown in color, with black lines in the middle. The hind wings are white. The larva is white or yellowish white in color with several bristly hairs and an orange-brown or red head.

Biology

The cream-colored eggs are laid either singly or in groups on the tender leaves, shoots, and petioles. The eggs are elongate and flat. A single female lays about 150-200 eggs The egg period varies from three to ten days. Eggs hatch in 3-5 days and young larvae bore into the stem and feed on the pith making longitudinal tunnels. They make silken cocoon within the feeding galleries and pupates in 7-10 days. They transform themselves into adults in 6-8 days.

Damage

Soon after hatching, the larva starts boring into the stem near ground level. Mostly they bore in the branching area or in leaf axils, and seal the entry holes with excretory materials. Larvae feed downward along the length of the main stem, which results in stunted growth or wilting and withering of the whole plant. The later stages of plant growth are most vulnerable to this insect.

Management

1. Remove and promptly destroy the infested plants.
2. Avoid ratoon cropping.
3. Protect the population of parasitoids such as *Pristomerus euzopherae.*
4. Light trap @ 1/ha to attract and kill adults.
5. Spray neem oil 2ml/lit, Imidacloprid 80.5 SC 0.6 ml/lit, Chlorpriphos 20 EC 2ml/lit, Thiamethoxam 25 WSG 0.6 mg/lit, Profenophos 2ml/lit.
6. Avoid using synthetic pyrethriods causing resurgence.

3. Spotted Leaf Beetle (or Hadda Beetle) *Henosepilachna vigintioctopunctata* (Fabr.) (Coccinellidae: Coleoptera)

Identification

The subfamily Epilachninae contains plant-feeding ladybird beetles because most other ladybird beetles are predators, not plant pests. These brownish or orangecolored, hemispherical beetles are larger than other ladybird species. *H. vigintioctopunctata* (in Latin, *viginti* means 20 and *octo* means 8) has 28 black spots on the forewing (elytra). *E. dodecastigma (dodeca* means 12 in Greek) has 12 black spots on the elytra. However, beetles with 14, 16, 18, 20, 22, 24 or 26 spots have been observed under field conditions, due to mating between females of *E. dodecastigma* and males of *H. vigintioctopunctata* (Lall and Mandal, 1958).

Biology

The female lays elongate, spindle-shaped yellowish eggs in groups of 10–20 on the under surface of leaves. About 120–180 eggs may be laid by a female. The egg period is 2–4 days. The yellowish spiny grubs become full grown in 10 -35 days and pupate on the leaf or stem. The pupa is hemispherical, yellowish with spines on the posterior part. The anterior portion being devoid of spines. Adults emerge in a week and live for a month feeding on leaves. The total life-history takes 17–50 days depending on weather conditions.

Damage

The grub and adult have chewing mouthparts. Hence, they scrape the chlorophyll from the epidermal layers of the leaves. The feeding results in a typical ladder-like window. The windows will dry and drop off, leaving holes in the leaves. In severe infestations, several windows coalesce together and lead to skeletonization and the formation of a papery structure on the leaf.

Management

1. Choose resistant or moderately resistant cultivars available in the region. Varieties such as Arka Shirish, Hissar Selection 14, and Shankar Vijay have been reported to be tolerant or resistant to Epilachna beetle, especially *E. vigintioctopunctata* (Parker *et al.,* 1995).

2. Protect the population of parasitoids such as *Pediobius foveolatus.*

3. In the initial stage, collection and destruction of affected leaves alongwith the eggs, grubs and adults.

4. Spray application of carbaryl 0.1 per cent or cypermethrin 0.025 per cent or profenofos 0.05 per cent or DDVP @ 1 ml/lit or Neem oil @ 3 ml/lit

4. Brinjal Lace-wing Bug, *Urentius histricellus (Tingidae, Hemiptera)*

Identification

Adults are straw coloured on dorsal side and black coloured on ventral side with sculptured (reticulated) forewing and Pronotum. Nymphs are yellowish white with prominent spines. Eggs are white nibble shaped.

Damage

Infestation starts after few weeks of transplantation. The caterpillars bore into the growing shoots or petioles large leaves and feed on internal tissues. As a result of damage, affected shoots wither and plants exhibit the symptoms of drooping. After fruit formation larva make their entry under the calyx when they are young. The holes later plugged with excreta leaving no visible sign of infestation. Large holes seen on the fruits are the exit holes.

Management

Spray methyl dematon 25 EC @ 1ml/lit, dimethoate 30 EC @ 1ml/lit

5. Leafhopper, *Amrasca devastans* Distant (Hemiptera: Cicadellidae)

Identification

The nymphs resemble the adults, but lack wings. Instead, they have slightly extended wing pads. They are pale green in color. They tend to move sideways when disturbed. Adults are wedge-shaped, pale green insects. They have fully developed wings with a prominent black spot on each forewing.

Damage

Both nymphs and adults suck the sap from the lower leaf surfaces through their piercing and sucking mouthparts. While sucking the plant sap, they also inject toxic saliva into the plant tissues, which leads to yellowing. When several insects suck the sap from the same leaf, yellow spots appear on the leaves, followed by crinkling, curling, bronzing, and drying, or "hopper burn". Leafhoppers also damage in okra, cotton, and potato.

Management

1. Choose tolerant or resistant cultivars with hairy leaves, as the length and density of the trichomes (hairs) repel leafhoppers such as Manjari Gota, Vaishali, Mukta Kesi, Round Green, and Kalyanipur T3 and Bangladeshi variety Bagun 6 are reported to be less susceptible or tolerant to damage.

2. Monitor the insects with yellow (570-580 nm) sticky traps.

3. Grow okra as a trap crop along the borders of an eggplant field; if pesticides are required to control an infestation, they can be applied to the trap crop.

4. Generalist predators such as ladybird beetles and green lacewings and Parasitoids such as *Anagrus flaveolus* Waterhouse and *Stethynium triclavatum* Enock are effective against leafhopper

5. Neem seed kernel extract (NSKE) @ 5 per cent can be sprayed.

6. Whitefly, *Bemisia tabaci* Gennadius (Hemiptera: Aleyrodidae)

Identification

The silverleaf whitefly is slightly smaller (about 0.96 mm in the female and 0.82 mm in the male) and slightly yellower than most other whitefly pests.The wings are covered with powdery wax and the body is light yellow in color. The head is broad at the antennae and narrow towards the mouth parts. The wings are held roof-like at about a 45 angle, whereas other whiteflies usually hold the wings nearly flat over the body. Hence, the silverleaf whitefly appears more slender than other common whiteflies. The first instar nymph has antennae, eyes, and three pairs of welldeveloped legs. The nymphs are flattened, oval-shaped, and greenish-yellow in color. The legs and antennae are atrophied during the next three instars and they are immobile during the remaining nymphal stages. The last nymphal stage has red eyes. This stage is sometimes referred to puparium, although insects of this order (Hemiptera) do not have a perfect pupal stage (incomplete metamorphosis).

Biology

The females mostly lay eggs near the veins on the underside of leaves. They prefer hairy leaf surfaces to lay more eggs. Each female can lay about 300 eggs in its lifetime. Eggs are small (about 0.25 mm), pear-shaped, and vertically attached to the leaf surface through a pedicel. Newly laid eggs are white and later turn brown. The eggs are not visible to the naked eye, and must be observed under a magnifying lens or microscope. Egg period is about three to five days during summer and 5 to 33 days in winter (David 2001). Upon hatching, the first instar larva (nymph) moves on the leaf surface to locate a suitable feeding site. Hence, it is commonly known as a "crawler." It then inserts its piercing and sucking mouthpart and begins sucking the plant sap from the phloem. Nymphal period is about 9 to 14 days during summer and 17 to 73 days in winter (David 2001). Adults emerge from puparia through a T-shaped slit, leaving behind empty pupal cases or exuviae. Adults live from one to three weeks.

Damage

Both the adults and nymphs suck the plant sap and reduce the vigor of the plant. In severe infestations, the leaves turn yellow and drop off. When the populations are high they secrete large quantities of honeydew, which favors the growth of sooty mould on leaf surfaces and reduces the photosynthetic efficiency of the plants.

Management

1. The field selected for eggplant or seedling production should be clean and not be located near any host plants and weeds.
2. Grow eggplant seedlings in insect-proof (50–64 mesh) net houses, net tunnels, greenhouses, or plastic houses.
3. If the seedlings are produced under open field conditions, use yellow sticky traps at the rate of 1-2 traps/50-100 m2 to trap the whiteflies. Hang the traps slightly above or at the canopy level for better trapping.
4. Maintain a high standard of weed control in seedling production areas and crop fields to reduce the availability of alternate host plants.
5. Plant fast-growing crops like maize, sorghum, or pearl millet in the border of the field to act as barriers to reduce whitefly infestations.
6. Reflective plastic or straw mulches may reduce landing of whiteflies on eggplant crops.
7. Spraying with 0.05 per cent dimethoate or 0.01 per cent imidacloprid or 0.1 per cent acetamiprid.
8. In case of severe infestation, two sprayings at 10–12 days interval with 0.03 per cent oxydemeton methyl or Emamectin benzoate 5 per cent SG@ 4 g/10 lit, Fipronil 5 per cent SC@1.5 ml/lit., Spinosad 45 per cent SC @3.2 ml/10 lit.

7. Aphids, *Aphis gossypii (Aphididae, Hemiptera)*

Biology

When first deposited, the eggs are yellow, but they soon become shiny black in color. The nymphs vary in color from tan to gray or green, and often are marked with dark head, thorax and wing pads, and with the distal portion of the abdomen dark green. The body is dull in color because it is dusted with wax secretions. The nymphal period averages about seven days. The wingless (apterous) parthenogenetic females are 1 to 2 mm in length. The body is quite variable in color: light green mottled with dark green is most common, but also occurring are whitish, yellow, pale green, and dark green forms. The legs are pale with the tips of the tibiae and tarsi black. The cornicles also are black. Small yellow forms apparently are produced in response to crowding or plant stress. Winged (alate) parthenogenetic females measure 1.1 to 1.7 mm in length. The head and thorax are black, and the abdomen yellowish green except for the tip of the abdomen, which is darker. The wing veins are brown. The egg-laying (oviparous) female is dark purplish green; the male is similar. The duration of the adult's reproductive period is about 15 days, and the post-reproductive period five days. These values vary considerably, mostly as a function of temperature. The optimal temperature for reproduction is reported to be about 21 to 27 degrees.C. *Viviparous* females produce a total of about 70 to 80 offspring at a rate of 4.3 per day.

Damage

Although *A. gossypii* is polyphagous, it prefers to feed on cotton and cucurbit vegetables; it is commonly known as "cotton aphid" or "melon aphid." Both the nymphs and adults possess piercing and sucking mouthparts. They occur in large numbers on the tender shoots and lower leaf surfaces, and suck the plant sap. Slightly infested leaves exhibit yellowing. Severe aphid infestations cause young leaves to curl and become deformed. Like whitefly, aphids also produce honeydew, which leads to the development of sooty mould.

Management

1. Although *A. gossypii* is a polyphagous insect, it overwhelmingly prefers to feed on cucurbits and cotton. Hence, the field selected for eggplant or seedling production should be located away from cucurbits and cotton.

2. Grow eggplant seedlings in insect-proof (50–64 mesh) net houses, net tunnels, greenhouses, or plastic houses to avoid early infestation.

3. The ladybird beetles *(Menochilus* sp. and *Coccinella* sp.) and green lacewings are efficient predators of aphids. Protect the population of these predators by avoiding the use of broadspectrum pesticides. Inundative release of ladybird beetles @ 200 pairs per ha at fortnightly intervals can suppress the aphid population.

4. *A.gossypii* can develop resistance to pesticides. Use only those pesticides that have been recommended.

5. Do not use the same compound or pesticide group continuously to avoid the development of pesticide resistance in insects.

6. Imidacloprid 17.8 per cent SL @3.0 ml/10 lit, Dimethoate 30 per cent EC @1.0 ml/lit., Emamectin benzoate 5 per cent SG@ 4 g/10 lit, Fipronil 5 per cent SC@1.5 ml/lit., Spinosad 45 per cent SC @3.2 ml/10 lit, Thiacloprid 21.7 per cent SC @6.0 ml/10 lit. Ethion 50 per cent EC @2.0 ml/lit., Oxydemeton –Methyl 25 per cent EC @1.0 ml/lit, Phosalone 35 per cent EC@2.0 ml/lit.

8. *Cestius (Hishimonus) phycitis* (Cicadellidae, Hemiptera)

Identification

Adults are small light brown leafhopper, measuring around 3 mm long.Males are little smaller in body length.Nymphs are creamy white, wingless and turn brownish with the advancement of age.They are found between the veins of leaves on the undersurface.

Damage

Reduction in leaf size and rosette appearance are the most prominent symptoms of little leaf disease of brinjal or eggplant. This disease is transmitted by a leafhopper (*Hishimonus phycitis*). In severe cases, affected plants do not bear any fruit, or, if formed, it becomes hard and tough.

Management

Spray application of Dimethoate 30 EC @ 2 ml/lit. or Methyl dematon 25 EC @ 2 ml/lit. or fipronil @ 2 ml/lit. with a waiting period of 15 days.

9. Blister Beetle, *Mylabris pustulata* Thunberg (Coleoptera: Meloidae)

Identification

The adult *Mylabris pustulata* is about 2.0-2.5 cm in length and bears red or reddish orange and black alternating bands on the forewing (elytra), Adults are rather soft-bodied, long-legged beetles with the head deflexed, fully exposed, and abruptly constricted behind to form an unusually narrow neck, the pronotum much narrower at the anterior end than the posterior and not carinate (keeled) laterally, the forecoxal cavities open behind, and each of the tarsal claws cleft into two blades. Body length generally ranges between 3/4 and 2 cm.

Biology

Eggs are laid in masses in the ground/hard soil in 20 batches. Females dig soil and lay 60 eggs at the same time of grass hopper. Incubation period is 20 – 22 days. Newly hatched larva is called as triangulin, very active on egg pods of grass hoppers. Larval Period is 10 – 30 days. Each larvae feeds upto 40 – 70 GH eggs. Pupation in soil upto 2 months. Female lays 10,000 to 20,000 eggs during life span and Over winters in egg stage.

Damage

The adult is the destructive stage. As the insects feed on the plants' reproductive parts, they can cause significant yield losses.

Management

1. Pick off beetles by hand (wear gloves or use insect nets) and destroy.

2. Spray during flowering stage of the crop with any of the following insecticides *viz.*, Cypermethrin or Fenvelerate or Dimethoate or Profenophos or Acephate.

10. Red Spider Mite, *Tetranychus urticae* Koch (Acarina: Tetranychidae)

Identification

T. urticae is commonly known as red spider mite or twospotted spider mite. They are minute in size, and vary in color (green, greenish yellow, brown, or orange red) with two dark spots on the body.

Biology

Eggs are round, white, or cream-colored egg period is two to four days. Upon hatching, it will pass through a larval stage and two nymphal stages (protonymph and deutonymph) before becoming adult. The lifecycle is completed in one to two weeks. There are several overlapping generations in a year. The adult lives up to three or four weeks.

Damage

Spider mites usually extract the cell contents from the leaves using their long, needle- like mouthparts. This results in reduced chlorophyll content in the leaves, leading to the formation of white or yellow speckles on the leaves. Under high population densities, the mites move to the tip of the leaf or top of the plant and congregate White and yellow speckles caused by spider mites. Webbing of leaves by spider mites using strands of silk to form a ball-like mass, which will be blown by winds to new leaves or plants, in a process known as "ballooning."

Management

1. Predatory mites such as *Phytoseiulus persimilis* and several species of *Amblyseius*, are more effective under protective structures and in high humidity.

2. Green lacewings *(Mallada basalis* Walker and *Chrysoperla carnea* Stephens) also are effective generalist predators of spider mites. A third instar grub of *C. carnea* could consume 25–30 spider mite adults per day (Hazarika *et al.,* 2001).

3. Spray acaricides and the macrocyclic lactones *(e.g.* avermectins and milbemycins) are effective. Use proper pesticide rotations.

12.2 Pests of Tomato

1. Tomato Fruit Borer, *Helicoverpa armigera* (Noctuidae: Lepidoptera)

Identification

Moths are medium sized, stout, ochreous with olive grey forewings in male and reddish brown forewings in female with dark brown circular spot in the centre and indistinct double waved antemedial lines. Hind wings are pale smoky-white with a broad blackish outer border.

Biology

The eggs are yellowish-white, ribbed and dome-shaped, 0.4–0.5 mm in diameter. Freshly hatched larvae are yellowish-white in colour but gradually change and acquire greenish tinge. Full grown caterpillars are 40–48 mm long, apple-green in colour with whitish and dark-grey broken longitudinal stripes. Full grown caterpillars drop down from the plants and burrow in the soil where they pupate. Pupae are dark-brown in colour, 11–14 mm long and have a sharp spine at the anal end. Fecundity of females is rather high ranging between 1200–1600 eggs. Incubation, larval and pupal stages last for 2–4, 15–24 and 10–14 days respectively and the entire life-cycle may be completed in 4–6 weeks.

Damage

Eggs are laid singly, generally on leaves and flowers but sometimes on fruits as well. On hatching, the young larvae feed on tender foliage; advanced stage larvae attack the fruits. They bore circular holes and thrust only a part of their body inside the fruit and eat the inner contents. If the fruit is bigger in size, it is only partly damaged by the caterpillar but later it is invariably invaded by fungi and bacteria and spoiled completely. The larvae move from one fruit to another and a single caterpillar may eat and destroy 2–8 fruits.

Management

1. Hand-picking of caterpillars and their mechanical destruction in the early stage of infestation can keep the population of this pest under check.
2. Spray application of HaNPV@ 250 larval equivalents/ha in 1125 lit. of water/ha when the larval population reaches 1 larva/m^2. (or) *Bacillus thuringiensis* var. *kurstaki* @ 1.5 kg in 1125 lit. of water/ha. at evening hours.
3. Installation of pheromone traps @5/ha coinciding with the initiation of flowering and destruction of collected moths.
4. Release of the egg parasitoid, *Trichogramma chilonis* or *T. brasiliensis* @ 1Lakh/ha coinciding with flower initiation at 15 days interval.
5. Planting of marigold as a trap crop (1: 16 ratio). Marigold is to be planted 20 days before tomato planting.
6. Erection of 20 bird perches/acre.

7. Spray Azadirachtin 1.0 per cent EC (10000 ppm) @2.0 ml/lit., Indoxacarb 14.5 per cent SC @8 ml/10 lit., Flubendiamide 20 WG@5 g/10 lit., Novaluron 10 per cent EC @7.5 ml/10 lit., Phosalone 35 per cent EC @13 ml/10 lit. Quinalphos 25 per cent EC @1.0 ml/lit.

2. Whitefly, *Bemisia tabaci* (Aleyrodidae: Hemiptera)

Identification

Adults are minute insects, about one mm long, covered completely with a white waxy bloom. Nymphs are oval, scale-like and greenish-white in colour.

Biology

Eggs are pear-shaped, light yellowish in colour, about 2 mm long and can be seen standing upright on leaves, being anchored by a tail-like appendage inserted into the stoma of leaves. On hatching, the nymphs crawl a little, settle down on a succulent spot on the same leaf and never move again during that stage. Incubation period is 3–5 days in summer extending up to 33 days during winter. Nymphal development takes 9–14 and 17–81 days in summer and winter, respectively and pupal period lasts for 2–8 days being longer during winter than in summer. A life-cycle may be completed in as little as 14 days or it may even be prolonged up to 107 days. There are about 12 overlapping generations in a year.

Damage

White, tiny, scale-like insects may be seen darting about near the plants or crowding in between the veins on ventral surface of leaves, sucking the sap from the infested parts. The pest is more active during the dry season and its activity decreases with the onset of rains. As a result of their feeding the affected parts become yellowish, the leaves wrinkle and curl downwards and are ultimately shed. Besides the feeding damage, these insects also exude honeydew which favours the development of sooty mould. In case of severe infestation, this black coating is so heavy that it interferes with the photosynthetic activity of the plant resulting in stunted growth. This whitefly also acts as a vector, transmitting the leaf curl virus.

Management

1. The field selected for tomato or seedling production should be clean and not be located near any host plants and weeds.

2. If the seedlings are produced under open field conditions, use yellow sticky traps at the rate of 1-2 traps/50-100 m2 to trap the whiteflies. Hang the traps slightly above or at the canopy level for better trapping.

3. Maintain a high standard of weed control in seedling production areas and crop fields to reduce the availability of alternate host plants.

4. Plant fast-growing crops like maize, sorghum, or pearl millet in the border of the field to act as barriers to reduce whitefly infestations.

5. Reflective plastic or straw mulches may reduce landing of whiteflies on tomato crops.

6. Spraying with 0.05 per cent dimethoate or 0.01 per cent imidacloprid or 0.1 per cent acetamiprid. Thiamethoxam 25 WSG 0.6 mg/lit, Profenophos 2ml/lit.

7. In case of severe infestation, two sprayings at 10–12 days interval with 0.03 per cent oxydemeton methyl or thiometon.

3. White Tailed Mealy Bug, *Ferrisia virgata* (Cockerell) (Coccidae: Homoptera)

Identification

Eggs of this mealy bug are pale-yellow, cylindrical and about 0.3 mm long. Freshly hatched crawlers are yellowish in colour and become pale white in 2–3 days. Adult females are apterous, long, slender, slightly oval (3.5 – 4.5 x 1.5 – 2.0 mm) covered with dusty white waxy secretion and having a pair of conspicuous long glossy wax tassels at the caudal end.

Biology

A single female lays 100–400 eggs which remain concealed under the female. Reproduction is sexual as well as parthenogenetic. Incubation period is 15 minutes to 4 hours and the immature stages may last for about 20–60 days in case of male and 19–47 days in case of females. Longevity of males is 1–3 days while the females live for 5–7 weeks.

Damage

Eggs are laid in clusters in cottony ovisac which remains concealed under the female. On hatching, the crawlers remain huddled together in cottony nest under the body of the mother. Later, these crawlers become active and wander about, moving swiftly till they find a succulent spot where they puncture the epidermis, inject their toxic saliva and start sucking the cell sap. The mechanical injury thus caused also serves as an entry for various disease producing organisms (bacteria and fungi). From 2nd instar onwards the nymphs secrete honeydew on which black sooty mould develop, which in turn hinders the photosynthetic activity of the plant resulting in stunted growth.

Management

1. Remove and destroy mechanically all the affected leaves and twigs in the early stages of infestation.

2. Spray application Phosphomidon 40 SL 2ml/lit, Imidacloprid 80.5 SC 0.6 ml/lit, Chlorpyriphos 20 EC 2ml/lit, Thiamethoxam 25 WSG 0.6 mg/lit, Profenophos 2ml/lit.

4. Serpentine Leaf Miner, *Liriomyza trifolii (Agromyzidae, Diptera)*

Identification

Egg: Eggs tend to be deposited in the middle of the plant; the adult seems to avoid immature leaves. The female deposits the eggs on the lower surface of the leaf, but they are inserted just below the epidermis. Eggs are oval in shape and small in size, measuring about 1.0 mm long and 0.2 mm wide. Initially they are clear but soon become creamy white in colour.

Larva: Body and mouth part size can be used to differentiate instars; the latter is particularly useful. For the first instar the mean and range of body and mouth parts (cephalopharyngeal skeleton) lengths are 0.39 (0.33 to 0.53) mm and 0.10 (0.08 to 0.11) mm, respectively. For the second instar the body and mouth parts measurements are 1.00 (0.55 to 1.21) mm and 0.17 (0.15 to 0.18) mm, respectively. For the third instar the body and mouth parts measurements are 1.99 (1.26 to 2.62) mm and 0.25 (0.22 to 0.31) mm, respectively. A fourth instar occurs between puparium formation and pupation, but this is a nonfeeding stage and is usually ignored by authors. The puparium is initially golden brown in color, but turns darker brown with time.

Adult: Adults are small, measuring less than 2 mm in length, with a wing length of 1.25 to 1.9 mm. The head is yellow with red eyes. The thorax and abdomen are mostly gray and black although the ventral surface and legs are yellow. The wings are transparent. Key characters that serve to differentiate this species from the vegetable, *Liriomyza sativae* Blanchard, are the matte, grayish black mesonotum and the yellow hind margins of the eyes. In vegetable leafminer the mesonotum is shining black and the hind margin of the eyes is black. The small size of this species serves to distinguish it from pea leafminer, *Liriomyza huidobrensis* (Blanchard), which has a wing length of 1.7 to 2.25 mm. Also, the yellow femora of American serpentine leafminer help to separate it from pea leafminer, which has darker femora.

Biology

Leafminers have a relatively short life cycle. The time required for a complete life cycle in warm environments is often 21 to 28 days, so numerous generations can occur annually in tropical climates. At 25°C the egg stage required 2.7 days for development; the three active larval instars required 1.4, 1.4, and 1.8 days, respectively; and the time spent in the puparium was 9.3 days. Also, there was an adult preovipostion period that averaged 1.3 days. The temperature threshold for development of the various stages is 6 to 10°C except that egg laying requires about 12°C.

Damage

Punctures caused by females during the feeding and oviposition processes can result in a stippled appearance on foliage, especially at the leaf tip and along the leaf margins.However, the major form of damage is the mining of leaves by larvae, which results in destruction of leaf mesophyll. The mine becomes noticeable about three to four days after oviposition and becomes larger in size as the larva matures. The pattern of mining is irregular. Both leaf mining and stippling can greatly depress the level of photosynthesis in the plant. Extensive mining also causes premature leaf drop, which can result in lack of shading and sun scalding of fruit. Wounding of the foliage also allows entry of bacterial and fungal diseases.

Management

Spray application of neem based formulations @ 3 ml/lit or Acephate @ 1ml/lit. or profenophos @ 2 ml/lit.

12.3 Pests of Bhendi

1. Okra Shoot and Fruit Borer, *Earias vittella* and *E. insulana* (Arctidae: Lepidoptera)

Identification

E. vittella: Eggs are spherical in shape, about half mm in diameter, light bluish-green in colour and beautifully sculptured with 26–32 longitudinal ridges; the alternate ridges project upwards to form a crown. Adults Fore wings buff coloured with a green wedge.

E. insulana: Larva: Brown coloured with dorsum showing a white median longtitudinal streak. All the abdominal segments have two pairs of fleshy tubercles (one dorsal and other lateral). Adults has completely green fore wings.

Biology

A female lays on an average 400 eggs (65–695). Incubation, larval and pupal periods last for 3–9, 9–20 (50–60 during winter) and 8–12 days respectively. A single life-cycle takes 22–25 days extending up to 74 days during winter and there may be 8–12 generations in a year. *Earias insulana* (Boisduval), is found damaging okra specially in drier regions.

Damage

Eggs are usually laid singly on buds and flowers and occasionally on fruits as well, but in absence of these parts *i.e.* at early stage of crop growth, the eggs are laid on shoot tips. When the crop is only a few weeks old, the freshly hatched larvae bore into tender shoots and tunnel downwards, these shoots wither, droop down and ultimately the growing points are killed, side shoots may arise giving the plants a bushy appearance. With the formation of buds, flowers and fruits, the caterpillars bore inside these and feed on inner tissues. They move from bud to bud and fruit to fruit thus causing damage to a number of fruiting bodies. The damaged buds and flowers wither and fall down without bearing any fruit whereas the affected fruits become deformed in shape and remain stunted in growth.

Management

1. Remove debris and all the alternate host plants from field; collect and destroy all the infested shoots and fruits.

2. Seed treatment with Imidacloprid 70 WS @ 10g/kg seed.

3. Spraying of Thiodocarp 75 WP @ 1 g/lit is recommended before fruiting.

4. Spray application of Carbaryl 50 per cent WP @ 3 g/lit or Profenophos 50 EC @ 2 ml/lit or Quinalphos 25 EC @ 2 ml/lit or Fenvalerate 20 EC @ 0.5ml/litre, Emamectin benzoate 5 per cent SG@3.0 g/10 lit,

Phosalone 35 per cent EC @1.5 ml/lit, starting from one month after planting at 10 days interval after harvesting the fruits.

2. Red Spider Mite, *Tetranychus cinnabarinus* (Tetranychidae: Arachnida)

Identification

Eggs are globular in shape, about 0.1 mm in diameter and whitish in colour. Larvae are about 0.2 mm in length and pinkish in colour. Nymphs are greenish-red in colour and about 3 mm in length. Larvae and nymphs look alike in shape but can be easily distinguished as larvae have 3 pairs of legs while nymphs and adults have 4 pairs of legs. There are only two nymphal stages – protonymphal and deutonymphal. Adults are ovate in shape, reddish-brown in colour and 0.4 mm (male)–0.5 mm (female) in length with four pairs of legs.

Biology

Eggs hatch in 4 to 7 days; larval development takes 3–5 days; protonymphal and deutonymphal stages last for 3–4 days each. Longevity of adult males and females is 4–9 and 9–18 days respectively. The females that are active during summer in northern India become active with the onset of monsoon and lay eggs parthenogenetically. These unfertilized eggs give rise to males only but the subsequent generations are sexual.

Damage

Colonies of mites comprising of eggs, nymphs and adults are found feeding on ventral surface of leaves under protective cover of fine silken webs. As a result of their feeding innumerable yellow spots appear on the dorsal surface of leaves and the affected leaves gradually start curling and finally get wrinkled and crumpled. This in turn affects the growth and fruit formation capacity of the plants.

Management

Dusting with sulphur dust or spray application with wettable sulphur powder.

Spraying dicofol (kelthane 18.5 EC) 1.5 ml/lit.

3. Cotton Jassid, *Amrasca devastans (=biguttula biguttula)* (Cicadellidae, Hemiptera)

As observed under "Pests of brinjal".

4. White Fly, *Bemisia tabaci (Aleurodidae, Hemiptera)*

Identification

Nymph is greenish yellow in colour and is present along with pupa on the under surface of the leaves. Adult is minute yellow colour bodied insects with white waxy covering.

Biology

The females mostly lay eggs near the veins on the underside of leaves. Each female can lay about 300 eggs in its lifetime. Egg period is about three to five days during summer and 5 to 33 days in winter. Nymphal period is about 9 to 14 days during summer and 17 to 73 days in winter. Adults emerge from puparia through a T-shaped slit, leaving behind empty pupal cases or exuviae. Adults live from one to three weeks.

Damage

Nymphs and adult remain on the undersurface of the leaf and suck the cell sap causing chlorotic spots and yellowing and drying of leaves. It is a vector of *"yellow vein mosaic virus"* (YVMV). The fruits turn white and hard which has no market value.

Management

1. Remove the wild Abelmoschus sp. from the crop bunds and destroy, since they are the carriers of YVMV.
2. Remove the infested crop and destroy.
3. Use of resistant genotypes, which are absolutely free from YVMV disease.
4. Use of yellow sticky traps coated with castor oil @ 5 per acre.
5. Spray application of methyl dematon 25 EC @ 2ml/lit. or dimethoate 30 EC @ 2 ml/lit. in the early stage and phasalone 35 EC @ 2ml/lit. in the later stages of crop.
6. In case of severe infestation, spraying of acephate 75 SP @ 1.5 g/lit.

5. Red Cotton Bug, *Dysdercus koengii (Pyrrhocoridae, Hemiptera)*

Identification

Nymphs are reddish in colour, with three black spots on the centre of the abdomen and three white spots on either margins of it. Adult body is predominantly red in colour, with white markings on the lateral side of the abdomen. Forewing is half reddish brown and half black in colour. There is a black spot on the reddish portion.

Damage

Nymph and adult suck the sap from the leaves and fruits. The plants become weak and stunted. The fruits curl up.

Management

1. Conservation of Reduvid, *Herpactor costalis* which is predaceous on this bugs.
2. Spray application of dimethoate 30 EC @ 2 ml/lit. or acephate 75 SP @ 1.5 g/lit.

Chapter 13

Pests of Cruciferous Crops

1. Diamond Back Moth, *Plutella xylostella (Plutellidae, Lepidoptera)*

Identification

The moth is grayish brown with narrow wings having pale white markings anteriorly which from diamond-like white patches dorsally when wings are folded over back at rest. Eggs are yellowish white, oval and flattened. Larva is greenish and short, tapering at both the ends.

Biology

It lays up to 57 eggs singly on the under surface of leaves along the veins. Egg period is 4 – 5 days. The larva is greenish with short thin hairs on the body. The larval period is 13 – 21 days. It pupates in a thin loose mesh of silken cocoon and the pupal period is 7 – 9 days. The period from egg to adult occupies 24–35 days.

Damage

The larvae bite holes in leaves and cause serious damage. The larvae cause blisters on the leaves which dry away in course of time. The typical symptoms are withered appearance of the leaves and skeletonisation of the plant.

Management

1. Mustard sown as trap crop twice *i.e.* 12 days preceding planting cabbage and again 40 days later controls DBM.
2. Remove and destroy plant remenent, stubbles, after harvest and plough the field.

3. Erection of Pheromone traps @ 12/ha

4. Crop rotation with cucurbits, beans, peas, tomato and melon

5. Larval parasitoid: *Diadegma semiclausm* @ 1,00000/ha (Hills – below 25 –27° C) Cotesia plutellae (plains) at 20000/ha release from 20 days after planting

6. *Bacillus thuringiensis* var *kurstaki* 2g/lit

7. Neem seed kernel extract 5 per cent

8. Cartap hydrochloride 0.5 per cent at 10,20 and 30 DAS, Spray with Azadirachtin 5 per cent Neem Extract Concentrate @ 5.0 ml/10 lit., Spinosad 2.5 per cent SC@ 1.2 ml/lit, Trichlorofon 50 per cent EC @ 1.0 ml/lit., Lufenuron 5.4 per cent EC@1.2 ml/lit.

9. Preservation of larval and pupal parasitoids *Apanteles ruficrus, A. plutellae, Brachymeria excarimata.*

2. Cabbage Butterfly, *Pieris brassicae* (Linnaeus) (Pieriidae: Lepidoptera)

Identification

Eggs are flask-shaped, about one mm long and yellowish in colour. Full grown caterpillars are 38–44 mm long, velvety bluish-green in colour with black dots and yellow dorsal and lateral stripes covered with white hair. Pupae are yellowish-green with black spots and dots. Adult butterflies have snow-white forewings with black distal margins more developed in females than in males; hind wings are also pure white with black apical spots. Wing expanse is 60–70 mm. Moths emerging in summer are larger in size than those of winter.

Biology

Eggs are laid in clusters under surface of the leaf. A single female lays only 2–3 egg-masses of 50–80 eggs each. Incubation, larval and pupal periods are on an average 3.2, 5.6 and 7.3 days during May extending upto 17.6, 40.7 and 28.8 days respectively in January. Generally there are two generations during winter (plains) and 4–5 in summer (hilly region).

Damage

On hatching, the young caterpillars feed gregariously on leaves for a couple of days, then disperse, spreading infestation to the adjacent plants and fields. As a result of their feeding the leaves are skeletonized, sometimes the caterpillars bore into the heads of cabbage and cauliflower.

Management

1. Pest can be checked by handpicking and mechanical destruction of caterpillars during early stage of attack when the caterpillars feed gregariously.

2. Conservation of parasitoid *Apanteles glomeratus.*

3. In case of widespread infestation spray with 0.05 per cent dichlorvos or 0.1 per cent Malathion.

3. Cabbage Head Borer, *Hellula undalis (Pyralidae, Lepidoptera)*

Identification

Larva is brown in colour with four longitudinal lines in the body. Adult is a small, pale, greyish brown moth, with wavy lines and a central elliptical marking in the fore wing. Hind wings are pale in colour.

Biology

The moth lays yellowish shiny eggs on the leaves and they hatch in about 4 days. The larva becomes full grown in about 9 days and pupates in soil. The pupal period lasts for 6 days.

Damage

The larvae in its first two instars, mines the leaves and renders it a papery white structure with excreta filled in it. The later stages feed on the leaves and shoots. Finally, they enter into cabbage heads, leaf petiole and stem. In severe case of attack the plants become weak and produce deformed heads.

Management

1. Collect and destroy mechanically caterpillars in the early stages of attack
2. Preservation of larval parasitoid, *Bracon hebator*
3. *Bacillus thuringiensis* @ 2g/lit at primordial stage
4. Cartap hydrochloride @ 500g/ha or malathion 50EC @500ml/ha

4. Cabbage Aphid, *Brevicoryne brassicae* (Aphididae: Homoptera)

Identification

Eggs when present are pale yellow with greenish tinge. Nymphs are 1.0–1.5 mm long and yellowish-green in colour while adults are 1.8–2.0 mm long and darker in colour than nymphs.

Biology

Reproduction is mostly viviparous parthenogenetic during summer and mild winter. However, during severe winter sexual reproduction may also occur. Eggs are laid during November – December. These hatch in 20–22 weeks. The nymphs mature in about 2 weeks and immediately start producing young ones, without mating. A single female may produce 40–45 young ones during her life time. The life cycle is completed in 11–45 days and as many as 21 generations have been recorded during a year when provided with favourable conditions.

Damage

Colonies of these insects are often found on tender shoots and as a result of sucking of vital sap from the tissues, the plant remain stunted in growth resulting in poor head formation. In the case of severe infestation plants may completely dry up and die away. When infestation occurs on seedlings, they loose their vigour, get distorted and become unfit for transplanting. The aphids also produce copious quantity of honeydew affected plant parts hindering the photosynthesis and adversely affecting the plant growth.

Management

1. Cut and destroy the infested shoots mechanically.

2. Spray with 0.025 per cent phosphamidon or methyl demeton or 0.01 per cent imidacloprid. Repeat the spraying after a fortnight

5. Cabbage Semilooper, *Trichoplusia ni (Noctuidae, Lepidoptera)*

Identification and Biology

Cabbage looper eggs are hemispherical in shape, with the flat side affixed to foliage. They are deposited singly on either the upper or lower surface of the leaf, although clusters of six to seven eggs are not uncommon. The eggs are yellowish white or greenish in color, bear longitudinal ridges, and measure about 0.6 mm in diameter and 0.4 mm in height. Eggs hatch in about two, three, and five days at 32, 27, and 20°C, respectively, but require nearly 10 days at 15°C. Young larvae initially are dusky white, but become pale green as they commence feeding on foliage. They are somewhat hairy initially, but the number of hairs decreases rapidly as larvae mature. Larvae have three pairs of prolegs, and crawl by arching their back to form a loop and then projecting the front section of the body forward. The mature larva is predominantly green, but is usually marked with a distinct white stripe on each side. Dorsally, the larva bears several narrow, faint white stripes clustered into two broad white bands. In some cases the mature larva is entirely green. The adult forewings of the cabbage looper moth are mottled gray-brown in color; the hind wings are light brown at the base, with the distal portions dark brown. The forewing bears silvery white spots centrally: a U-shaped mark and a circle or dot that are often connected. The forewing spots, although slightly variable, serve to distinguish cabbage looper from most other crop-feeding noctuid moths. The moths have a wingspan of 33 to 38 mm.

Damage

Cabbage loopers are leaf feeders, and in the first three instars they confine their feeding to the lower leaf surface, leaving the upper surface intact. The fourth and fifth instars chew large holes, and usually do not feed at the leaf margin. In the case of cabbage, however, they feed not only on the wrapper leaves, but also may bore into the developing head. Larvae consume three times their weight in plant material daily. Feeding sites are marked by large accumulations of sticky, wet fecal material. Despite their voracious appetite, larvae are not always as destructive as presumed.

Biology

Development time (egg to adult) requires 18 to 25 days when insects are held at 32 to 21°C, respectively. so at least one generation per month could be completed successfully under favorable weather conditions.

Management

1. Pest can be checked by handpicking and mechanical destruction of caterpillars during early stage of attack when the caterpillars feed gregariously.

2. Conservation of parasitoid *Apanteles glomeratus*.

3. In case of widespread infestation spray with 0.05 per cent dichlorvos or 0.1 per cent Malathion.

6. Painted Bug, *Bagrada cruciferarum* (Pentatomidae: Hemiptera)

Identification

Eggs are oval in shape about one mm long, pale yellow when freshly laid gradually becoming pinkish-orange. Nymphs are beautifully patterned with a mixture of black, white and orange colour, 1.5–4.5 mm long depending on their age. Adults are also black and orange colour bugs similar in colour pattern as nymphs – that's why they have earned the common name of painted bugs. Males are 6–7 mm long and females 7–8 mm.

Biology

Eggs are laid singly or in batches of 2–12 on leaves, stems and flower buds. The mating takes place 2–6 days after the final nymphal moult and the oviposition commences a week after first mating and may continue intermittently throughout the life span of the female. A single female may lay as many as 230 eggs @ 15–20 eggs per day. Eggs and nymphal duration is recorded as 2–5 and 18–20 days respectively. A single life-cycle is completed in 3–4 weeks and adults live for 16 -18 days with 6–8 generations in a year.

Damage

Both nymphs and adults suck cell sap from tender plant parts causing yellowing of leaves which gradually dry up and ultimately fall down exposing the plants to secondary invasion of bacteria and fungi. The plants wilt and wither affecting adversely the yield both quantitatively and qualitatively.

Management

1. Clean cultivation by removing weeds harbouring this pest is imperative for avoiding infestation of these bugs.

2. In case of heavy infestation, spray with 0.05 per cent dichlorvos or 0.05 per cent phosalone. Atleast 7–10 days waiting period should be there between treatment and harvest.

7. Mustard Sawfly, *Athalia lugens proxima* Kulg (Tenthredinidae: Hymenoptera)

Identification

Newly hatched grubs are 2–3 mm long, smooth, cylindrical and greenish-grey in colour; full grown ones are cylindrical in shape, 16–20 mm long and greenish-black in colour. They look and behave like caterpillars but have 8 pairs of prolegs. Adults are 8–12 mm long, having dark head and thorax, orange coloured abdomen and translucent smoky wings with black veins. Females have a strong saw-like ovipositor – hence it has been given the popular name sawfly.

Biology

A female lays on an average 35 eggs (20–150). Egg period is 6–8 days. Grub development takes 21–31 days. Pre-pupal and pupal periods last for 3–4 and 7–10 days respectively. Severe winter is passed in pupal stage and lasts for about 14 weeks. In Northern India there are three generations during cold season. In South India where there is no severe winter, the pest undergoes as many as 10 overlapping generations in a year.

Damage

Eggs are laid singly, mostly during day time and inserted into leaf tissues near the periphery of leaves. On hatching the grubs nibble the margins of tender leaves but later on bite holes in the leaves. Grubs are diurnal in habit and feed generally during early morning and evening hours. With slight disturbance they fall on the soil and feign death.

Management

1. Hand-picking of grubs which are not active during dawn and dusk if the area under crop is limited. Spraying with 0.2 per cent carbaryl.

2. Dusting with 5 per cent Malathion is also effective.

8. Tobacco Caterpillar, *Spodoptera litura (Noctuidae, Lepidoptera)*

Identification

The egg is round and dirty white. The body of the newly emerged larva is cylindrical; head size is wider than the body while the abdomen tapers towards thecaudalregion.But during the 2nd instar until pupation, the body turns wider than the head. The true legs and pro-legs are distinct. The newly emerged larva is whitish, then turns yellow green an hour after with a pattern of red, yellow, and green lines from the head to the anal region. As the larva grew bigger, the body turns brown with 3 thin yellow lines down the back. The newly emerged larva singled out the leaf veins during feeding while the big larva includes them. Bigger larva is an excellent feeder and active at night. The larva hides in the soil at day time and ceaseless when about to molt. The adult female and male are hairy. The female is pale brown while the male is darker. The female is bigger with a stout abdomen while the male is narrower and tapering towards the tip.Adult moth fore wings have golden and greyish white patches.with wavy white markings.Hindwings white with a brown patch along the margin.

Biology

Females lay 1000-2000 eggs in egg masses of 100- 300 on the lower leaf surface of the host plant. The masses are covered by hair-like scales from the end of the insect's abdomen. Fecundity is adversely affected by high temperature and low humidity (about 960 eggs laid at 30°C and 90 per cent RH and 145 eggs at 35°C and 30 per cent RH). The eggs hatch in about 4 days in warm conditions, or up to 11-12 days in winter. The larvae pass through six instars in 15-23 days at 25-26°C.

Damage

Larva is nocturnal and heavily defoliates the freshly emerged leaves.

Management

1. Conservation of larval parasitoids *viz., Cosmopolites* sp., *Eriborous* sp., *Rogas* sp.
2. Collection and destruction of egg masses and larvae.
3. Spray application of Fipronil 80 WG @ 75 g ai/ha, Spinosad 2.5 SC @ 15 g ai/ha. Or cypermethrin @ 30 g a.i. or fenvalerate @ 50 g a.i. or deltamethrin @ 10 g a.i. or cartap hydrochloride @ 175 g a.i./ha once at primordial initiation (22 days after planting) and repeated either thrice at 7 days interval or twice at 10 days interval or *Bacillus thuringiensis* var. kurstaki @ 1 kg/ha.

9. Green Peach Aphid, *Myzus persicae*

Identification

The eggs measure about 0.6 mm long and 0.3 mm wide, and are elliptical in shape. Eggs initially are yellow or green, but soon turn black. Mortality in the egg stage sometimes is quite high. Nymphs initially are greenish, but soon turn yellowish, greatly resembling viviparous (parthenogenetic, nymph-producing) adults.

Damage

Green peach aphids can attain very high densities on young plant tissue, causing water stress, wilting, and reduced growth rate of the plant. Prolonged aphid infestation can cause appreciable reduction in yield of root crops and foliage crops. However, green peach aphid does not seem to produce the high volume of honeydew observed with some other species of aphids. Blemishes to the plant tissue, usually in the form of yellow spots, may result from aphid feeding. Leaf distortions are not common except on the primary host. In cabbage and cauliflower, poor head formation occurs. In severe cases of attack, crinkling and cupping of the leaves occur.Aphids also excrete copius amount of honey dew in which sooty mould fungus develops resulting in contamination of harvested products.

Management

1. Conservation of predators, *Coccinella septumpunctata, Menochilus sexmaculata,* syrphids and *Chrysoperla carnea.*
2. Spray 300 ml of Metasystox 25EC (Methyl demeton) or 200 ml of Rogar 30EC (Dimethoate) or 75 ml of Dimecran 85SL (phosphamidon) in 100 liters of water.

Chapter 14

Pests of Cucurbits

1. Fruitflies, *Bactrocera cucurbitae* Coq. and *B. ciliatus* Loew (Tephritidae: Diptera)

Identification

Adult with a dark spot in each antennal furrow; facial spot round to elongate. Frons - 2-3 pairs frontal setae; 1 pair orbital setae.Thorax: Predominant colour of scutum red-brown. Postpronotal (=humeral) lobe entirely pale (yellow or orange). Notopleuron yellow. Scutum with parallel sided lateral postsutural vittae (yellow/orange stripes) which extend anterior to suture and posteriorly to level of the intra-alar setae. Medial vitta present; not extended anterior to suture. Scutellum yellow, except for narrow basal band. Anepisternal stripe not reaching anterior notopleural seta. Yellow marking on both anatergite and katatergite. Postpronotal lobe (=humerus) without a seta. Notopleuron with anterior seta. Scutum with or without anterior supra-alar setae; with prescutellar acrostichal setae. Scutellum rarely (5 per cent) with basal as well as apical pair of setae.The eggs of *Bactrocera* Size, 0.8 mm long, 0.2 mm wide, with the micropyle protruding slightly at the anterior end. The chorion is reticulate (requires scanning electron microscope examination). White to yellow-white in colour.Third instar larva: Large, length 9.0-11.0 mm; width 1.0-2.0 mm.Head: Stomal sensory organ small, completely surrounded by 6-7 large preoral lobes, some bearing serrated edges similar to oral ridges; oral ridges with 17-23 rows of moderately long, uniform, bluntly rounded teeth; accessory plates numerous, with serrated edges and interlocking with oral ridges; mouthhooks large, heavily sclerotized, each with a small, but well-defined preapical tooth. Puparium is

Barrel-shaped with most larval features unrecognisable, the exception being the anterior and posterior spiracles which are little changed by pupariation. White to yellow-brown in colour. Usually about 60-80 per cent length of larva.

Biology

In nature the population is generally low during dry weather and increases rapidly with adequate rainfall. Pre-oviposition, egg, maggot and pupal periods last for 9 - 21, 1 - 1½, 3 - 9 and 6 - 8 days respectively. During winter the larval and pupal stages are extended up to 3 and 4 weeks, respectively. A single life-cycle is completed in 10 - 18 days but it takes 12 - 13 weeks to complete a single life-cycle in winter. Adult longevity is 2 - 5 months; females live longer than males. Pupae are 5 - 8 mm long, barrel shaped and brown to ochraceous in colour. The fully fed maggots come out of the fallen fruits and pupate 10 - 15 cm deep in the soil. Where the fruits do not fall, the maggots pupate inside the fruits (which are not common) or come out, drop down and pupate in the soil.

Damage

The female flies puncture the soft and tender fruits with their stout and hard ovipositor and lay eggs below the epidermis @ 4–10 eggs per fruit each time. A single female can lay about 200 eggs in her life span of 8–10 weeks. A puncture made by one female is often used by others also for ovipositing and a single fruit may have more than one puncture made by one or more females. On hatching the maggots feed inside on the pulp of fruits and the infested fruits can be identified by the presence of brown resinous juice which oozes out of the punctures made by the flies for oviposition. These punctures also serve as an entry for various bacteria and fungi; as a result, the infested fruits start rotting, get distorted and malformed in shape and fall off from the plants pre-maturely.

Management

1. To check the damage by these flies, fruits should be harvested before they start ripening.
2. All the fallen and infested fruits should be collected and destroyed to prevent the carry over of the pest.
3. The flies when they congregate and rest on the under surface of large leaves of ribbed gourd may be controlled by spray application of cypermethin 0.025 per cent.
4. Spray application of three to five rounds of profenofos 0.05 per cent or fenthion 0.1 per cent or carbaryl 0.1 per cent at intervals of 15 days commencing from flowering may be useful.
5. Frequent raking of the soil under the vines or ploughing the infested fields after the crop is harvested can help in killing the pupae.
6. Baits prepared with 10 per cent ripe banana, 10 per cent jaggery mixed with 0.1 per cent Malathion or 1g carbofuran used in bait traps was found effective or this bait mixture is to be applied as 200 spot splashes per hectare on the undersurface of cucurbit leaves.

7. Use of 0.4 ml methyl eugenol with 1ml of dichlorvos in bait traps was also found effective.

2. Snake Gourd Semilooper, *Anadevidia (=Plusia) peponis* F. (Noctuidae: Lepidoptera)

Identification

The brownish moth has shiny brown fore wings. The caterpillar is greenish with white longitudinal lines and black tubercles with thin hairs arising on them.

Biology

The female moth lays white spherical eggs on the under surface of leaves. The semilooper caterpillar with humped last abdominal segment measures 35–40mm long. It pupates in a thin silken cocoon in leaf fold. The pupa is greenish but turns dark brown before the emergence of the adult moth. Egg, larval and pupal periods last for 4–5, 24–30 and 7–8 days. Its life-history occupies about six weeks.

Damage

The caterpillar cut the leaf partly and rolls it and lives inside the roll. It is a specific pest of snake gourd and larvae defoliate the plants entirely if infestation is serious.

Management

1. The larvae when found in small numbers may be hand-picked and destroyed.
2. Encourage activity of *Apanteles taragamae, A. plusiae*
3. Spray application of 0.02 per cent carbaryl or or Malathion @ 2 ml/lit.

3. Pumpkin Beteles, *Raphidopalpa foveicollis* (red beetle), *Aulacophora cincta* (Grey with black border) and *A. lewisii* (*=A. intermedia*) (Blue beetle) (Galerucinae: Coleoptera)

Identification

Adult beetles of R. *foevicollis* are 6 – 8 mm long having glistening yellowish red to yellowish brown elytra that are uniformly covered with fine punctures. Adults of *A. cincta* are similar in size and appearance except that the colour of elytra is greyish yellow to brownish grey with distinct palm margin all around. Adults of *A. lewisii* are slightly smaller (5 – 6 mm) with blackish blue elytra.

Biology

The yellowish pink spherical eggs are laid in the soil which turns orange after two days and a beetle may lay 150–300 eggs. The egg period is 5–8 days. The grubs become full grown in 13–25 days and pupate in the soil. The prepupal period is 2–5 days. The pupal period ranges from 7–17 days. Total life cycle occupies 32–65 days. In a year there may be 5–8 generations of the insect.

Damage

The beetles bite irregular holes on leaves and also feed on flowers. They prefer young seedlings and tender leaves and the damage at this stage may kill the seedlings. The roots as well as the stem and fruits that come in contact with the soil are damaged by the grubs. Pumpkin is preferred by *R. foveicollis* and sponge gourd by *A. lewisi.* Both the species feed also on snake gourd, pumpkin, cucumber, melon and ribbed gourd.

Mangement

1. Cultural practices like clean cultivation and early sowing reduces pest damage.
2. After harvesting deep ploughing of infested field to kill the grub in the soil.
3. Spray application of 0.2 per cent carbaryl or dusting with 5 per cent carbaryl dust or malathion 50 EC @ 500 ml or dimethoate 30 EC 500 ml or methyl demeton 25 EC@ 500 ml/ha.

4. Pumpkin Leaf Caterpillar, *Diaphania indica* (Pyralidae, Lepidoptera)

Identification

The wingspan is about 30 mm. Adults have translucent whitish wings with broad dark brown borders. The body is whitish below, and brown on top of head and thorax as well as the end of the abdomen. There is a tuft of light brown "hairs" on the tip of the abdomen, vestigial in the male but well-developed in the female. It is formed by long scales which are carried in a pocket on each side of the 7th abdominal segment, from where they can be everted to form the tufts. Unfertilized females are often seen sitting around with the tuft fully spread, forming two flower-like clumps of scales, which move slowly to spread their pheromones.Larva is elongate, bright green in colour with two narrow ends, white longitudinal stripes dorsally. Adult is a medium sized moth having white transparent wings with big brown marginal patches. Females have orange coloured hairs at the anal end.

Damage

Larvae scrap the chlorophyll and webs leaves together and feed within. In addition feeds on flowers and occasionally bores into developing fruits.

Management

Spray application of malathion 50 EC @ 500 ml or dimethoate 30 EC 500 ml Quinolphos 25 EC @ 2 ml/lit.

5. Coccinia Gall Fly, *Neolasioptera cephalandrae* (Cecidomydae: Diptera)

Identification

Adults are small dark brown coloured fly.

Damage

Gall formation due to attack of the larvae of *Neolasioptera cephalandrae* is common on the stem of *Coccinia indica*. Isolated patches of secondary meristematic centres develop throughout the proliferated tissue and they often differentiate into vascular tissues (either xylem or phloem or both). The larvae bore through the stem and live in lysigenous cavities in the ground tissue. Proliferation of cells of the secondary meristematic centres results in gall formation and the normal vascular pattern is disturbed and vascular bundles crushed. Development of mechanical tissues in the affected region is suppressed. Additional cambial layers with no definite pattern of orientation also develop in the ground tissue and these produce patches of xylem tissue. Larval cavities are irregular and are surrounded by a nutritive tissue rich in protein and polysaccharides.

Management

1. Clipping of the galls and destroy.
2. Spray application of 0.2 per cent carbaryl or dusting with 5 per cent carbaryl dust or malathion 50 EC @ 500 ml or dimethoate 30 EC 500 ml or methyl demeton 25 EC@ 500 ml/ha.

6. Serpentine Leaf Miner, *Liriomyza trifolii (Agromyzidae, Diptera)*

Identification

Eggs are small, creamy white in colour and oval shaped. The maggots are apodous, bright yellow to yellow green in color, measuring 1/6 inch in length. Pupa is distinctly segmented and yellowish brown in colour. Adult fly is yellowish black coloured fly measuring about 1/12 inch of length.

Biology

Eggs are laid in upper or lower surface of the leaf. Eggs hatch in 2-4 days. Larva feeds between the epidermal layers. Larval stage is completed in 6-7 days. Larva on maturity, makes a longitudinal slit in the leaf, comes out and pupates in the leaves or in the soil.

Damage

Larva mines and feeds on the chlorophyll in between the epidermal layers of the leaves. Characteristic symptom of serpentine mining is observed. Severely affected leaves dry and drop. Large population of the pest destroys leaves in large extent and affects the growth of plants.

Management

1. Collect and destroy mined leaves.
2. Spray application of neem based formulations or Acephate @ 1ml/lit. or profenophos @ 2 ml/lit
3. Spray NSKE 5 per cent

7. Green Peach Aphid, *Myzus persicae*

Identification

Nymphs and adults are yellow in colour. In adults, both winged and wingless forms are present.

Damage

Nymphs and adults are found in colonies of hundred in tender portions of the plant. They suck the sap from tissues leading to curling and crinkling of leaves. In case of severe attack, the leaves curl up, fade gradually and dry up. Sooty mould develops due to the excretion of honey dew giving the plants a dark appearance.

Management

1. Conservation of predators, *Coccinella septumpunctata, Menochilus sexmaculata,* syrphids and *Chrysoperla carnea.*
2. Spray 300 ml of Metasystox 25EC (Methyl demeton) or 200 ml of Rogar 30EC (Dimethoate) or 75 ml of Dimecran 85SL (phosphamidon) in 100 liters of water.

Chapter 15

Pests of Peas and Beans

1. Pea Pod Borer, *Etiella zinckenella (Phycitidae, Lepidoptera)*

Identification
Caterpillar is green in colour with with 5 black spots. Adults are greyish brown with distinct pale white band along the front margin of forewings.

Biology
Eggs are laid singly or in groups of 2- 27 on pea pod. Larval development is completed in 12.6 days. Pupation takes place in soil. The pest overwinters in pupal stage and the adults emerge in February–March Adults live for 35.8 days. There are 5 generations in a year.

Damage
Immediately after hatching, the larvae feed on the floral parts and green pods but later bore into the pods and feed on the seeds within. Generally one larva is seen per pod.

Management
Spray application with fenthion [0.05 per cent], Chlorpyriphos [0.02 per cent] and phosphomidon [0.03 per cent], malathion 50 EC @ 500 ml or dimethoate 30 EC 500 ml is effective in controlling the pest.

2. Pea Leaf Miner, *Phytomyza* (Chromatomyia) *horticola (Agromyzidae, Diptera)*

Identification
Adults are small (1.5-2.0 mm long), mostly shiny black except for yellow

on the scutellum, sides of thorax, and middle portion of head; two pairs of reclinate orbital setae; and costal vein extending to M_{1+2}. Wing length: 2.2 - 2.7 mm. Frons can vary in color. It is at most completely yellow but sometimes exhibit some brownish darkenings. Margins of frons not distinctly darkened. Arista appears to be divided into two sections with the apical part distinctly thinner than the basal part.The larva forms a narrow, linear mine on the upper or lower leaf surface. Pupation is in the mine.

Biology

Eggs are translucent white and the incubation period is 1 -2 days. The larval duration of instars is about 6 days. Pupal stage occupies 9 days. A female on an average lays 300 eggs during its life span of 26 days.

Damage

Larva bore between the epidermis, form galleries leaving intact the epidermal layer and feed on leaf tissues. Flies puncture leaves for both feeding and oviposition; punctures may be numerous enough to greatly reduce photosynthesis and may kill young plants. Unsightly mines and punctures further reduce the value of plants.

Management

Sprayingf of carbaryl [0.1 per cent], monocrotophos [0.07 per cent], Dimethoate [300 g ai/ha],Methyl demeton [500 g ai/ha] and the spray can be repeated after 15 if the attack persists.

3. Cut Worms

a. Common Cut Worm, *Agrotis segetum (Noctuidae, Lepidoptera)*
b. Black Cut Worm, *Agrotis ipsilon (Noctuidae, Lepidoptera)*
Identification

A. segetum: Larva is black colour with a brown head. Triangular spots are present in the spiracular region. After five instars, they pupate in earthern cocoon. Adult moth is stout and has wavy lines and spots on brownish fore wings. Males have bipectinate antenna while females have filiform antenna.

A. ipsilon: Larva is black with pale mid dorsal stripes. Skin with coarse granules interspread with small granules. There are five larval instars. Adult fore wing is pale brown in colour with dark purplish brown patch along coastal and towards the base. Hind wings white with brown tinge. Male has bipectinate antenna whereas female has filiform antenna.

Biology

The cut worms are active from October to april in the plains and during summer in mountains regions of the country. Females lay creamy white dome shaped eggs [700 eggs during its life span] in clusters on the under surface of the leaves or in soil. The eggs hatch in 3-6 days. There are 6 larval instars. The larval stage lasts for a month. Pupation takes place in the earthern chambers in the soil. Pupal period is 10-15 days. Life cycle is completed in 8 weeks.

Damage

The larva hide during day time in cracks and crevices of soil, become active at dusk, cut the tender stem of young and growing plants near the ground level. Infested field looks like a cattle grazed field. They feed from dusk to dawn and cut many plants than they consume.

Management

1. Cutworm larvae can be controlled to some extent by flooding the field.
2. Burning the heaps of grass in the affected fields containing the Cutworm larvae.
3. Application of baits formulations containing 0.025 per cent quinalphos applied at 15-20 kg/ha by broadcasting method or trichlorfon baited with mixture of bran, water, and sugar, broadcasted at the rate of 30 kg/ha is very effective in controlling the pest.

4. Pea Stem Fly, *Ophiomyia phaseoli* (Agromyzidae, Diptera)

Biology

The incubation period varied from 2-3 days under laboratory conditions and 2-4 days in the field. There are 3 larval instars. The larval period varied from 7-9 days in March-April and from 9-11 days in July-Sept. in the laboratory, while under field conditions, it varied from 8-10 days in March-April and from 9-11 days in July-Sept. Pupation occurred in the stem near the surface of the soil where the larva had been feeding. The pupal period varied from 7-9 days in March-May and 8-10 days in July-Sept. in the laboratory. Under field conditions, it varied from 8-9 days in March-May and 9-12 days in July-Sept. The fly completes its life cycle in about 3 wk.

Damage:

Larvae feeding soon after hatching produce numerous larval mines which are better seen on the underside of the leaves just under the epidermis, and appear as silvery, curved stripes; on the upper side of the leaf only a few tunnels are visible. Later, both egg holes and larval mines turn dark brown and are clearly visible. In cases of severe attack, infested leaves become blotchy and later hang down. These leaves may dry out and may even be shed. When mature plants become infested, insect damage is confined to the leaf petioles, which become swollen and at times the leaves may wilt.

The developing larvae in second and third instar mine downward into the cortex just underneath the epidermis. The third instar continues to feed downwards into the tap root and returns to pupate still inside the stem, close to the soil surface. The feeding tunnels are clearly visible on the stems. If the *O. phaseoli* larvae population is high, larval feeding leads to destruction of the cortex tissue around the root-shoot junction. This initially leads to yellowing of the leaves, stunting of plant growth and even plant mortality. If the damage is less severe, the root-shoot junction area appears swollen. In some cases the host plant produces adventitious roots above this swollen area on the stem.

Management

1. Remove and destroy the affected branches at the initial stages of attack.
2. Spray application Phosphomidon 40 SL 2ml/lit, Imidacloprid 80.5 SC 0.6 ml/lit, Chlorpriphos 20 EC 2ml/lit, Thiamethoxam 25 WSG 0.6 mg/lit, Profenophos 2ml/lit.

5. Pea Blue Butterfly, *Lampides boeticus* [Lycaenidae: Lepidoptera]

Identification

Pea blue butterflies have a wingspan of 30 mm. The male´s wings are pale blue on top, whereas females are blue with wide dark brown edges. Underneath, both sexes are pale brown with white markings and have a distinctive broad white band across the outer sector of both wings. Both sexes have a small, thin tail on each hindwing. There are two small black eye spots, ringed with orange, at the base of each tail, on the underside of the wings. Larvae reach 10 mm in length. Larvae are variable in colour ranging from a uniform pale cream to pale pink with darker pink markings. The pale brown head is usually retracted out of site when the larvae are disturbed. Younger larvae may be pale pink with darker pink markings. The eggs are flattened, white and only 0.2 mm across. Pupae are mottled brown with a length of 8 mm.

Biology

Female lays eggs on buds, flowers and green pods which are round and light green in colour. Eggs hatch in 4- 6 days. The larva passes through 5 instars. The larval period varies from 22 to 35 days. Pupation takes place in the leaves or in the infested pod itself. The pupal stage lasts for 6 to 7 days. Adults live for 2-5 days.

Damage

The larvae eat out buds and flowers. small pods may also be damaged.

Management

The pest spraying can be controlled by carbaryl [0.1 per cent], monocrotophos [0.07 per cent], Dimethoate [300 g ai/ha],Methyl demeton [500 g ai/ha] and the spray can be repeated after 15 if the attack persists.

6. Gram Pod Borer, *Helicoverpa armigera* (Noctuidae: Lepidoptera)

Identification

Moths are medium sized, stout, ochreous with olive grey forewings in male and reddish brown forewings in female with dark brown circular spot in the centre and indistinct double waved antemedial lines. Hind wings are pale smoky-white with a broad blackish outer border.

Biology

The eggs are yellowish-white, ribbed and dome-shaped, 0.4–0.5 mm in diameter. Freshly hatched larvae are yellowish-white in colour but gradually change and acquire greenish tinge. Full grown caterpillars are 40–48 mm long, apple-green in colour with whitish and dark-grey broken longitudinal stripes. Full grown caterpillars drop down from the plants and burrow in the soil where they pupate. Pupae are dark-brown in colour, 11–14 mm long and have a sharp spine at the anal end. Fecundity of females is rather high ranging between 1200–1600 eggs. Incubation, larval and pupal stages last for 2–4, 15–24 and 10–14 days respectively and the entire life-cycle may be completed in 4–6 weeks.

Damage

Eggs are laid singly, generally on leaves and flowers but sometimes on fruits as well. On hatching, the young larvae feed on tender foliage; advanced stage larvae attack the fruits. They bore circular holes and thrust only a part of their body inside the fruit and eat the inner contents. If the fruit is bigger in size, it is only partly damaged by the caterpillar but later it is invariably invaded by fungi and bacteria and spoiled completely. The larvae move from one fruit to another and a single caterpillar may eat and destroy 2–8 fruits.

Management

1. Hand-picking of caterpillars and their mechanical destruction in the early stage of infestation can keep the population of this pest under check.
2. Spray *Ha*NPV @250 LE/ha. Spray application of HaNPV@ 250 larval equivalents/ha in 1125 lit. of water/ha when the larval population reaches 1 larva/m². (or) *Bacillus thuringiensis* var. *kurstaki* @ 1.5 kg in 1125 lit. of water/ha. at evening hours.
3. Installation of pheromone traps @5/ha coinciding with the initiation of flowering and destruction of collected moths.
4. Release of the egg parasitoid, *Trichogramma chilonis* or *T. brasiliensis* @ 1Lakh/ha coinciding with flower initiation at 15 days interval.
5. Planting of marigold as a trap crop (1: 16 ratio).
6. Erection of 20 bord perches/acre
7. In case of severe attack, 5 per cent dust or 0.2 per cent spray of carbaryl or 0.005 per cent cypermethrin has been found to be effective.

7. Blister Beetle, *Mylabris pustulata* Thunberg (Coleoptera: Meloidae)

Identification

The adult *Mylabris pustulata* is about 2.0-2.5 cm in length and bears red or reddish orange and black alternating bands on the forewing (elytra), Adults are rather soft-bodied, long-legged beetles with the head deflexed, fully exposed, and abruptly constricted behind to form an unusually narrow neck, the pronotum much narrower at the anterior end than the posterior and not carinate (keeled) laterally, the forecoxal cavities open behind, and each of the tarsal claws cleft into two blades. Body length generally ranges between 3/4 and 2 cm.

Biology

Eggs are laid in masses in the ground/hard soil in 20 batches. Females dig soil and lay 60 eggs at the same time of grass hopper. Incubation period is 20 – 22 days. Newly hatched larva is called as triangulin, very active on egg pods of grass hoppers. Larval Period is 10 – 30 days. Each larvae feeds upto 40 – 70 GH eggs. Pupation in soil upto 2 months. Female lays 10,000 to 20,000 eggs during life span and Over winters in egg stage.

Damage

The adult is the destructive stage. As the insects feed on the plants' reproductive parts, they can cause significant yield losses.

Management

1. Pick off beetles by hand (wear gloves or use insect nets) and destroy.
2. Spray during flowering stage of the crop with any of the following insecticides *viz.*, Cypermethrin [30 g ai/ha] or Fenvelerate [75g ai/ha]or Dimethoate [0.03 per cent].

8. White Fly, *Bemisia tabaci (Aleurodidae, Hemiptera)*

Identification

Nymph is greenish yellow in colour and is present along with pupa on the under surface of the leaves. Adult is yellow colour bodied insects with white waxy covering.

Biology

The females mostly lay eggs near the veins on the underside of leaves. Each female can lay about 300 eggs in its lifetime. Egg period is about three to five days during summer and 5 to 33 days in winter. Nymphal period is about 9 to 14 days during summer and 17 to 73 days in winter. Adults emerge from puparia through a T-shaped slit, leaving behind empty pupal cases or exuviae.

Damage

Nymphs and adult remain on the undersurface of the leaf and suck the cell sap causing chlorotic spots and yellowing and drying of leaves. It is a vector of many viruses and transmits numbet of viral dideases. Females are more efficient vectors than males. Heavy infestation leads to death of the seedlings, wilting of older plants, shedding of leaves and flowers. The honey dew excreted by the insects interferes with the photosynthetic activity in the leaves

Management

1. Alternate hosts like brinjal, tomato, sunflower and tobacco may be avoided.
2. Judicious application of recommended doses of fertilizers, particularly nitrogenous fertilizers may be followed.
3. Remove the infested crop and destroy.
4. Use of yellow sticky traps coated with castor oil @ 5 per acre.

5. Preservation and use of natural enemies like coccinellids, chrysopids and spiders, Encarsia sp and microbials like *Beavaria bassiana.*

6. Spray application of methyl dematon 25 EC @ 2ml/lit. or dimethoate 30 EC @ 2 ml/lit. in the early stage and phasalone 35 EC @ 2ml/lit. in the later stages of crop. In case of severe infestation, spraying of acephate 75 SP @ 1.5 g/lit.

9. Bean Aphid, *Aphis craccivora (Aphididae, Hemiptera)*

Identification
Nymphs are dark brown coloured whereas adults are black coloured aphids.

Biology
Adults reproduce parthenogenetically and viviparously. There are four nymphal instars. The total nymphal period is about 6 days. Adults live for 10 days of which 8 days are utilized by the females to lay young ones. An average of 36.5 nymphs are laid by the female during its life time. Under favaourable environmental conditions, the life cycle is completed in about 15 days.

Damage
Nymphs and adults suck the sap of the plants tender shoots, leaves and inflorescence stalk. They are covered with dark coloured aphids. Infestation in early stages cause stunting of plants and reduction in vigour. When the attack occurs at the time of flowering and pod formation, the yield is reduced considerably.

Management
1. Cultural control can be effective through early planting and close planting.
2. Conserve coccinellids, carabids aand syrphids are the most important predators.
3. Spray application of methyl dematon 25 EC @ 2ml/lit. or dimethoate 30 EC @ 2 ml/lit. or phasalone 35 EC @ 2ml/lit or acephate 75 SP @ 1.5 g/lit.

Chapter 16

Pests of Tuber Crop

1. Potato Tuber Moth, *Phthorimaea operculella (Gelechidae, Lepidoptera)*

Identification

Adult is a small brown moth (approximately 0.95cm long, with an approximately 1.30cm wingspan) whose larvae feed on solanaceous plants. The adult has a a silvery-gray body, with brown forewings and white hind wings. Eggs are spherical, less than 0.1 cm, and iridescent white upon oviposition. As the embryo develops, the color darkens from yellow to brown. Larvae molt four times before pupation. The first instar is 0.3 to 0.4 cm and is capable of traveling up at least 74 cm from the site of oviposition to suitable plant material.The head capsule is dark brown-to-black. The neonate body is light brown and shifts to green or pink in later instars. Fourth instar males (approximately 0.95 cm) can be distinguished from female larvae by two yellowish testes, which can be observed through the larval cuticle in the 5th and 6th abdominal segments. Larvae spend their entire lifetime in a single gallery. After 4th instars cease feeding, the larva drops from the mine and wanders on the soil surface or leaf litter until pupation. Prior to pupation, the larvae create a silk pupal case.

Biology

The adult lays about 200 eggs singly on under surface of leaves or on exposed tubers. Eggs hatch in 3 days. Larval period lasts for 10-15 days. Larvae pupate in soil or leaf litter.Pupal period ranges from 5-8 days. Adult longevity is 10-20 days. Eight generations are completed per year/storage season.

Damage

The larvae, also known as potato tubeworms, are the most damaging to potato plants. Larvae grow up to 1/2 inch and are white or yellow with brown heads. They feed on their host plants for up to two weeks, boring tunnels in the tubers, depositing excrement and rendering the potatoes unfit for human consumption. The larvae may also feed on the leaves of the potato plant, damaging leaf tissue and inhibiting the plant's growth.In tubers, the typical symptom is presence of irregularly shaped galleries with excrement near tuber eyes. It is an important pest in the field and storage.

Management

1. Select healthy tubers.
2. Avoid shallow planting of tubers. Plant the tubers to a depth at 10–15 cm deep.
3. Adopt intercropping with chillies, onion or peas.
4. Collect and destroy all the infested tubers from the field.
5. Crop sanitation.
6. Earthing up of the crop helps in minimising infestation.
7. Install pheromone traps at 15/ha.
8. Do not leave the harvested tubers in the field overnight.
9. Cover the tuber with 2.5 cm thick layer of sand.
10. Release egg larval parasitoid: *Chelonus blackburnii* @ 30,000/ha twice – 40 and 70 day after planting.
11. Spray *Bacillus thuringiensis* @ 1 kg/ha at 10 days interval.
12. Spray NSKE @ 5 per cent or quinalphos 25 EC @ 2ml/lit of water to manage foliar damage.
13. Fumigation with Carbon di sulphide/methyl bromide/CCl4 @ 2.5-5 kg/1000 cu. m. for 3 hrs will bring about control of the pest.
14. Application of carbaryl 4 D @ 3 g/lit. in the field to protect the tubers.
15. Spray application of Malathion 50 EC @ 3 ml/lit on gunny bags that are used for storing tuber.

2. Cut Worms

a. Common Cut Worm, *Agrotis segetum (Noctuidae, Lepidoptera)*

b. Black Cut Worm, *Agrotis ipsilon (Noctuidae, Lepidoptera)*

Identification

A. segetum: Larva is black colour with a brown head. Triangular spots are present in the spiracular region. After five instars, they pupate in earthern cocoon. Adult moth is stout and has wavy lines and spots on brownish fore wings. Males have bipectinate antenna while females have filiform antenna.

A. ipsilon: Larva is black with pale mid dorsal stripes. Skin with coarse granules interspread with small granules. There are five larval instars. Adult fore wing is pale brown in colour with dark purplish brown patch along coastal and towards the base. Hind wings white with brown tinge. Male has bipectinate antenna whereas female has filiform antenna.

Biology

The cut worms are active from October to april in the plains and during summer in mountains regions of the country. Females lay creamy white dome shaped eggs [700 eggs during its life span] in clusters on the under surface of the leaves or in soil. The eggs hatch in 3-6 days. There are 6 larval instars. The larval stage lasts for a month. Pupation takes place in the earthern chambers in the soil. Pupal period is 10-15 days. Life cycle is completed in 8 weeks.

Damage

The larva hide during day time in cracks and crevices of soil, become active at dusk, cut the tender stem of young and growing plants near the ground level. Infested field looks like a cattle grazed field. They feed from dusk to dawn and cut many plants than they consume.

Management

1. Cutworm larvae can be controlled to some extent by flooding the field.
2. Hand –pick and destroy the larvae – morning and evening hours on cracks and crevices in the field.
3. Set up light trap @ 1/ha.
4. Pheromone traps @ 12/ha to attract male moths.
5. Plough the soil during summer months to expose larvae and pupae to avian predators.
6. Burning the heaps of grass in the affected fields containing the Cutworm larvae.
7. Application of baits formulations containing 0.025 per cent quinalphos applied at 15 and 20 kh/ha by broadcasting method or trichlorfon baited with mixture of bran, water, and sugar, broadcasted at the rate of 30 kg/ha is very effective in controlling the pest.
8. Spray insecticides like chlorpyriphos 20EC @ 1 lit/ha or neem oil @ 3 per cent.

3. Bihar Hairy Caterpillar, *Spilosoma (Diacrisia) obliqua (Arctiidae, Lepidoptera)*

Identification

Larva crimson colored with black head with broad transverse band and tufts of yellow hairs that are dark at both ends. Pupa forms a thin silken cocoon by interwoven shed hairs of the larvae.

Biology

Female lays around 50-100 eggs on lower surface of the leaves. Pupation takes place under dry debris, foliage and soil

Damage

1. Young larvae feed gregariously on the under surface of the leaves. Early instars skelatonise the leaves gregariously. Feed on leaves and cause loss by way of defoliation. Leaf Skeletonisation.

2. Defoliation leaving only stem and petioles.Similar to Red headed hairy caterpillar, goat or cattle grazed appearance is seen.In severe cases, only stems are left behind.

Management

Conservation of parasitoids like *Apanteles obliquae.*

The pest spraying can be controlled by carbaryl [0.1 per cent], monocrotophos [0.07 per cent], Dimethoate [300 g ai/ha], Methyl demeton [500 g ai/ha] and the spray can be repeated after 15 if the attack persists.

4. Spotted Leaf Beetle (or Hadda Beetle) *Henosepilachna vigintioctopunctata* (Fabr.) (Coccinellidae: Coleoptera)

Identification

The subfamily Epilachninae contains plant-feeding ladybird beetles because most other ladybird beetles are predators, not plant pests. These brownish or orangecolored, hemispherical beetles are larger than other ladybird species. *H. vigintioctopunctata* (in Latin, *viginti* means 20 and *octo* means 8) has 28 black spots on the forewing (elytra). *E. dodecastigma (dodeca* means 12 in Greek) has 12 black spots on the elytra. However, beetles with 14, 16, 18, 20, 22, 24 or 26 spots have been observed under field conditions, due to mating between females of *E. dodecastigma* and males of *H. vigintioctopunctata* (Lall and Mandal, 1958).

Biology

The female lays elongate, spindle-shaped yellowish eggs in groups of 10–20 on the under surface of leaves. About 120–180 eggs may be laid by a female. The egg period is 2–4 days. The yellowish spiny grubs become full grown in 10 -35 days and pupate on the leaf or stem. The pupa is hemispherical, yellowish with spines on the posterior part. The anterior portion being devoid of spines. Adults emerge in a week and live for a month feeding on leaves. The total life-history takes 17–50 days depending on weather conditions.

Damage

The grub and adult have chewing mouthparts. Hence, they scrape the chlorophyll from the epidermal layers of the leaves. The feeding results in a typical ladder-like window. The windows will dry and drop off, leaving holes in the leaves. In severe infestations, several windows coalesce together and lead to skeletonization and the formation of a papery structure on the leaf.

Management

1. Preservation of Eulophid Egg parasitoids, *Tetrastichus ovularum,* Chalcid parasitoid *Ugna menoni, Pedobius foveolatus.*

2. Protect the population of parasitoids such as *Pediobius foveolatus.*

3. In the initial stage, collection and destruction of affected leaves alongwith the eggs, grubs and adults.

4. Spray application of carbaryl 0.1 per cent or cypermethrin 0.025 per cent or profenofos 0.05 per cent or DDVP @ 1 ml/lit or Neem oil @ 3 ml/lit.

5. Sweet Potato Weevil, *Cylas formicarius (Apionidae, Coleoptera)*

Identification and Biology

A complete life cycle requires one to two months, with 35 to 40 days.Adults do not undergo a period of diapause in the winter, but seek shelter and remain inactive until the weather is favorable. All stages can be found throughout the year if suitable host material is available.

Eggs are deposited in small cavities created by the female with her mouthparts in the sweet potato root or stem. The female deposits a single egg at a time, and seals the egg within the oviposition cavity with a plug of fecal material, making it difficult to observe the egg. Most eggs tend to be deposited near the juncture of the stem and root (tuber). Sometimes the adult will crawl down cracks in the soil to access tubers for oviposition, in preference to depositing eggs in stem tissue. The egg is oval in shape and creamy white in color. Its size is reported to be about 0.7 mm in length and 0.5 mm in width. Duration of the egg stage varies from about five to six days during the summer to about 11 to 12 days during colder weather. Females apparently produce two to four eggs per day, or 75 to 90 eggs during their life span of about 30 days. Under laboratory conditions, however, mean fecundity of 122 and 50 to 250 eggs per female has been reported.

When the egg hatches the larva usually burrows directly into the tuber or stem of the plant. Those hatching in the stem usually burrow down into the tuber. The larva is legless, white in color, and displays three instars. Temperature is the principal factor affecting larval development rate, with larval development (not including the prepupal period) occurring in about 10 and 35 days at 30° and 24° C, respectively. The larva creates winding tunnels packed with fecal material as it feeds and grows.

The mature larva creates a small pupal chamber in the tuber or stem. Duration of the pupal stage averages 7 to 10 days, but in cool weather it may be extended to up to 28 days.

The adult emerges from the pupation site by chewing a hole through the exterior of the plant tissue, The body, legs, and head are long and thin, giving it an ant-like appearance. The head is black, the antennae, thorax and legs orange to reddish brown, and the abdomen and elytra are metallic blue. The snout is slightly curved and about as long as the thorax; the antennae are attached at about the mid point on the snout. The beetle appears smooth and shiny, but close examination shows a layer of short hairs. The adult measures 5.5 to 8.0 mm in length. Under laboratory conditions at 15 °C, adults can live over 200 days if provided with food and about 30 days if starved. In contrast, their longevity decreases to about three months if held at 30° C with food, and eight days without food. Adults are secretive, often feeding on the lower surface of leaves, and are not readily noticed. The adult is quick to feign death if disturbed. Adults can fly, but seem to do so rarely and in short, low flights. However, because they are active mostly at night, their dispersive abilities are probably underestimated. Females feed for a day or more before becoming sexually active, but commence oviposition shortly after mating; the average preoviposition period is seven days.

Damages

Sweetpotato weevil is often considered to be the most serious pest of sweet potato, with reports of losses ranging from five to 97 per cent in areas where the weevil occurs. There is a positive relationship between vine damage or weevil density, and tuber damage. However, the plants exhibited some compensatory ability, with the relationship between vine damage and yield non-linear, and sometimes not significant. It is a pest in the field and storage.

A symptom of infestation by sweetpotato weevil is yellowing of the vines, but a heavy infestation is usually necessary before this is apparent. Thus, incipient problems are easily overlooked, and damage not apparent until tubers are harvested. The principal form of damage to sweet potato is mining of the tubers by larvae. The infested tuber is often riddled with cavities, spongy in appearance, and dark in color. In addition to damage caused directly by tunneling, larvae cause damage indirectly by facilitating entry of soil-borne pathogens. Even low levels of feeding induce a chemical reaction that imparts a bitter taste and terpene odor to the tubers. Larvae also mine the vine of the plant, causing it to darken, crack, or collapse. The adult may feed on the tubers, creating numerous small holes that measure about the length of its head. The adult generally has limited access to the tubers, however, so damage by this stage is less severe than by larvae. Adult feeding on the foliage seldom is of consequence.

Management

1. Crop rotation.
2. Crop sanitation.
3. Remove rotting tubers or infected tubers.
4. Harvest the crop immediately after maturity.
5. Flooding of infested fields for at least 48 hours after completing harvest drowns weevils induces rotting of the leftover plant materials and thereby reduces weevil densities.
6. Immersing preplanting material in Chlorpyriphos 3 ml/lit + Carbandazim @ 2 g/lit.
7. Instal pheromone trap @ 1/100 sq.m. to trap male weevil.

8. Earthing up operation to be done at 60 DAP.

9. Spray 2.5 kg carbaryl and 2.5 lit of Malathion 50 EC in 625 lit of water per ha.

10. Soil application of phorate 10 G @ 1 kg a.i./ha at 15 and 45 DAP or Carbofuran 3 G @ 0.5 kg a.i./ha at 30 and 60 DAP are effective in reducing weevil infestation.

6. Vine Borer, *Omphisa anastomosalis (Pyraustidae, Lepidoptera)*

Identification

The adult wingspan is about 33 mm with reddish-brown body and reddish brown markings/dark wavy lines on white wings. Full-grown larvae are 25–30 mm long and pale yellowish white. The pupa is about 16 mm long and 3 mm wide and nearly cylindrical. It is formed in a slight cocoon in the larval tunnel in the vine. The pupal period lasts 12–16 days.

Damage

The larvae feed on *Ipomoea batatas* and other Convolvulaceae species. They bores into the main stem and sometimes penetrate the storage roots. The larvae create large tunnels causing hollow cavities in the stem. Infested plants usually have a pile of frass that can be found close to the attacked stem.Symptoms are slight swelling on the basal portion of the stem.

Management

Spray application of Fenvalerate 0.4 per cent D @ 20 kg/ha.

7. Sweet Potato Hopper, *Exitianus indicus (Cicadellidae, Hemiptera)*

Identification

This leaf beetle has black back ground color of its protonum and elytra with metallic green to golden marks on it. The shell of the pronotum and elytra is transparrent and extended over its body. The size is about 5mm.

Biology

Tortoise beetles overwinter as adults in dry protected areas such as under bark or leaf trash. They emerge in May and June and feed on various weeds of the morning glory family until sweet potato plants are available. Eggs are deposited on the leaf undersides, hatching in 7–10 days. Larvae and adults feed on the underside of leaves, but eat entirely through the foliage. Larvae feed for 3–4 weeks, then attach themselves to leaves by their anal end, and pupate for 1–2 weeks to emerge as adult beetles. New Jersey has several generations per year.

Damage:

Occasionally these beetles are abundant enough to cause serious defoliation to sweet potato foliage, but usually they are just a curiosity. However, when the beetle population is high, leaves may become riddled with many small holes. Newly set plants are most susceptible, and can be severely damaged or killed by feeding of this pest. The grubs and adults of these beetle bites large round holes in leaves and eventually skeletonising the leaves.

Management

Spray application of Fenitrothion 50 EC @ 2 ml/lit, Spray Azadirachtin 1.0 per cent EC (10000 ppm) @2.0 ml/lit, Quinalphos 25 per cent EC @1.0 ml/lit.

Chapter 17

Pests of Perrenial and Leafy Vegetable Crops

1. Moringa Hairy Caterpillar, *Eupterote mollifera (Eupterotidae, Lepidoptera)*

Identification and Biology

Adults are large moths with yellowish wings having faded lines. Moths emerge with inception of monsoon and lay eggs in groups on leaves and tender stems. Egg period is 6 days. Matured larva are brownish and densely hairy. Hairs are annoying to handle. Pupation takes place in soil.

Damage

Larvae feed gregariously by scrapping bark and defoliating foliage. Severe invasion results in absolute defoliation of the tree.

Management

1. Collect and destroy egg masses and caterpillars.
2. Use light traps to attract and kill adults @ 1/ha.
3. The larva in groups are killed by burning torch.
4. Spray application of Quinalphos 25 EC @ 2 ml/lit. Azadirachtin 1.0 per cent EC (10000 ppm) @2.0 ml/lit., Chlorantraniliprole 18.5 SC @ 60 ml/ha, Emamectin benzoate 5 per cent SG @4 g/10 lit.

2. Moringa Leaf Eating Caterpillar, *Noorda bliteallis* (*Crambidae, Lepidoptera*)

Identification

Caterpillar is brown colour with a noticeable mid dorsal stripe and a black head. Adult is tiny moth with dark brown fore wings and white hind wings with brown border.

Damage

Caterpillar webs the leaves and feed on the leaflets further reducing them into papery structures.

Management

1. Deep Plough the soil around the trees to expose and kill pupae.
2. Spray application of Malathion 50 EC @ 2 ml/lit.,Quinalphos 25 EC @ 2ml/lit. or insecticides like carbaryl 50 WP @ 1g/lit.

3. Moringa Bud Midge, *Contarinia moringae (Cecidomyiidae, Diptera)*

Identification

Adult fly is a small brownish in colour .

Damage

Adult Female lays eggs in batches on anthers of the bud. Maggot nourish on the internal contents of the flower buds and causes shedding of flower buds.

Management

Spray application of Malathion 50 EC @ 2 ml/lit., Fipronil 5 per cent SC@1.5 ml/lit.,Dichlorvos 76 WSC @ 1 ml/lit, Imidacloprid 17.8 per cent SL @3.0 ml/ 10 lit, Dimethoate 30 per cent EC @1.0 ml/lit., Emamectin benzoate 5 per cent SG@ 4 g/10 lit.

4. Moringa Pod Fly, *Gitona distigma (Drosophilidae, Diptera)*

Identification

Adult fly is a minute yellow in colour with red eyes.

Damage

Injury is caused by maggot, drying and splitting of fruits from tip, gummy exudates oozing from fruits are the main symptoms.

Management

1. Raking up the soil under the trees or plough the infested field to destroy puparia and application of folidol dust @ 25 kg/ha.
2. Poison baits containing attractants like citronella oil, eucalyptus oil, acetic acid and lactic acid.
3. Spray application of Phasalone 35 EC @ 2 ml/lit at the time of flowering.

4. At fruit formation stage, spray Dichlorvos 76 WSC @ 1 ml/lit. If infestation is severe, spraying should be repeated at 25 days interval.

5. Amaranthus Caterpillar, *Hymenia recurvalis (Pyraustidae, Lepidoptera)*

Identification

Wingspan of both sexes 22-24 mm, forewings are 9-11 mm long, deep brown with broad white median band; hindwings deep brown with a broad white median bar. Mature larva is 15 mm long, head light yellowish-brown with many brown spots, body appears green with transparent epidermis.

Damage

Larvae skeletonize leaves of leafy vegetables.Affected portion dry up.

Management

Spray application of Dichlorvos 75 WSC @1.5 ml/lit.

6. Amarantus Weevil, *Hypolixus truncatulus (Curculionidae, Coleoptera)*

Identification

Grub is apodous stout, "C" shaped, white in colour. Adult weevil with conspicuous snout, grey in colour.

Damage

Larvae bore into the stems of amaranthus and girdle the stems. In affected plants, gall like swellings can be observed on the stem which is scarred with longitudinal white markings. Inside the stem, tunneling by stout creamy white larva can be seen with frass. In severe cases, plants show yellowing of leaves and wilting.

Management

Spray application of Spray Azadirachtin 1.0 per cent EC (10000 ppm) @2.0 ml/lit, Fipronil 5 per cent SC@1.5 ml/lit., Spinosad 45 per cent SC @3.2 ml/10 lit.

Chapter 18

Pests of Chillies

1. Chilli Thrips, *Scirtothrips dorsalis (Thripidae, Thysanoptera)*

Identification

Chilli thrips are pale colored and the lengths of their first and second instar larvae and the pupae are 0.37-0.39, 0.68-0.71 and 0.78-0.80 mm, respectively. Adults are about 1.2 mm long with dark fringed wings and dark spots forming incomplete stripes which appear dorsally on the abdomen.There are numerous microtrichia and dark transverse antecostal ridges on the abdominal tergites as well as sternites.

Biology

The life cycle stages of *Scirtothrips dorsalis* include egg, first and second instar larvae, prepupa, pupa and adult. Gravid females insert the eggs inside plant tissues above the soil surface. The eggs are microscopic (0.075 mm long and 0.070 mm wide), kidney-shaped and creamy white in color. The eggs hatch between two to seven days, depending upon temperature. Larvae and adults tend to gather near the mid-vein or borders of the host leaf.The two larval stages are completed in eight to ten days and the pupal stage lasts for 2.6-3.3 days. The life span of chilli thrips is influenced by the host plant species. Unlike other thrips, pupae of chilli thrips are generally found on leaves, leaf litter or on the axils of leaves, in curled leaves or under the calyces of flowers and fruits.Reproduction in thrips is generally sexual, parthenogenesis is also present.

Damage

Thrips possesses lacerating and sucking mouthparts and cause damage by extracting the contents of individual epidermal cells leading to necrosis of tissue.

This changes the tissue color from silvery to brown or black. Chilli thrips create damaging feeding scars, distortions of leaves, and discolorations of buds, flowers and young fruits by feeding on the meristems of the host plant's terminals and on other tender parts above the soil surface. *Scirtothrips dorsalis* has not been reported feeding on mature host tissues. According to Sanap and Nawale (1987), adult and nymphs of *Scirtothrips dorsalis* suck the cell sap of leaves, causing rolling of the leaf upward and leaf size reduction. For example, a heavy infestation of *Scirtothrips dorsalis* in pepper plants changes the appearance of the plant to what is called "chilli leaf curl." Appearance of discolored or disfigured plant parts suggests the presence of *Scirtothrips dorsalis*.A severe infestation of chilli thrips makes the tender leaves and buds brittle, resulting in complete defoliation and total crop loss. Infested fruits develop corky tissues.

The infested leaves curling upward, crumbling and shedding Infested buds turning brittle with petiole becoming brown and dropping down, affected fruits showing light brown scars.

Management

1. Inter crop with *Sesbania grandiflora* to provide shade which regulate the thrips population
2. Do not grow chilli after sorghum – more susceptible to thrips
3. Do not follow chilli and onion mixed crop – both the crops attacked by thrips
4. Sprinkle water over the seedlings to check the multiplication of thrips
5. Treat seeds with imidacloprid 70 per cent WS @ 12 g/kg of seed
6. Apply carbofuran 3 per cent G @ 33 kg/ha or phorate 10 per cent G @ 10 kg/ha or spray any one of the following insecticide
7. Imidacloprid 17.8 per cent SL @3.0 ml/10 lit, Dimethoate 30 per cent EC @1.0 ml/lit., Emamectin benzoate 5 per cent SG@ 4 g/10 lit, Fipronil 5 per cent SC@1.5 ml/lit., Spinosad 45 per cent SC @3.2 ml/10 lit, Thiacloprid 21.7 per cent SC @6.0 ml/10 lit. Ethion 50 per cent EC @2.0 ml/lit., Oxydemeton –Methyl 25 per cent EC @1.0 ml/lit, Phosalone 35 per cent EC@2.0 ml/lit.
8. Thrips, *Franklinothrips vespiformis* and *Erythrothrips asiaticus* are predaceous on the insect.

2. Fruit Borers, *Spodoptera litura, Helicoverpa armigera, Utethesia pulchella (Noctuidae, Lepidoptera)*

Damage

1. Irregular hole is seen in the fruit due to the feeding of *S. litura*.
2. Round neat hole is seen on the fruit due to the feeding of *H. armigera*.
3. Faded pericarp in typical fashion with seeds intact due to the infestation of *U. pulchella*.

Management
1. Remove the affected plants.
2. Spray application of Endosulphan 35 EC/Malathion 50 EC/ Chlorpyriphos 20 EC @ 2 ml/lit. or thiodicarb 200g/ac.
3. For *Spodoptera*, spray application of poison baits 5 kg of rice bran + 500 g Carbaryl/500 ml Chlorpyriphos + 500 g jaggery + sufficient amount of water is made into small balls and is applied in the evening.

3. Chilli Mites, *Polyphagotarsonemus latus (Tarsanomidae, Acari)*

Identification and Biology

Adult broad mites are elliptically shaped, but slightly wider at the front than the rear. Females are about 1.5 mm long and males are slightly shorter and more broad. Live specimens are light, translucent yellowish green. A pale white stripe runs longitudinally down the back of the female. Dead specimens are yellowish brown. They have 4 pairs of whitish legs, but the fourth pair of the female adult is greatly reduced. Females live for about 10 days and lay an average of 2 to 5 eggs per day (20 - 50 eggs per female).Without fertilization, females produce eggs which result in only male progeny.

Broad mite eggs are oval in shape and slightly flattened.The exposed translucent surface is covered with five or six rows of white bumps called tubercles. Eggs are about 3/100 inch (0.7 mm) long and can be discerned with a 14X handlens. They are usually laid singly on the undersides of new growth leaves.On fruit eggs are laid on the protected surface or in the depressions of the fruit.Eggs usually hatch in 2 to 3 days.

Larvae are very small, pear-shaped and have three pairs of legs. Just after hatching the larvae are translucent, but females become yellowish green or dark green in color and males yellowish brown.The larvae feed for 1 to 3 days before going into the resting pupal stage.

The pupal stage of this mite is a resting period in which there is no feeding. Sexes are similar in appearance, except for the fourth pair of legs. On males the fourth pair of legs are enlarged; on females, the fourth pair of legs are reduced and whip-like. The pupal lasts 2 to 3 days.

Damage

Mites feed by piercing plant cells and sucking up the sap that oozes from the wound.Reduction in photosynthesis and instability of water balance are some the damaging effects to plants. Feeding damage also causes terminal leaves and flower buds to become cupped and distorted. As a result of feeding injury, corky brown areas appear between the main veins on the underside on the leaf. Young foliage sometimes becomes rust colored and nearly always is deformed. Blooms abort, and the plant growth is stunted. Damaged leaves often turn dark green. *P. latus:* is a vector of leaf curl [Murda disease].

Management

1. Avoid use of excess nitrogen.
2. Spray application of Dichofol (Kelthane) @ 5 ml/lit, wettable sulphur @ 3 g/lit, micronized sulphur @ 2.5 g/lit.
3. Encourage the activity of predatory mite: *Amblyseius ovalis*
4. Apply phorate 10 per cent G @ 10 kg/ha or spray any one of the following insecticide:

Insecticide	*Dose*
Buprofezin 25 per cent SC	8.0 ml/10 lit.
Chlorfenapyr 10 per cent SC	1.5 ml/lit.
Diafenthiuron 50 per cent WP	8.0 g/10 lit.
Dimethoate 30 per cent EC	1.0 ml/lit.
Ethion 50 per cent EC	2.0 ml/lit.
Fenazaquin 10 per cent EC	2.0 ml/lit.
Fenpyroximate 5 per cent EC	1.0 ml/lit.
Hexythiazox 5.45 per cent EC	8.0 ml/10 lit.
Milbemectin 1 per cent EC	6.5 ml/10 lit.
Oxydemeton –Methyl 25 per cent EC	2.0 ml/lit.
Phosalone 35 per cent EC	1.3 ml/lit.
Propargite 57 per cent EC	2.5 ml/lit.
Quinalphos 25 per cent EC	1.5 ml/lit.
Spiromesifen 22.9 per cent SC	5.0 ml/10 lit.

4.Green Peach Aphid: *Myzus persicae*

Identification and Biology

Nymphs are greenish, but later turn yellowish. Adults are yellowish green in colour. Adults live for 2-3 weeks and produce 8-22 nymphs per day.Both winged and wingless forms breed Parthenogenetically. The Nymphal period lasts for about 7-9 days. It has 12-14 generations per year.

Damage

Appear on the tender shoots, leaves and on the lower surface of the leaves.Suck the sap and reduce the vigour of the plant.Secrete sweet substances which attracts ants and develops sooty mould. The pods that develop black colour due to sooty mould lose quality and fetch low price.The yields are also reduced by aphids directly and more through the spread of virus diseases acting as vectors indirectly.The infested plants turn pale with sicky appearance. The leaves curled and crinkled.

Management

1. Treat seeds with imidacloprid 70 per cent WS @12 g/kg of seed.

2. Apply phorate 10 per cent G @ 10 kg/ha or spray any one of the following insecticide Carbosulfan 25 per cent EC @1.0 ml/lit., Fipronil 5 per cent SC @1.0 ml/lit., Imidacloprid 17.8 per cent SL@ 3.5 ml/10 lit, Oxydemeton –Methyl 25 per cent EC @ 1.6 ml/lit., Phosalone 35 per cent EC @2.0 ml/lit., Quinalphos 25 per cent EC @1.0 ml/lit.

5. Chilli Blossom Midge, *Asphondylia capsici*, Cecidomyiidae: Diptera

Identification

Fly is dark reddish brown mosquito like midge.

Damage

Fly lays eggs in flower buds. Maggot is tiny pale orange colored and feeds on the floral parts leading to poor development of fruits. The ovary is distorted into gall like structure of varied shape.

Management

Foliarspray with triazophos 2 ml/l or carbosulfan 2 ml/l followed by chlorpyriphos 2.0 ml/l one week later is found effective.

Chapter 19

Pests of Onion

1. Onion Fly, *Delia antiqua (Anthomyiidae, Diptera)*

Identification and Biology

Maggots are small, creamy white in colour. Adult fly is slightly smaller than the common house fly, slender, greyish, large winged and bristly.

Biology

Onion flies are slightly smaller than houseflies. They have longer legs, are more slender, and overlap their wings when at rest.The onion fly deposits white elongated eggs about 1/25 inch in length on the soil near the stem and occasionally on the young leaves and neck of the onion plant. Eggs hatch into maggots 2-3 days after being laid.The legless maggots age tapered, creamy-white in color, and reach a length of about 1/3 inch. Maggots develop through three larval stages in 2 to 4 weeks depending on the temperature. Most newly hatched larvae crawl below the soil surface and feed upon the roots or burrow into the basal plate of the bulbs.When full-sized, the maggot leaves the bulb and enters the soil to pupate at a depth of 1-4 inches. The pupa is chestnut brown and 1/3 inch long. First and second generation pupae remain in the soil for 2-4 weeks before adult emergence. Larvae of the third generation develop into pupae and pass the winter in that stag. Flies emerging the following spring constitute the spring flight.

Damage

The larvae damage bulb onions, garlic and the bulbs of flowering plants. The first generation of larvae is the most harmful because it extends over a long period owing to the females' longevity and occurs when the host plants are small. Seedlings of onion and leek can be severely affected as can thinned-out onions.Less damage occurs in wet and cold springs as this delays the development of the larvae. When plants are attacked, the leaves start to turn yellow and the bulbs rot quickly, especially in damp conditions. Control measures include crop rotation, the use of seed dressings, survey and removal of infested plants, and autumn

Management
1. Seed treatment with imidacloprid @ 3g/kg seed.
2. Early sowing or planting
3. Deep summer ploughing of the ground to destroy the pupae.[1]
4. Regular crop rotation should be followed
5. Application of 10 kg of Carbaryl 4 G to the soil followed by light irrigation.

2. Onion Thrips, *Thrips tabaci (Thripidae, Thysanoptera)*

Identification
Thrips are very small, slender insects that are best seen with a hand lens. mature onion thrips are about 0.05 inch (1.3 mm) long and flower thrips are slightly larger at 0.06 inch (1.5 mm) long. The most distinctive characteristic of thrips are two pairs of wings that are fringed with long hairs. Adults are pale yellow to light brown in color. The immature stages have the same body shape as adults but are lighter in color and are wingless. When viewed under a microscope, western flower can be distinguished from onion thrips by its red eyes and 8-segmented antennae, while onion thrips' eyes are gray and its antennae are 7-segmented.

Damage
Thrips are the most common and serious insect pest of onions, and are found wherever onions are grown in California. High populations of thrips can reduce both yield and keeping quality of onions. Thrips are most damaging when they feed during the early bulbing stage of plant development. Scarring of leaves is a serious problem on green onions. Thrips have rasping-sucking mouthparts and feed by rasping the surface of the leaves and sucking up the liberated plant fluid. They feed under the leaf folds and in the protected inner leaves near the bulb. When population levels are high, thrips can also be found feeding on exposed leaf surfaces. Both adults and nymphs cause damage. When foliage is severely damaged, the entire field takes on a silvery appearance. Severe scarring also creates an entry point for foliar leaf diseases.

Management
1. Clean cultivation, regular hoeing.
2. Flooding of infested field will check the thrips population.
3. Spray dimethoate @ 0.06 per cent, profenofos @ 0.05 per cent or Fipronil 2 ml/lit.

Chapter 20

Pests of Mango

1. Mango Hoppers, *Amritodus atkinsoni*, I*dioscopus clypealis*, and *I. niveosparsus* (Cicadellidae: Homoptera)

Damage

The hoppers are found in abundance during November – February synchronizing with the flowering of mango trees. During the remaining part of the year they occur in small numbers inside barks or on leaves of mango. Both the nymphs and adults suck the sap from the inflorescence in large numbers causing withering and shedding of flower buds and flowers which result in heavy loss ranging from 25–60 per cent due to poor fruit setting. The honey dew excreted by them affords conditions for development of sooty mould. Egg laying also inflicts injury to the inflorescence.

Biology

I. niveosparsus (L.) is slightly smaller with three spots on the scutellum and prominent white band across its light brown wings. *I. clypealis* (L.) is the smallest with two spots on the scutellum and dark spots on the vertex and is light brown in colour. *A. atkinsoni* (L.) is the largest and light brown having two spots on the scutellum. The female hopper inserts the eggs into flower buds and the inflorescence stalk. The nymphs hatch out in 4 to 7 days. Freshly hatched nymphs are wedge shaped and whitish in colour with two small red eyes. Gradually with each moulting, the colour changes to yellow, yellowish green, green and ultimately greenish brown. The period from egg to adult takes about 12–17 days and during a flowering season two or more broods of the pest may occur.

Management

1. First spray of imidacloprid (0.005 per cent, 0.3 ml per litre of water), acephate 75 SP@ 1g/lit or phosalone 35 EC@ 1.5 ml/li should be done at early stages of panicle formation, if hopper population is more than 5 per panicle.

2. The second spray of thiamethoxam (0.005 per cent, *i.e.*, 0.2 g per litre of water) or acephate (1.5 g per liter of water) should be carried out after fruit set.

3. If substantial hopper population still persists, third spray of carbaryl (0.15 per cent, *i.e.*, 3 g per liter of water) should be done before maturity of fruits.

4. Synthetic pyrethroids such as cypermethrin, permethrin, fenvalerate and deltamethrin should not be sprayed in mango as they cause resurgence and are harmful to human health.

5. Do not spray if more than 50 per cent flowering has already occured because it will affect the pollinator activity leading to low fruit set.

6. Good orchard management practices such as keeping the orchard clean, regular ploughing, removal of weeds and pruning of overcrowded and overlapping branches in the month of December will reduce the hopper population.

2. Mango Nut Weevil, *Sternochetus mangiferae* (Curculionidae: Coleoptera)

Damage

Eggs are laid singly on the epicarp of partially developed fruits or under the rind of ripening fruits. The grubs as soon as they hatch out from the eggs tunnel in a zigzag manner through the pulp, endocarp and the seed coat and finally reach the cotyledons. As the fruit develops the tunnels get closed up. The grubs feed on the cotyledons and destroy them. The adults which emerge from the pupae also feed on the developing seed and this may hasten the maturity of infested fruits. The adults hibernate in between the crevices on the tree trunks. The weevil attacks only mango.

Biology

The dark brownish stout weevil measures about 6 mm long. The female scoop out the surface of the developing fruit (till it is half ripe) and deposits the eggs singly. On a fruit 12–36 eggs may be deposited. However, finally only a maximum of about 7 weevils can be noticed in a highly susceptible variety. The fluid that oozes from the fruit covers the egg. The incubation period is 7 days. The grub is apodous, fleshy, light yellow with a dark head and pupates inside the nut itself. It emerges as adult in 7 days. The total life cycle from egg to adult occupies 40–50 days. The weevils hibernate from July – August till next fruiting season. There is only one generation in a year.

Management

1. Avoid close planting of mango trees. Plough the soil under the trees. Maintain field sanitation. Collect and destroy the affected fruits at weekly intervals.

2. Dip the fruits in hot water at 50 degree celsius for two hours to kill the grubs. This treatment will not harm the fruit pulp.

3. Treat the bark with kerosene suspension or diazinon 0.05 per cent to kill the hiding adults before the start of fruiting season.

4. Apply carbaryl at 4g/lit alternated with monocrotophos at 1.5ml ml/lit at 22 days interval from the beginning of March to the end of May.

5. The damage due to nut weevil can be minimized appreciably by spraying deltamethrin 0.025 per cent thrice at 15 days interval at 45 days after fruit set.

6. Destruction of affected fruits and digging of soil to expose hibernating weevils.

3. Mango Stem Borer, *Batocera rufomaculata* (Cerambycidae: Coleoptera)

Damage

Eggs are laid singly either in the slits of tree trunks or in the cavities in the main branches and stem which are covered with a viscous fluid. The grubs feed by tunneling through the bark of branches and cause wilting. Though it is an occasional pest of importance, in case of severe attack the trees succumb. Normally, the attack goes unnoticed till the branch start drying up. Sometimes sap and frass may be seen exuding from the bore holes.

Biology

The adult beetle has two pink dots and lateral spines on thorax and measures about 50 mm long. The eggs laid singly on the bark or in crevices on tree trunk or branches hatch in about 1–2 weeks. The grub feeds for 3–6 months and pupates inside the tunnel itself. The adult emerges in about 4–9 months.

Management

1. Exclude alternative host trees such as silk cotton, from mango orchards.

2. Prune off the infested twigs of tree and destroy the late instar larvae of the insect.

3. Grow less susceptible varieties such as Neelam, Himayudin, Panchavarma. Varieties such as Banganapalli, Mulgoa, Gundur, Jehangir, Rumani, Padiri and Amlet are more susceptible.

4. Swab coal tar + Kerosene (1: 2) on the basal part of the trunk up to 3 feet high after scraping the loose bark in order to deter the females from laying eggs.

5. Put Carbofuron 3G at the rate of 5 g per hole and then plug it with copper oxychloride paste or plug it with mud. Close the holes by the cotton plugs dipped in chloroform or kerosene.

6. Inject the tree with 0.2 ml methyl parathion.

7. Apply few drops of Carbon disulfide or fenvolarate or dichlorovos into the holes and plug the same.

8. Keep aluminium phosphide tablets in the holes and plug them.

9. If infestation is severe then apply the copper oxychloride paste on the trunk of the tree.

4. The Mango Fruit Borer, *Deanolis albizonalis* (Hampson)

Biology

The infestation of the pest begins in the first week of February when the fruits are in pea size and continue to persist in the field till 3ʳᵈ week of May. The damage progressively increases up to a maximum of 44 per cent as the mango season advances. The pest has a shorter life cycle of about one month with three to four generations during a mango season. The adult moth is medium sized, brown in colour with prominent snout. Round and cream coloured eggs are laid on the fruit surface or fruit stalks. The larva is cream in colour with brown head having pink bands on the dorsal side. Mean larval period is 12.4 days. Larvae are nocturnal, migratory and gregarious in behaviour. The full grown larva pupates in dead wood on the tree or cracks and crevices of the bark and pupal period completes in 13.5 days. The pupa was greyish brown and obtect. Pre-pupae of the last brood enter into diapause at the end of the mango season, *i.e.*, in the month of May.

Damage

The bunch of dried fruits with a black entry hole of larvae at beak region is the typical symptom. Actual feeding site is kernel part, after exhausting the feed in small fruits, larvae migrate to other fresh fruits of the same or nearby bunch thereby causing extensive damage. Generally, in most of the infested fruits, 4 – 6 larvae/fruit are found.

Management

1. Collection and destruction of dried branches/dead wood in the mango orchards during off-season.

2. Collection and destruction of infested fruits regularly starting from the beginning of pest appearance to prevent the spread of pest infestation.

3. Collection and destruction of diapausing pre pupae from the cracks, crevices of the bark during off-season.

4. The mango fruit borer larvae/pupae/diapausing larvae can be trapped by using artificial niches on the trees during the season and destroyed.

5. Swabbing of the chemical chlorpyriphos 20 per cent EC 20 per cent EC @ 5 ml/lit. of water would kill the prepupae/pupae population under the bark and helps in reduction of fruit damage.

6. Spraying relatively safer insecticides such as chlorpyriphos 20 per cent EC (2.5 ml/1 litre of water), dichlorvos 76 per cent EC (1.5 ml/1 litre of water) or carbaryl 50 per cent W. P. (3 gm/1 litre of water) in the second fortnight of January coinciding with the moth emergence/ hatching of eggs of first brood in the gardens where the pest incidence was severe in previous year.

7. Application of dichlorvos 76 per cent EC (1.5 ml/1 litre of water) or chlorpyriphos 20 per cent EC (2.5 ml/1 litre of water), or carbaryl 50 per cent W.P. (3 gm/1 litre of water) or neem oil 3 ml + chlorpyriphos 20 per cent EC 1 ml/1 litre of water late in the evening (as the pest is nocturnal) when the fruits are of marble size to coincide with the migration of the larvae.

5. Fruit Fly *Bactrocera dorsalis* Hendel, (Tephritidae: Diptera)

Damage
The female flies lay eggs just below the fruit epidermis (1–4 mm deep). On hatching the maggots feed on pulp of those fruits. As a result a brown patch appears around the place of oviposition and the infested fruits start rotting. These affected fruits drop down prematurely and the maggots come out from these fallen fruits to pupate in the soil. Semi ripe fruits are attacked usually by April-May. Sometimes it becomes serious.

Biology
The adult fly is light brown with transparent wings. Adult flies are very conspicuous. These are about 7 mm long, with hyaline wings (expanse: 13–15 mm), thorax ferrugineous without yellow middle stripe, legs yellow, abdomen conical in shape and dark brown in colour. Preoviposition period is 2–5 days. A single female can lay 150–200 eggs (average 50) in about a month. The eggs are laid in clusters of 2–15 eggs and these hatches in 2–3 days during March and 1–1½ days during April. Maggot duration is 6 days in summer and extends up to 19 days with the fall in temperature. Pupation usually takes place 80–160 mm below the soil surface and pupal period ranges from 6 days (summer)–44 days (winter).

Management
1. Harvest the fruits before ripening.

2. To check the carry over pest, collect and destroy all fallen and attacked fruits.

3. Plough around the trees during winter to expose and kill the pupae.

4. The adult flies may be trapped and killed by poison-baiting or bait-spray (20 gm Malathion, 50 per cent wettable powder in 2 liters of water for baiting and 20 litres of water for spraying.

5. Spraying with 0.3 per cent oxydemeton methyl or 0.03 per cent phosphamidon or 0.06 per cent dimethoate or 0.2 per cent carbaryl.

6. Mango Mealy-Bug, *Drosicha mangiferae* (Green) (Coccidae: Homoptera)

Damage

These are large, fleshy, flat-bodied creatures measuring about 1.5 cm in length and a little less than a centimeter in width, covered with ashy-white mealy powder and crawling up or down the tree-trunks or on the ground round the tree-base or even invading the houses if the mango trees are near about. These mango mealy-bugs are also referred to as the giant mealy-bugs. They suck the plant-sap and although their name seems to suggest that they are specific pests of mango only, their list of food plants includes at least 62 species of trees, shrubs and herbs. When they are in large number they devitalize the plant and they produce honeydew which encourages growth of sooty mould, giving a very unhealthy look to the plant as a whole. At times, they are found clustering in masses on young shoots, like fungus outgrowths.

Biology

There is a well-established sexual dimorphism in the adult stage which is generally found during the midsummer period, *i.e.* from April to June. Adult females are wingless and large-bodied. The male is a winged creature with only one pair of wings and a very delicate reddish body which flies actively and fertilizes the females. The adult gravid females after fertilization crawl down along the tree-trunk to the ground where they lay eggs at depths of about 5–15 cm and in clusters of 300–400 eggs each. The oviposition is generally confined to an area near and around the base of the tree. These activities of migration from the tree downwards to the ground and oviposition in the soil are generally confined to the months of April, May and June. The males die soon after mating and the females soon after oviposition. The eggs laid in the soil take quite a few months before they hatch and is influenced by the temperature and moisture conditions of the soil. Hatching can be as early as November of the same year or as late as March of the succeeding year. The young nymphs soon after hatching crawl about in search of some suitable food-plant on which, if found, they spend some time. Thereafter, they begin their ascent along the tree-trunks and this upward migration lasts for several weeks. On reaching the fresh growths, the nymphs congregate there and begin to suck the plant-sap. They moult thrice during their nymphal period which lasts about three months or more, depending on the environmental temperature. Thereafter, the nymphs developing into males undergo some sort of pupation and transform themselves into winged adults and the female-producing nymphs do not undergo any appreciable change except in size. Thus, there is only one generation during the year. Unlike many other coccids, the nymphs of this pest do not remain stationary although they are sluggish.

Management

1. Raking the soil under the tree to a depth of 15 cm after the month of May to expose the eggs.

2. Destruction of weed host *Clerodendron inflortunatum* and grasses, by ploughing during June-July.

3. Nymphs can be prevented from climbing the tree by applying a sticky band made of a 20 cm wide polythene sheet (400 gauge) above the ground level on the trunk, during December.

4. It is secured to the stem with a jute thread and grease is applied on the lower edge of the band.

5. The nymphs congregate in the lower end of the band and they can be killed by insecticidal spray.

6. Spray application of chlorpyriphos 0.05 per cent with due precaution, on the colonies settled on the shoots.

7. Release of Australian ladybird beetle, *Cryptolaemus montrouzieri* @ 10/tree.

7. Mango Leaf Gall Midge, *Procystiphora mangiferae* (Diptera: cecidomyiidae)

Damage

Midges are very small flies, 1-2 mm in length. The female lays eggs into the tissue of young leaves leaving a small reddish spot. The leaf tissue under the red spot becomes swollen and soft. Gall formation begins within seven days and attains a maximum diameter of 3-4 mm. Adults usually emerge from the underside of the leaf leaving the pupal skin protruding from the emergence hole. Due to the attack of unopened flower buds, they fail to open and drop down. When the inflorescence stalk is attacked, the inflorescence becomes stunted and malformed. Mango leaf gall midge is spread by wind currents and movement of infested plant material.

Biology

The light orange coloured fly lays the eggs inside immature blossoms. The maggots that hatch out from the eggs feed on stalks of stamens, anthers, ovary, etc. Only one maggot is found in each bud and it pupates inside the bud itself. The life-cycle from egg to adult occupies 12-14 days).

Management

1. Stem injection by making 5 – 10 cm deep holes in the main branches with dimethoate or monocrotophos @ 0.5 ml a.i./cm circumference gave effective control of the pest.

2. Single spray of 2, 4-D @150 mg/l in October resulted in opening of galls causing 90 per cent autocidal mortality of the nymphs.

8. Red Tree Ant, *Oecophylla smaragdina* Fb. (Formicidae: Hymenoptera)

Damage

The ants web and stitch together a few leaves usually at the top of the branches and build their nests on citrus, jack-fruit, jamun, litchi, mango, sapota etc. The ants do not cause any direct injury or loss to the tree. Indirectly, the damage is caused by protecting aphids and scale insects from being preyed upon by their parasitoids and predators and also carries the nymphs of aphids, mealy bugs and scale insects from tree to tree thus spreading the infection of these noxious pests. Besides, being very ferocious, they also prove to be a nuisance to the persons who climb the trees and other workers around, who often get badly bitten by these ants.

Biology

Eggs are oval in shape and whitish in colour. Larvae are also whitish, 1.2– 1.4 mm long when freshly formed while full grown ones are 9–11 mm long. Pupae are also pure white in colour. Adults are light orange red in colour; the workers are 14–18 mm long, wingless and infertile. Sexually functional males and females are winged and usually mate outside the nests in the course of their nuptial flights. The fertilized females, also known as Queens, shed their wings at the time of nest formation. Egg, larval and pupal periods occupy 4–8, 10–17 and 5–7 days respectively.

Management

1. Spraying with cypermethrin 0.025 per cent
2. Removal and destruction of nests mechanically.

9. Mango Leaf Webber, *Orthaga exvinacea* (Pyralidae Lepidoptera)

Symptoms

The larvae web the leaves and terminal shoots into clusters. A webbed cluster of leaves may harbour several larvae in the initial stage. The larvae are initially gregarious and feed by scraping the leaf surface. Late-instar larvae feed individually on the whole leaf lamina leaving only the midrib. As a consequence of severe feeding, clusters of webbed leaves become dry and brown in colour. With severe infestation, the shoots become dry and photosynthesis is severely hampered. Trees which bear clusters of affected leaves present a sickly appearance and can be seen from a distance due to the brown, dried, clustered leaves.

Biology

The medium sized moth with brownish wings has wavy lines on fore wings. It lays 30 – 50 yellowish green eggs singly near the leaf viens. The incubation period is 4 days. The pale greenish caterpillar with brown head pupates in soil or in ab leaf. The adult emerges in 11-14 days.

Management

1. Cultural control: Mango orchards should be inspected at least once a month for webbed leaves and shoots. Infested clusters should be systemically pruned and destroyed along with the larvae of *Orthaga* spp. Webbed, infested leaves which have fallen to the ground under infested trees should also be collected and destroyed as they harbour mature larvae or pupae. Soil under trees should be ploughed to expose hibernating larvae during winter. Raking the soil around the base of the trees in January and spraying insecticides like chloropyriphos at 2ml or acephate at 1.5gm or monocrotophos at 2 ml + 1ml diclorovos per litre of water immediately after disturbance of webs have shown to control the pest.

Biological Control

The release of egg parasitoids such as *Trichogramma chilonis* or *T. pretiosum* may be effective. Beauveria bassiana should be sprayed two or three times during June-July (the period of high humidity) to achieve good control. Parasites like *Brachyameria eascus* are found to effectively control the pest.

Chapter 21

Pests of Citrus

1. Lemon Butterfly, *Papilio demoleus* (Papilionidae: Lepidoptera)

Symptoms

The caterpillars feed on the leaves and especially young seedlings and trees are seriously affected. Complete defoliation occurs in severe attack. It is present throughout the year.

Biology

The butterfly lays yellow, spherical eggs, scattered singly on the tender foliage. They hatch in 4 or 5 days. The larva is dark brown with irregular whitish patches, resembling the excreta of birds. After about 10 days it turns green with white and pink lines. On disturbance the caterpillar everts out two orange coloured osmeteria with a characteristic smell. Full grown caterpillar is stout and about 40 mm in length. Larval period is 15 – 25 days. Pupate on the plant itself as green chrysalis attached to the leaf or twig by a silken girdle. Pupal period is 10 days.

Management

1. Hand pick the larvae and destroy.
2. First instar–Spraying of 1ml DDVP (Nuvan).
3. Field release of parasitoids *Trichogramme evanescens* and *Telenomus* sp. on eggs.
4. *Conserve Brachymeria sp* on larvae and *Pterolus* sp. on pupae.

2. Leaf Miner, *Phyllocnistis citrella* St. (Gracillaridae: Lepidoptera)

Symptoms

The minute caterpillars mine into the leaf tissues of tender leaves and feed on them leaving the outer tissue intact. Leaves become crinkled with whitish lines. In case of severe infestation leaves turn pale, curl badly and dry off.

Biology

The tiny moth lays the eggs singly on the undersurface of the leaves near the mid-rib. On hatching, the caterpillar enters tissue and starts mining between the two layers. It is thin and yellowish green in colour. In about one or two weeks apodous caterpillars become full grown and pupate inside the mine. The pupal period is 5 – 7 days. The adult is a tiny moth, grayish in colour with a wing expanse of 5 mm only.

Management

1. Spray dimethoate 0.03 per cent or methyl demeton 0.025 per cent. or imidacloprid 0.01 per cent.
2. Spray NSKE 5 per cent or Neem oil 3 per cent.

2. Fruit Sucking Moths, *Othreis fullonica* and *O. materna* (Noctuidae: Lepidoptera)

Damage

The two species of moths attack the fruits during nights. They pierce the rind and suck the juice through the long proboscis. This puncture causes the fruit to rot. It is only the moths that are destructive to citrus fruits. The moths are distinguished by having particularly well developed proboscis with dentate tips with which they are able to pierce the ripening fruits. The moths are nocturnal in habit and may be seen flying about in orchards after dusk especially during rainy season. The damaged fruits soon start rotting as the punctured regions are easily infected with bacteria and fungi and ultimately the fruits drop prematurely.

Biology

The adults of *O. fullonica* has pale orange brown body with forewings dark grayish and the hind wings orange red with two black curved patches. The adults of *O. materna* has pale greenish gray upper wings with pale white markings and the lower wings with a marginal dark brown region mixed with white spots and a circular dark spot. The moth lays 200 – 300 eggs on a weed, *Tinospora cardifolia*. Egg period is 8–10 days. The caterpillar is a semilooper, dark brown with yellow and red spots. Full grown caterpillars are 50–60 mm long, stout, velvety-blue with yellow patterns on dorsal and lateral sides and having a hump at anal end. Pupation takes place in a transparent pale whitish silken cover enclosed in leaf fold. Larval period is 28–35 days.

Pupal period is 14–18 days. Pupates on the leaves itself.

Management

1. Systematic destruction of weed host plants like *Tinospora cardifolia* and *coccules pendules* on which the caterpillars feed in the vicinity of orchards helps to check the pest population.
2. Bagging of fruits has been suggested.
3. Creating smoke in the orchards after sunset may keep the pest at bay.
4. Spraying oil emulsions once in 10 days to act as a deterrent.
5. Poison baiting (20 g malathion 50 per cent W.P. + 200 g gur or molasses in 2 litres of water) has been found quite affective.
6. Growing tomato as a trap crop in the orchards to attract the moths.

4. Citrus Psyllid *Diaphorina citri*

Identification

Eggs are almond shaped yellow colour eggs, present on the bolds of half opened leaves. Nymphs are flattish, oval in shape and light orange colour. Adult are small, brown in colour and brown colour band present on the half of fore wing.

Symptoms

Both nymphs and adults suck sap from the plants and injection of toxic saliva. Nymphs are more destructive, crowd on the terminal shoots, buds and tender leaves. Excrete honeydew–growth of sooty moulds. Affected plant parts dry and die away. It is transmits the "Greening" virus.

Management

1. Collect and destroy the damaged plant parts
2. Spraying with systemic insecticides at flush growth periods
3. Spray malathion 0.05 per cent or carbaryl 0.1 per cent or methyl parathion 0.05 per cent.
4. Encourage the activities natural enemies such as Syrphids and Chrysopids.

5. Citrus Rust Mite (Mangu Mite) *Phyllocoptruta oleivora* (Acari: eriophyidae)

Identification

The adult body is elongated and wedge-shaped, approximately three times longer than wide. Magnification is required to see CRM as the adult is extremely small, about 0.15 mm long with the immature stages being slightly smaller. Color ranges from light yellow to straw. CRM can be found any time during the year with peak populations usually occurring during June and July. CRM prefer outer canopy fruit exposed to sunlight but avoids the most sun-exposed portion of the fruit. Populations of CRM can develop quickly under ideal conditions with a female laying up to 20 to 30 eggs over a 20-day period. Eggs are spherical, clear, and found along leaf midribs or clustered in fruit depressions.

Damage

Citrus rust mite (CRM) damage epidermal cells of plant leaves, fruit, and green twigs of all citrus varieties using piercing-sucking mouthparts. The rust mite feeds on the outside exposed surface of fruit that is 0.5 inch (1.3 cm) or larger. Feeding destroys rind cells and the surface becomes silvery on lemons, rust brown on mature oranges, or black on green oranges. Rust mite damage is similar to broad mite damage, except that somewhat larger fruit are affected. Most rust mite damage occurs from late spring to late summer.

Management

Mites increase their reproduction on water-stressed trees. Good irrigation reduces red mite outbreaks. Use of acaricides like fenpropathrin @ 2ml/lit, Dicofol 18.5 E @ 2 ml/lit and Ethion 50 EC @ 2 ml/lit.

6. Shoot and Bark Eating Caterpillar, *Indarbella tetraonis* [Metarbelida: Lepidoptera]

Identification

The pale brown moth has wavy grey marks on thw wings. The female oviposits under loose barks of tree. The full grown caterpillar is pale brown with dark head and measures 50-60 mm long. Pupation takes place inside the bore hole.

Damage

The presence of winding galleries of frassy web on the stem and near forks or angles of branches indicate infestation by the pest. The caterpillar remains in a small tunnel at the axils of branches and moves about concealed inside the silken gallery and feed on the bark by scrapping. Only one larva is seen in a gallery and heavy infestation retards the growth of trees and ultimately the yield of fruits.

Management

Removal of frassy galleries found on the trunks and injecting chlorpyriphos 0.1 per cent or profenophos 0.1 per cent into the bore holes and closing them afford effective control of the pest.

7. Citrus Black Fly, *Aleurocanthus woglumi* [Homoptera: Aleyrodidae]

Biology

The female blackfly lays batches of eggs in a spiral pattern on the undersides of leaves. The eggs are golden-brown but darken before hatching which happens in seven to ten days. The nymph moults three times. The first instars is a brown, elongated oval shape, about 0.30 mm long with two transparent filaments curling back over the body. The second instar is a darker brown and has short spines on its body. It measures about 0.40 mm long. The third instar is glossy black with many stout spines and measures about 0.87 mm long by 0.74 mm wide. The larval stages last three to nine weeks. The pupa is ovate, black with short bristles and a marginal fringe of waxy secretion. The adult when it first emerges has a pale yellow head, whitish legs and reddish-brown eyes. Soon afterwards it darkens, developing a fine covering of waxy powder which gives it a slate blue appearance. The wings are angled and held in a tented fashion. The whole life cycle takes from 45 to 130 days, depending largely on the temperature.

Damage

The principal harm done by the citrus blackfly is the sucking of the tree's sap which deprives it of both water and nutrients. The excretion of honeydew coats the leaf surfaces and encourages the growth of sooty mould fungus. This can severely impair both leaf respiration and photosynthesis. The combination of all these factors causes a decline in the health and vigour of the tree and a reduction in fruit yield.

Management

The citrus blackfly has a number of natural enemies. The most effective agents for controlling it in Florida are the parasitic wasps, *Encarsia perplexa* and *Amitus hesperidum*. The former has a lower rate of reproduction than does *A. hesperidum* but is better able to search out suitable hosts. The latter is well synchronized with its host as adult female wasps are ready to lay their eggs at about the same time as suitable larval stages of the blackfly are present. These species have been used in biological control of the pest. For example, both species were introduced into Hawaii in 1999 after discovery of the presence of the citrus blackfly the previous year. Both wasps have succeeded in establishing themselves and are helping to reduce the damage done by the pest.

The use of insecticides *viz.*, Bifenthrin, Acephate, fenpropathrin, imidachloprid ect may help to control infestations temporarily but this is not advised because of the adverse effects on the environment and any existing predators. A sufficiency of water and appropriate applications of fertiliser will encourage growth and minimise damage.

Chapter 22

Pests of Guava, Sapota, Banana and Pomogranate

1. Tea Mosquito Bug, *Helopeltis antonii* S. (Miridae: Hemiptera)

Damage

The nymphs and adults of the tea-mosquito bug *Helopeltis antonii* (Miridae) cause corky scabs formation on fruits. The blisters are formed due to the toxic substance injected by the bugs. Due to the attack the fruits become unsuitable for marketing.

Biology

It is a slender insect 6–8 mm in length with a yellowish brown head and abdomen, a dark red thorax and long dark appendages. The adult lays elongate and sausage shaped eggs that possess two filamentous long processes which remain jutting out from the tender plant tissue in which the eggs are bedded by the female. Incubation period varies between 5–27 days. The newly hatched nymphs resemble spider in general appearance with elongate appendages. They undergo five moults and complete one generation in two weeks in June and eight weeks or more in cold weather.

Management

Periodical spray application of Malathion 0.1 per cent minimizes the damage.

2. Guava Fruit Fly, *Bactrocera correcta* (Bezzi) (Diptera: Tephritidae)

Identification

Bactrocera correcta is a brightly-colored brown and yellow fly approximately 6.0 millimeters (mm) in length. The wings are clear with a light brown band along the leading edge and a spot at the tip. The band along the leading edge has a clear gap before the wing tip. The top of the body of both sexes are entirely yellow and the legs mostly yellow.

Damage

The female flies puncture the soft and tender fruits with their stout and hard ovipositor and lay eggs below the epidermis. On hatching the maggots feed inside on the pulp of fruits and the infested fruits can be identified by the presence of brown resinous juice which oozes out of the punctures made by the flies for oviposition. These punctures also serve as an entry for various bacteria and fungi; as a result, the infested fruits start rotting, get distorted and malformed in shape and fall off from the plants pre-maturely.

Management

1. To check the damage by these flies, fruits should be harvested before they start ripening.
2. All the fallen and infested fruits should be collected and destroyed to prevent the carry over of the pest.
3. The flies when they congregate and rest on the under surface of large leaves of ribbed gourd may be controlled by spray application of cypermethin 0.025 per cent.
4. Frequent raking of the soil under the vines or ploughing the infested fields after the crop is harvested can help in killing the pupae.
5. Use of 0.4 ml methyl eugenol with 1ml of dichlorvos in bait traps was also found effective.

3. Gauva Fruit Borer, *Conogethes punctiferalis (Pyraustidae, Lepidoptera)*

Identification

Larva is pale greenish with pinkish warts dorsally and fine hairs arising on warts of the body and dark head and prothoracic shield. It grows up to 20-24 mm long. Adult is a medium sized yellow moth with small, black spots on the wings.

Damage

The larva of *C. punctiferalis* bores into the fruits and make them unfit for consumption and marketing.

Management

1. Collect and destroy damaged fruits.
2. Clean cultivation as weed plants serve as alternate hosts.

3. Use light trap @ 1/ha to monitor the activity of adults.

4. Insecticides: malathion 50 EC 0.1 per cent two rounds, one at flower formation and next at fruit set.

4. Sapota Leaf Webber *Nephopteryx eugraphella* Rag. (Phycitidae: Lepidoptera)

Damage
The caterpillar webs and feeds on the leaves. It also feeds on flower buds and fruits and 3–4 larvae may occur together occasionally. Leaves of *Mimusops elengi* and cured tobacco are also infested by the larvae.

Biology
The grayish moth lays pale yellow eggs on leaves singly or in groups of two or three. The pinkish larva measuring 25 mm long has close-set of longitudinal lines on the dorsal surface. Pupation takes place in the leaf web itself. The total life-cycle occupies 32–45 days, the egg, larval and pupal periods respectively being 3 -5, 17–32 and 7–11 days.

Management
1. Remove and destruct the infested fruits from the orchard.

2. Collect and remove the dried clusters of leaf web.

3. Spray phosalone 35 EC 2 ml/lit or phosphamidon 40 SL 2 ml/lit or NSKE 5 per cent.

4. Spray application of cypermethrin 0.025 per cent affords protection.

5. Hairy Caterpillar *Metanastria hyrtaca* C. (Lasiocampidae: Lepidoptera)

Damage
The caterpillars feed on leaves voraciously and defoliate the trees.

Biology
The female moth has grayish brown wings and is stout. The male is smaller and has a white spot in the centre of a black patch on forewing. The antenna is pectinate. The moth oviposits on leaves or twigs in rows or groups and lays about 140 eggs. The incubation period is 9–12 days. The long stout grayish hairy caterpillar measuring about 65 mm long has black head and median dorsal brownish band extending to second abdominal segment. The larval stage occupies 45–60 days. Pupation takes place on tree trunks in a cocoon of silk and body hairs. Pupal stage occupies 7–10 days.

Management
1. Field sanitation.

2. Free from weeds and debris.

3. Collect and destroy the egg mass.

4. Burning the groups of larvae found on tree trunks with torches.

5. Spray chlorpyriphos 20EC or phosalone 2 ml/l.

6. Spraying of cypermethrin 0.025 per cent.

7. Dusting carbaryl 10 D on the trunk and branches (around the tree 4 feet).

8. Field release of chalcidid wasp, *Brachymeria* sp.

6. Banana Rhizome Weevil *Cosmopolites sordidus* G. (Curculionidae: Coleoptera)

Damage

The dark weevil oviposits in the root stock or leaf sheath just above the ground level. The grubs and adults bore into the rhizome and cause stunting of rhizome development. If the infestation occurs on a mature rhizome, damage symptoms appear through the reduction in the leaf number, bunch size and the fruit number. Most damage is done by extensive tunneling of the larvae in the corn, thus weakening the plant and causing blow-down by even slight winds.

Biology

Adults lay eggs in between leaf sheaths and stems as well as around the corn, often in an enlarged cell-like compartment in the tissue. Eggs are laid singly and the newly hatched larvae bore into the corm. The egg, larval and pupal stages are completed in 5–7, 15–20 and 6–8 days, respectively. Adults can live over two years without food.

Management

1. Select healthy sucker and plant.

2. Do not take regular crop in the same field to avoid initial infestation.

3. Ensure clean cultivation.

4. Deep ploughing before planting to expose the weevils to sun and predators.

5. Setting traps in the field using length-wise split pseudostem of 50cm length. Adults attracted to it during nights may be collected and destroyed.

6. Removal of pseudo stems below ground level.

7. Trimming the rhizome.

8. Use cosmolure trap at 5/ha.

9. Stem injection with monocrotophos Or Drenching with chlorpyriphos 0.1 per cent emulsion in the soil before planting.

7. Banana Stem Weevil *Odoiporus longicollis* Oliver (Curculionidae: Coleoptera)

Damage

Infestation of the weevil starts in 5 month old plants. Early symptoms of the infestation are the presence of small pinhead sized holes on the stem, fibrous extrusions from bases of leaf petiole and exudation of a gummy substance from the holes on the pseudostem. In advanced stages of infestation, the stem when split open will show extensive tunneling both in the leaf sheath and in the true stem. Rotting occurs and foul odour is emitted due to secondary infection of pathogens. When the true stem and peduncle are tunnelled after flowering, the fruits do not develop properly, become dehydrated with premature ripening of the bunch. Weakening of the stem by larval tunneling often result in breakage by wind. The estimated yield loss due to this pest is between 10 – 90 per cent depending on the growth stage in which the infestation occurs and it is the highest in 5 months old crop.

Biology

The adult weevils are black-coloured and measures 23–39 mm. Red coloured morphs are also encountered. All life stages of the weevil are present throughout the year. Adults are strong fliers and this way they spread quickly from field to field. The pre-oviposition period is 15–30 days. The adults mate throughout the day and night and after a single mating lay an average number of one egg per day for 9 days. Gravid females lay yellowish white, 3.14 x 1.1 mm sized elliptical eggs through ovipositional slits cut by the rostrum on the outer epidermal layer of the leaf sheath of the pseudostem. The incubation period ranges between 3–8 days. The larvae are fleshy, yellowish white and apodous and they pass through 5 instars. Pupation takes place in a fibrous cocoon and the pupate are exarate. The total life cycle from egg to adult stage is competed in 44 days.

Management

1. Field sanitation by removing and destroying the affected plants alongwith rhizome and also the destruction of pseudostem and rhizome of harvested plants is the most important method.
2. Application of carbofuran 3g @ 30g/plant at planting and @ 15g/plant at 60th and 90th day after planting.
3. Spray application of quinalphos 0.05 per cent or chlorpyriphos 0.03 per cent or carbaryl 0.2 per cent at planting. In case of severe infestation spraying may be repeated after 3 weeks.

8. Banana Aphid – *Pentalonia nigronervosa* (Coquerel) (Aphididae: Homoptera)

Damage

Nymphs and adults suck the sap causing deformation of plants. The leaves become curled and shriveled and in case of severe infestation young plants are killed. Feeding also results in honey dew secretion on which the sooty mould grows resulting in decrease of photosynthetic activity and vigour of the plant. It is a vector of the "bunchy top disease" in banana and "Katte disease" in cardamom.

Biology

They reproduce parthenogenetically giving rise to nymphs which complete their life cycle in 9–16 days. The first, second, third and fourth nymphal stages are completed in 2–4, 3–4, 2–4 and 2–4 days, respectively.

Management

1. Application of 25g of phorate 10G or 20g of carbofuran 3G/plant 20 days after planting around the rhizome in the soil.

2. Application of 12.5g phorate 10G or 10g of carbofuran 3G/plant in the leaf axils or 25g phorate 10G or 20g carbofuran 3G/plant in the soil 75 days after planting which may be repeated 165 days after planting.

Pests of Pomegranate

10. Fruit Borer, *Conogethes punctiferalis*

Damage

Caterpillar bores into young fruits and Feeds on internal contents (pulp and seeds). Fruits Dry up and fall off in without ripening.

Identification

Larva is Pale greenish with pink tinge and fine hairs with dark head and prothoracic shield. Adult is Yellowish moth with black spots on the wing and body.

Management

1. Collect and destroy damaged fruits.
2. Clean cultivation as weed plants serve as alternate hosts.
3. Use light trap @ 1/ha to monitor the activity of adults.
4. Insecticides: malathion 50 EC 0.1 per cent or dimethoate 30 EC 0.06 per cent, two rounds, one at flower formation and next at fruit set.

Chapter 23

Pests of Papaya and Grapevine

1. Papaya Mealy Bug, *Paracoccus marginatus*

Identification

Nymphs and adults female are greenish-yellow in colour with yellowish body fluid. Some of the nymphs have a reddish appearance. Females are also without dorsal stripes and dusted with mealy wax not thick enough to hide body colour on the dorsum and without discrete bare areas on the dorsum. Adult females are about 2-3 mm long *i.e.* about the size of the pink hibiscus mealybug. Ovisacs are most commonly produced beneath the body but sometimes behind the body of the female. The body is fringed with many short waxy filaments; the caudal filaments are about one forth of the body length. Various stages of the papaya mealybug can be seen in a colony. Adult males are smaller, have a reddish body and white wings and two caudal (tail) filaments. Papaya mealybug secretes a sticky wax, which may partially cover or fully envelop the entire colony. The mealybug also produces sticky, sweet, straw-coloured honeydew that ants feed upon. The body of *Paracoccus* species often turns black in alcohol and goes dark-brown/black in potassium hydroxide.

Adult Male

The adult male of papaya mealybug differs from the other instars by having a distinct aedeagus with ventral lobes that are broad and cylindrical in dorsal-ventral view, lateral pore clusters, a heavily sclerotized thorax and head, and well-developed wings. Antennae are 10-segmented with bristle shaped and fleshy setae. Eighth abdominal tergite usually without setae.

Second-Instar Female

The second-instar female of papaya mealybug differs from other mealybugs by lacking oral-collar tubular ducts and multilocular pores, and by having 5 setae on the third antennal segment.

Third-Instar Female

The third instar females can be distinguished from all other instars by having 6 or 7 segmented antennae. When 6-segmented with hind tibia divided by hind tarsus the length is usually 1.2 µm, and about 9 setae on the hind tibia. Multilocular pores absent, without vulva.

Adult Female

Field features – body greenish-yellow, dusted with mealy wax not thick enough to hide body colour, without discrete bare areas on dorsum, with many short waxy filaments around the body margin. Ovisacs developed beneath and behind adult female. Slide-mounted characters – adult females can be distinguished from all other instars by having multilocular pores, translucent pores on the hind coxa and a vulva.

Biology

There are at least three female instars and four male instars. Eggs are pale yellowish-green in colour and are laid in ovisacs that are situated either at the tip of or beneath the abdomen.

Damage

1. Leaf damage – Curling, crinkling, rosetting, twisting and general leaf distortion; reduced in leaf size and surface area.
2. Stem and shoot damage–Shoots and young stems may be distorted and malformed; arrested growth at the shoot terminals lead to shortened internodes and rosetting at the shoot tip.
3. Flower damage–Flowers may be distorted and fail to open; where they open, petals may be twisted and/or malformed or show various types of blemishes. Premature flower drop and poor fruit set may occur.
4. Fruit damage–Fruit blemish and sooty mould may reduce the marketability and market value of fruits. Fruits may fail to develop normally and may be unusually small. Such fruits eventually shrivel and drop.

Management

A. Cultural Control

1. Localized quarantine should be employed to avoid moving infested plants or plant material from place to place and outside of the infested area.
2. Wash plants with mild soapy water.
3. Dislodge mealybugs physically by hosing down plants frequently.
4. Use high pressure hose to wash produce clean of mealybugs.

5. Spray with insecticidal soap.

6. Control ants, which feed on the honeydew produced by the mealybugs. Ants also protect mealybugs from natural enemies and transport them from place to place.

7. Prune and burn infested shoots and branches. Use pruning shears or secateurs for softer stems. A sharp machete/cutlass or hand saw should be used to make a clean cut of branches.

8. Remove and burn dry crinkled, older leaves with attached mealybug colonies.

9. Do not use infested plant material as mulch this should be removed from the field and burned Cultural control may be difficult when the mealybug attacks several host plants at the same time.

B. Biological Control

1. **Parasitoids:** Several primary parasitoids have been identified as natural enemies of *P. marginatus*. These included: *Anagyrus loecki* Noyes and Menezes, *Apoanagyrus californicus, Acerophagus* sp. and *Pseudophycus* sp.

   In India, Papaya mealy bug *was* successfully managed by the hymenopteran parasitoid *Acerophagous papaya* that was Imported and introduced from Mexico in 2010-11 by NBAII Bangalore which stands as a latest successful example of Classical Biological Control in India.

2. **Predators:** A number of predators attack *Paracoccus marginatus*. Among them are the larvae and adults of the introduced Australlian ladybird beetle *Cryptolaemus montrouzieri*. indigenous *Scymnus* spp., the Cecidomyiidae *Diadiplosis coccidarum* (Cockerel), chrysopid (lace wing) larvae and adults and Syrphid (hover fly) larvae. Predators such as ladybird beetles and lace wing flies are effective in rapidly reducing high populations of mealybugs.

C. Chemical Control

Apply chemical control judiciously to avoid killing natural enemies *i.e.* parasites and predators.

1. Chemicals give good short-term control but chemical control is difficult and requires repeated application of the insecticide. Long-term chemical treatment is therefore not advised.

2. Spray or drench the roots when necessary with imidacloprid (Attack, Admire) every 5-6 weeks.

3. This insecticide when used as a root drench is compatible with natural enemies.

4. Spray more frequently (only when absolutely necessary) with other insecticides such as Neemex, Carbaryl (sevin), white (mineral) oils.

5. A sticker should always be used with the insecticide when spraying for mealybugs.

2. White Fly in Papaya, *Bemisia tabaci*

Damage

Nymphs and adults suck the sap from undersurface of the leaves resulting in Yellowing of leaves.

Identification

Egg are pear shaped, light yellowish. Nymph are Oval, scale-like, greenish white and settle down on a succulent part of leaves. Adult are White, tiny, scale-like adults.

Management

1. Field sanitation
2. Removal of host plants
3. Installation of yellow sticky traps
4. Spray application of imidacloprid 200SL at 0.01 per cent or triazophos 40EC at 0.06 per cent during heavy infestation.
5. Spray neem oil 3 per cent or NSKE 5 per cent
6. Release of predators *viz.*, Coccinellid predator, *Cryptolaemus montrouzieri*.
7. Release of parasitoids *viz.*, *Encarsia haitierrsis* and *E. guadeloupae*.

3. Fruit Fly in Papaya, *Bactrocera (Dacus) dorsalis*

Damage

Maggots puncture into semi-ripe fruits with decayed spots. Oozing of fluid is observed and brownish rotten patches on fruits. Dropping of fruits are noticed.

Identification

Larva is yellowish apodous maggots and adults are light brown with transparent wing.

Management

1. Collect fallen infested fruits and dispose them by dumping in a pit and covering with soil.
2. Provide summer ploughing to expose the pupa.
3. Monitor the activity of flies with methyl eugenol sex lure traps.
4. Heavy application of dust and sprays of pyrethrum or BHC.
5. Spray fenthion 100 EC 2 ml/lit or malathion 50 EC 2ml/lit.
6. Field release of natural enemies *Opius compensates* and *Spalangia philippines*.

Bait Preparation
1. Prepare bait with methyl eugenol 1 per cent solution mixed with malathion 0.1 per cent.
2. Take 10 ml of this mixture per trap and keep them in 25 different places in one hectare between 6 a.m. and 8 a.m.
3. 250 ml capacity wide mouthed bottle fitted with hanging device at its neck.
4. Change the solution at fortnightly interval from March to July.

5. Grapevine Flea Beetle, *Scelodonta strigicollis,* Chrysomelidae: Coleoptera

Identification
It is the most destructive pest of grapevine all over India. Adult is a shiny flea beetle with a metallic bronze colour and six black patches on the elytra and is 4.5 mm long. The adults are very destructive during Sep–Nov particularly when the vines put forth new flush after pruning.

Damage
The beetles feed on the sprouting buds and eat them completely without allowing them to develop. They feed on mature leaves cutting elongated holes on the leaf laminalike shot holes. The damage results in Complete fed sprouting buds. Shot holes (rectangular cuttings) on mature leaves. Adult beetles hibernate during winter under tree bark and become active from March till November. Adults have characteristic habit of falling down and feigning death when disturbed.

Biology
The females lay eggs about one month after emergence and continue from middle of March to middle of October. Eggs are laid beneath the bark in groups of 20-40. A female lays about 220-569 eggs in 10-14 installments during its life of 8 -12 months. Egg period is 4 days. On hatching small, dirty white grubs drop down to the wa ter basin and burrow into the soil and feed on the cortical layer of roots not causing any appreciable damage. Larval period is 6 -7 weeks. Pupation takes place in an earthen cell and the pupal period is 7-11 days. Total life history takes 52 -54 days.

Management
1. Adult beetles may be collected and killed.
2. Removal of loose bark in rainy season after pruning to expose and eliminate eggs and adults found underneath.
3. First spraying when buds swell in early morning or evening hours to kill beetles and second spray after 10 days with monocrotophos 1.6 ml/l or carbaryl 3.0 g/l or imidachloprid 0.3 ml/l or quinalphos 2ml/l.

6. Grapevine Thrips, *Rhipiphorothrips cruentatus*, Thripidae: Thysanoptera

Biology

Most destructive pest of grapevine India. It also feeds on rose, jasmine, cashew and other fruit trees. Adults are minute, pale, blackish brown, found on the underside of leaves. Reproduction is either with or without fertilization. Fertilized eggs give rise to female and unfertilized ones to male. Adults appear in March and lay eggs on the underside of leaves by making slits in leaf tissue, placing one egg in each slit. About 50 eggs are laid by each female. The egg is dirty white and bean shaped. Eggs hatch in 3 -8 days, Young nymphs on hatching feed on the undersurface of leaves. Nymphal period is 9 -20 days Pupation on leaves and pupae possess locomotion and crawls when disturbed. Pupal period is 2 -5 days.

Damage

Both the nymphs and adults lacerate tender foliage and suck the oozing sap. The attacked leaves appear silvery initially and later turn brown and give withered appearance, curl up and drop off the plants. Severely affected vines do not bear fruits. If fruits are attacked, they develop corky layer on the fruits and turn brown. Infestation results in Silvery patches on the affected leaves. Brown corky patches on fruits (scab).

Biology

1. Removal of weeds in and around garden.
2. Cutting of infested branches and burning.
3. Spraying dimethoate2ml/l or methyl demeton 2.0 ml/l or thiamethoxam 0.25 g/l.

7. Grapevine Mealy Bug, *Maconellicoccus hirsutus*, Pseudococcidae: Hemiptera

Damage

It is a serious pest on grapevine varieties having compact fruit bunches like Thompson seedless. Anab – e – shahi with loose bunches is less infested. Clusters of mealy bugs with white mealy mass suck the sap from fruits making berries or fruits unfit for consumption. They also feed on stems and foliage resulting in Sooty mould development that affects photosynthesis and final yield. Malformation of growi ng shoots and leaves and sooty mould are the common.

Management

1. Clearing mealy bug clusters on stem using gunny cloth.
2. Releasing 8-10 *Cryptolaemus montrouzieri* (Australian lady bird beetle)/each tree.
3. Removal of loose bark and paste mixture of carbaryl 6 g + Copper oxy chloride 10 g + neem oil 1ml+ gum 1ml on the stem and branches.
4. Spraying of dichlorvos 2.0 ml/l or methomyl 1.0 g/l have been found effective.

5. Applying sticky bands like greeze or sticky tapes around stem, stalks of branches to prevent crawlers from reaching young shoots.

6. Dipping grape bunches in a solution of DDVP 1.5ml + soap 2.5g +water 1 litre for 30 seconds.

8. Grapevine Stem Girdler, *Sthenias grisator* [Cerambycidae: Coleoptera]

Damage

Besides grapevine, this insect also infests apple, citrus, mango. It is a medium sized, stout beetle which girdles (ringing) the vine as a pre – ovipositional operation resulting in drying up of regions beyond the cut. Eggs are inserted under the bark in cuts made by the beetle on the girdled vines. 1-4 eggs are laid at one place. Egg period is 8 days. The grub tunnels in to the wood and completes its life cycle within the stem. Pupation takes place within the tunnel.

Management

Cutting attached branches below girdling point and burning, applying dichlorvos in the holes or placing half a tablet of aluminium phosphide in to the hole and closing it with mud are recommended.

9. Grapevine Stem Borer, *Coelosterna scabrator,* Cerambycidae: Coleoptera

Damage

The grubs *C. scrabator* bores in to stem and branches and causes drying and withering of affected branches. Initially reddish sap oozes from wounds, chewed particles of wood are seen on the ground just below the site of damage.

Management

Removal of loose bark in pre monsoon period, later painting bark with Chlorpyriphos is recommended. Applying dichlorvos in the holes and closing it with mud or placing half tablet of aluminium phosphide in to the hole and closing it.

Chapter 24

Pests of Ber, Custard Apple and Tamarind

1. Ber Fruit Fly, *Carpomyia vesuviana,* Tephritidae: Diptera

Identification

Adult is a small black spotted fly with banded wings.

Biology

Two to three generations are completed from November to April. Eggs are laid in cavities made on the fruit with the ovipositor. Up to 22 eggs are laid by a female either singly or in groups of 2-4. Incubation period is 2-3 days. As many as 18 maggots may infest a single fruit. Maggots are white tapering anteriorly. Larval period is 7-10 days. The full grown maggot falls to ground to pupate in soil 5 to 7.5 cm deep for periods varying from 14 to 300 days depending upon the climate.

Damage

Oviposition punctures made by the flies on the fruits give them rough appearance. The punctures appear as black spots in depression later on. Maggots bore into the pulp forming reddish brown galleries. The damage fruits rots that turns dark brown and smells offensively.

Management

    1. Harvest the fruits before ripening.

    2. To check the carry over pest, collect and destroy all fallen and attacked fruits.

3. Plough around the trees during winter to expose and kill the pupae.

4. The adult flies may be trapped and killed by poison-baiting or bait-spray (20 gm Malathion, 50 per cent wettable powder in 2 liters of water for baiting and 20 litres of water for spraying.

5. Spraying with 0.3 per cent oxydemeton methyl or 0.03 per cent phosphamidon or 0.06 per cent dimethoate or 0.2 per cent carbaryl.

2. Ber Fruit Borer, *Meridarchis scyrodes,* Carposinidae: Lepidoptera

Damage

It is distributed all over the country. Adult is a small dark brown moth. Eggs are laid on young fruits. Larvae are reddish. The larva bores into the fruit feeding on the pulp and accumulating faecal frass within. Up to 40 per cent of the fruits are damaged during July and August. Pupation takes place in the soil.

Management

1. Collection and destruction of affected fruits.

2. Raking up of soil in tree basins in summer.

3. Spraying 2 – 3 times at 10 day interval from pea sized fruit stage with malathion @2ml/lit or polytrin 1ml/l.

3. Ber Fruit Weevil, *Aubeus himalayanus,* Curculionidae: Coleoptera

Damage

Grub feeds on the seed and adult feeds on the fruit. The fruit loses its shape. Fruits become round and fruit stalk bulges. The adult also sometimes feed on the seed and comes out of the fruit.

Management

Spray monocrotophos 1.6 ml/l or deltamethrin 2 ml/l at 15 day interval from maturity of fruit till harvest.

PESTS OF CUSTARD APPLE

1. Mealybug, *Ferrisia virgata,* Pseudococcidae: Hemiptera

Identification

Female bug is apterous with two long prominent waxy filaments at the posterior end and a number of waxy hairs over the body covered with waxy powder. In the posterior end of the body, the dorsum has a prominent blackish patch. It has the habit of encircling itself by secreting thin glassy threads of wax specially when its population is less.

Biology

Reproduction takes place both sexually and parthenogenitically, the latter being more common. Mating takes place only once and lasts for about 12-23 minutes. The female lays the eggs in groups which lie under its body. Fecundity ranges from 109 to 185 during aoviposition period of 20-29 days. Incubation period is about 3-4 hours. Male and female nymphs moult 3–4 times respectively and their development period varies from 26 to 47 and 31 to 57 days. Longevity of female is 36-53 days and that of male is only 1-3 days.

Damage

Nymphs and adults remain clustering upon the terminal shoots, leaves and fruits and suck the sap which results in Yellowing, withering and drying of plants or shedding of fruits *etc.* Formation of sooty mould due to honey dew excretion. In dry weather they may move down below ground and inhabit the roots.

Management

1. Pruning and destruction of the infested twigs.
2. The branches that are touching the ground to be cut and destroyed
3. Periodical raking of basins and application of balanced dose of fertilizers especially.
4. Arranging the polythene sheet around the stem.
5. Predators *Chrysoperla carnea, Cryptolaemus montrouzieri, Pullus* sp. suppress the mealy bug.
6. Spray dichlorvos 1.0 ml/l or acephate 1.5 g/l. Imidacloprid 17.8 per cent SL @3.0 ml/10 lit, Dimethoate 30 per cent EC @1.0 ml/lit., Spinosad 45 per cent SC @3.2 ml/10 lit.

PESTS OF TAMARIND

1. Tamarind Fruit Borer: *Phycita orthoclina* [Phycitidae: Lepidoptera]

Identification

Moths are small, having elongate forewings. The hind wings are broad, bear hairs on dorsal side. Full gron larva is cylindrical, pink and measures 14 mm.

Biology

The female lays about 200 eggs on pods, cracks and crevices. The larva emerges in 4-5 days and enter into fruit pulp and feeds inside making silken web. Larval period is 27–40 DAYS. The full grown larva makes a silken cocoon inside the infested pod and pupates there. The moth emerges in 6-8 days.

Damage

The larva feeds on the pulp and their castings, excrements, webbings render fruits unfit for consumption.

Management

Spray dichlorvos 1.0 ml/l or acephate 1.5 g/l.

Chapter 25

Pests of Apple, Peach, Pear, and Plum

1. San Jose Scale *Quadraspidiotus perniciosus* (Comstock) (= *Aspidiotus perniciosus* Comstock) (Diaspididae: Homoptera)

Damage

These tiny insects suck the sap; as a result, the young plants in the nursery become weak and ultimately die away. The leaves, twigs, fruits and sometimes even the entire bark may be seen covered with ashy-grey scales which can be easily scraped off exposing the orange coloured individuals beneath. The affected fruits present pink coloured areas around the scales and the market value of such fruits is reduced.

Biology

The nymphs hibernate from December to March, resume their activity around end of March and mature in about 4 weeks. The second stage lasts 10–12 days. The males (winged) fertilize the females (wingless) and die. The females do not lay eggs but produce young ones; 300–400 per female, the eggs mature into tiny nymphs in the female ovisac in about a month, during April–May. Nymphal period varies between 40–50 days. The number of generations in a year depends mainly on elevation and climatic conditions. There are six to seven overlapping generations in a year. The 1st instar crawlers are the main dispersal phase and are carried for a few kilometers by the wind.

Management

1. Select nursery stock free scale infestation.

2. Fumigate nursery stock with HCN gas or methyl bromide.

3. Spray phosalone 50 EC 0.05 per cent or fenitrothion 50 EC 0.05 per cent.

4. Winter spray with diesel oil emulsion at 8-12 l/tree (diesel oil 4.5 l, soap 1 kg, water 54 -72 l).

5. Encourage the activity of parasitoids: *Encarsia perniciosi* and *Aspidiotophagus* sp.

6. Field release of coccinellid, *Chilocorus circumdatus* predator.

2. Wooly Aphid *Eriosoma lanigerum* (Hausman) (Aphididae: Homoptera)

Damage

It attacks primarily the underground roots but winged form also attacks trunk, branches, stems, twigs, leaf petioles and fruit stalks. Upward and downward migrations are accentuated during hottest and coldest seasons respectively. Maximum migration from roots to aerial parts takes place in May and in the opposite direction during December–January. Due to the desapping caused by this pest, the affected trees present a sickly appearance, lose vigour and the growth of these trees as also their fruiting capacity are adversely affected. In case of young tees, the roots disintegrate to such an extent that these trees are easily blown over by even moderately strong winds.

Biology

The pest overwinters either as egg or young nymph on the roots of the host tree. The eggs hatch and the nymphs mature during spring. The reproduction during this period is parthenogenetical as well as viviparous. A single female produces 30–116 young ones in her life time. New nymphs soon settle down in batches and start sucking the plant sap. Within 24 hours, these nymphs begin to secrete wooly filaments of wax over their bodies – hence the name, wooly aphid. Nymphal period is about 11 days in June which gradually increases with fall in temperature and becomes 93 days by December. During summer and early monsoon months there is rapid multiplication both on stems and roots and considerable dispersal of pest. The winged adults fly away while the wingless forms are blown off by the wind. With the advent of winter, the sexual forms appear, mate and lay eggs; while the immature nymphs on the trees descend and enter the root zone for hibernation.

Management

1. Use resistant root stocks M 778, M 779, MM 14, MM 110, MM 112.

2. Soil application (80–100 mm deep) of dimethoate or thiometon granules @ 15g/tree during spring and summer against the root forms.

3. Foliar spraying with 0.03 per cent dimethoate or phosphamidon or oxydemeton methyl during March – April (spring) and again in June.

4. The aphid population can also be effectively checked by an exotic parasitoid, *Aphelinus mali* Hald.

5. Predators: *Chilomenus bijugus* and *Coccinella septumpunctata.*

3. Apple Codling Moth *Cydia pomonella* (Tortricidae: Lepidoptera)

Damage

Eggs are laid singly on leaves, blossoms and fruits. The freshly hatched caterpillars feed on leaves for a while, then burrow inside the fruits and feed on the pulp. The entry holes become quite conspicuous as these are filled with dry brown frass and are surrounded by a dark reddish ring. The infested apples become brighter in colour than those that are not infested and also ripe prematurely. The fruits that are attacked early in the season often drop down before the crop is ready for harvest.

Biology

Eggs are flattened and white in colour. Full grown caterpillars are 16–22 mm long and pinkish in colour. Moths are greenish to dark brown with chocolate-brown or copper coloured circular markings near the tip of forewings. The colour pattern resembles bark of the tree trunk which makes the moths quite inconspicuous. Wing expanse is 18–24 mm. Egg, larval, and pupal periods are 4–12, 28–35 and 8–14 days respectively. The caterpillars of third brood over-winter by forming thick silken cocoons in which they pass the winter under loose scales of the bark of the host trees. When spring comes, the larvae become pupae inside these cocoons and the moths emerge from the cocoons during March – April.

Management

1. Field sanitation.

2. Collect and destroy the infested fruits and cocoons.

3. Banding–corrugated cardboard bands to tree trunks.

4. Use sex pheromone trap.

5. Mass trap males with codling moth lure traps.

6. Release egg parasitoids, *Trichogramma embryophagum* @2000/tree

7. Strict domestic quarantine is to be followed by screening of consignments of fruits to prevent the spread of the insect from Ladak to other apple growing regions.

8. Application of 0.2 per cent Pyrethrum extract is also helpful in checking the pest infestation. The protective treatment may be applied about ten days before ripening of the fruits.

9. Apply diazinon, acetempride (4 application per season) during egg laying stage. Spray DDVP 0.04 per cent.

10. Apply Virosoft CP4 Granulovirus.

5. Peach Leaf Curl Aphid *Brachycaudus helichrysi* (Aphididae: Homoptera)

Damage

Nymphs and adults suck the cell sap from leaves, petioles, blossoms and fruits. Affected leaves turn pale and curl up; blossoms wither and fruits do not develop into normal size and drop prematurely.

Biology

The alternate host in cooler region is *Golden rod, Erigeron canadensis* Linnaeus while in plains the aphid breeds on a weed, *Ageratum conyzoides* Linnaeus. Eggs are cylindrical, 0.6 mm long, light green in colour when freshly laid, later turning shiny black. Nymphs are dark green in colour and the adults that feed on leaves are green while those that feed on bark are chocolate coloured. Reproduction is sexual as well as parthenogenetic. Sexual forms appear early in November in cooler regions and lay eggs which hatch in March. At lower altitudes there is no egg-laying and over-wintering is in adult stage. With the rise in temperature, there is rapid multiplication (parthenogenetically). A single female gives birth to about 50 young ones in her life time of two weeks and each of these takes about 10 days to mature during March – April and start reproducing. All these young ones are apterous, viviparous females. After producing 3–4 asexual generations the aphids migrate to pass summer on its alternate host. The migration takes place during mid May in plains and around July in cooler regions.

Management

1. Spray with 0.03 per cent dimethoate or oxydemeton methyl or phosphamidon or quinalphos or 0.04 per cent diazinon or dichlorvos just before flowering (pink bud stage) and again after 7–10 days.
2. Another one or two sprayings should be given when the fruit is pea-sized. In higher hills only one pre-bloom spraying is sufficient while in mid and lower hills, in addition to pre-bloom spray, a post-bloom spray (8–10 days after petal fall) is also necessary.

6. Plum Weevil, *Conotrachelus nenuphar*

Damage

A female curculio uses a number of hosts to lay her eggs, including plums, peaches, apples, pears, and other stone fruits. After the female has chosen a suitable host, she will build an egg chamber under the fruit skin to receive the egg. She then turns around and places the egg in the cavity. Next, she slices a curved slit underneath the egg cavity to leave the egg in a flap of flesh. Eggs that do not hatch are killed from the pressure presented by the growth of the host fruit, resulting in crescent-shaped scars visible on the outside of the fruit. Plum curculio beetles can cause irreparable damage to a fruit harvest. In badly damaged fruit, one can identify large scars and bumps due to feeding. Most internally damaged fruit (through burrowing into the fruit) drops prematurely.

Biology

Plum curculio larva are typically 6 to 9 mm long when fully grown. After such, the beetle reaches the pupal stage measuring about 5 to 7 mm, all adult characteristics are visible in this stage prior to transformation. Adult plum curculio are about 4 to 6 mm and have a small, rough, snout colored with black, gray, and brown specks. Four pairs of ridges cover the wings; however, because of the middle humps it only appears to have two ridges.

Management

Pest control of the plum curculio is fairly easy to maintain. Application of proper insecticide during the pink and petal-fall stages of apples, also the petal-fall and shuck-split stages in peaches and cherries is usually enough to reduce plum curculio damage to a minimum. A recommended preventative measure is destroying the fallen, damaged host fruits before the adults emerge. Pest management at petal fall is of particular importance.

7. Cottony Cushion Scale in Apple, *Icerya purchase*

Identification

Female with a cottony ovisac and nymphs are Pinkish crawler, long antenna and group of hairs.

Damage

Nymph and adults suck the sap from leaves and twigs resulting in Yellowing of leaves.

Management

1. Select healthy and pest free rootstock.
2. Collect and destroy the infested plant parts.
3. Spray application of neem oil 2 per cent, NSKE 5 per cent.
4. Spray application of chlorpyriphos 20 EC 0.04 per cent with sticking agent.
5. Field release of some coccinelid predators and *Chilocorus nigritus*.

8. Stem Borer in Apple, *Apriona cinera*

Identification

Grub is Creamy yellow with dark brown, flat head and Adult is Ashy grey beetle with numerous black tubercles at the base of elytra.

Damage

Grub provide circular holes and mass of excreta. Chewed up wood particles protruding out. Bark gnawed and leaves defoliated. Shoots with circuitous galleries and the infested trees stunted.

Management

1. Prune the branches containing grubs before they entered the tree trunk.
2. Inject 10 ml of monocrotophos and plug with wet clay to kill the grub.

9. Peach Borer, *Sphenoptera lafertei*

Damage

Beetle feed on leaves and Young grubs bore into and feed inside bark and make irregular galleries.Gum globules seen at the points of entrance on bark.

Management

1. Collect and destroy the damaged shoots and branches.
2. Swab trunk with carbaryl 50 WP at 0.2 per cent.
3. Spray malathion 0.1 per cent.
4. Field release of some natural enemies like gray field ant, *Formica aerata,* chalcid wasps, *Copidosoma (=Paralitomastix) varicornis* and *Hyperteles lividus.*
5. Spray application of *Bacillus thuringiensis* (BT) products have been effective if applied when larvae are first noticed and before they tunnel into twigs, buds or fruit.
6. Spray with Imidacloprid 17.8 per cent SL @3.0 ml/10 lit, Emamectin benzoate 5 per cent SG@ 4 g/10 lit, Spinosad 45 per cent SC @3.2 ml/10 lit.

10. Pear Stem Borer, *Sahydrassus (=Phassus) malabaricus*

Identification

Larva is Stout caterpillar and Adult is Big, brownish white moth.

Damage

Bore hole at the base of the stem with circular mat covering. Wilting of tree.

Management

1. Remove and destroy damaged branches and tree.
2. Use light trap 1/ha to attract and kill the adults.
3. Iron hooking – locate the bore hole and kill the grub.
4. Inject or pour 10 ml of monocrotophos + 10 ml of water in bore hole plug with mud.
5. Drench the stem with Chlorpyriphos 20 EC @ 2.5 l in 250 l of water.

11.Green Peach Aphid, *Myzus persicae*

Identification

Adults are pale yellowish green or pink with three dark lines on abdomen.

Damage

Nymph and adults suck the sap from leaves, new shoots, fruits and flower buds. Infested leaves become pitted and curled. Young fruits are shriveled and drop pre maturely. Fruit setting affected in severe infestation. Its transmits plum pox virus.

Management

1. Remove and destroy the damaged plant parts along with nymphs and adults.

2. Encourage parasitoid, *Aphelinus mali* and predators, *Coccinella septumpunctata* and *Bacillus eucharis.*

3. Spray dimethoate 0.03 per cent or methyl demeton 0.025 per cent.

12. Peach Twig Borer, *Anarsia lineatella*

Identification

Matured larvae is reddish brown in colour and has a black colour head. Adult is brown in colour and black colour line present on the both forewings and hind wings.

Damage

Caterpillar damages twigs of peach, plum and apricot. Eggs are laid on shoots, twigs and fruits. After egg hatching the larvae bores into pith of twigs and fruits. Fruit damage is more serious.

Management

1. Collect and destroy the damaged shoots and branches.

2. Swab trunk with carbaryl 50 WP at 0.2 per cent.

3. Spray malathion 0.1 per cent.

13. Bark Borer *Indarbella tetraonis*

Damage

Make tunnels in the main trunk and branches. Larvae construct loose irregular webbing of silken threads. Deterioration of vitality. Reduction in yield.

Management

1. Keep orchard clean.

2. Collect loose and damaged bark and destroy.

3. Kill larvae by inserting iron spike or wire into hole.

4. Spot application of 10 ml monocrotophos or fenthion in 1 lit of water.

14. Mealy Bug, *Ferrisia virgata*

Damage

Covers tender growing points with white mass. Suck the sap. Vitality reduced.

Management

1. Early detection of mealy bugs–presence of ants – indicator.

2. Cutting of infested twigs and leaves and burying them.

3. Field release of green lace wing *Chrysoperla carnea,* Several species of ladybird beetles such as *Chilocorus* sp. *Cryptolaemus montrouzieri* are efficient predator.

4. Spray with 0.03 per cent dimethoate or oxydemeton methyl or phosphamidon or quinalphos or 0.04 per cent diazinon or dichlorvos.

Chapter 26

Pests of Coconut

1. Rhinoceros Beetle, *Oryctes rhinoceros* L.

Identification

Adult is dark brown to black in colour, shiny with 35-50 mm long and 14-21 mm wide. Male beetle is with prominent horn on the head and with bare pygidium. Female beetle is with small horn and with dense reddish brown hair on pygidium. Egg is white to dirty white or white brown, tough, 3-4 mm long and 2-3 mm wide. It is fairly hard-shelled, and when dropped on the ground bounces like a table tennis ball. Larva is yellowish white to ash grey in colour with brown hair (this color is due to the color of the food in side the abdomen), head and legs are brown. Hind part of the abdomen is greyish blue, body is curved, 6-7 mm long on emergence and 75-125 mm long and 20-30 mm wide when fully grown. The full-grown larva is stout, fleshy, voraciously feeding at all the times. The body is long and somewhat cylindrical, but strongly arched convexly above and concavely beneath, so that the head may touch the caudal end of the body forming a ring. The larva passes through 3 instars during its developmental period. Before pupation and formation of the cocoon, the larva assumes a dull cream color, stops feeding and becomes inactive. Before pupating, the larva makes cocoons in the soil or in the trunks of palms. The cocoons are oval, internally smooth, brittle and hard in the soil. Pupa is uniformly brown, convex dorsally, 40-70 mm long and 20-25 mm wide. A conspicuous horn in male pupa is present.

Biology

Life cycle varied from a minimum of 3 to 9 months and a maximum of 11 – 14 months. Life cycle occupied on an average 171 (101-260) days, which comprised a mean egg period of 10 (8-14), larval 130 (74-191) in three instars and pupal 20 (14 to 29) days. The adult beetles emerging from the breeding sites, rest for 5 to 26 days and then fly to the palm crown for feeding. Fecundity per female was 108 (range 48-152) and longevity of the female is 142 (66-240) days. Male lived for about 120 days. Peak oviposition period occurred in February-April and September-October. Eggs are laid singly at a depth of 20 cm in the substratum. The average incubation period was 11.4 days. The three larval instars on an average took 13.5, 22.2 and 124.5 days, respectively. The mean pupal period was 15.4 days. The average longevity of adult was 55.5 days for male and 64.7 days in case of female. The mean total life cycle recorded was 187.7 days.

Damage

The adult beetle damages the unopened leaves and spathes. The beetle causes up to 10 per cent loss in yield. Generally palm aged 3 to 10 years are more prone to infestation by the beetle. More over, injured portion may attract pests such as red palm weevil and also pave way for fungal infection (bud rot). Repeated infestation to the growing points may eventually lead to the death of seedlings. The affected frond, when fully opened, shows characteristic geometric cuts. Holes are present on the unopened spindle leaves, unopened spathes show round to oblong holes. Usually one or two leaves show the cut marks but under severe conditions all the leaves and tender portion of the crown shows injuries by the beetle; chewed fibrous material is present in the crown or near the bored hole.

Managment

a. Sanitational

Plant and field sanitation is one of the important methods to ward off the beetle population. Remove and destroy all possible decaying debris and dead palms from the plantation since these act as prolific breeding grounds for the beetle.

b. Mechanical

The adult beetles may be extracted using curved beetle hooks from the palm crown particularly during the peak period of population build up. The holes made by the beetles are to be filled with a mixture of neem seed kernel powder 100g + 150g sand.

c. Prophylactic Leaf Axil Filling

To protect the young palms from rhinoceros beetle attack, the innermost 2-3 leaf axils may be filled with a mixture of Neem seed kernel powder 100g + fine sand (150g) per palm during May, September and December.

d. Biological

A. Pathogens: Baculovirus of *Oryctes* and the green muscardine fungus cause diseases to the immature and adult stages of the beetle.

The viral pathogen Baculovirus of Oryctes: The adult beetles are inoculated either by feeding the viral inoculam or by allowing the insect to crawl over the viral suspension. Release of such baculovirus inoculated beetles has to be done @ 10-15 beetles/ha of the plantation periodically.

The entomo pathogen Metarrhizium anisopliae (Metch.) Sorokin: This fungus could be mass cultured in rhinoceros grubs or broken maize grain or coconut water or on cassava chips and rice bran supplemented with a nitrogen source. Periodical spraying of fungus on manure heaps is to be done during monsoon season. The fungus can be inoculated @ 5 X 10^{11} spores/m^3 of the breeding material.

B. Predators: Insect predators are frequently observed in the breeding materials of the beetle. *Santalus Parallelusm, Pheropsophus occipitalis, Harpalus* sp., *Scarites* sp., and *Agrypnus* sp. are some of the dominant predators. The immature and adult stages of exotic predator, *Platymeris laevicollis* consume eggs and early instar larvae of the rhinoceros beetle. Ectoparasitic mites were observed in association with different stages of the beetle. Some species particularly Laelaphid mites feed and destroy the eggs. Rhabditid nematodes were isolated from the beetle. The nematode-cum-bacterium culture (DD-136) killed the immature stages of the beetle. The nematodes survived in the field for one month, and caused 60-100 per cent mortality of the grubs.

e. Periodical Spraying of Insecticide

Carbaryl 50 per cent w. p @ 3gms/1 lt. of water on the farmyard manure heaps may be done for the control of immature stages of the pest.

f. Pheromone Traps

Studies with *Oryctes rhinoceros* have helped to identify an active compound *i.e.* aggregation male pheromone (ethyl-4-methyl-octanoate) and opens up new prospects for controlling the pest effectively. Rhinolures used for trapping rhinoceros beetles were effective up to 60 days and establishment of these traps.

2. Red Palm Weevil, *Rhynchophorus ferrugineus* Oliver

Identification

Eggs are creamy white, long, oval and slightly broader at one end. It measures 2.62 mm in length and 1.12 mm in width. Eggs are translucent, smooth and shining. Newly hatched grub is yellowish white with pale brown mouth parts. Full grown grub is stout, fleshy, apodous, having a conical body bulged in the middle and tapering at the ends. Head is pointed downwards, brownish with strongly chitinised mouth parts. The prothorax is large with a dark color than the rest of the body. Rest of the body segments are nonchitinised with a wrinkled skin covering. The full grown grub has on an average 50 mm length and 20 mm in width in the middle. Pupa is cream colored in the early days but latter on turns brown. It measures on an average 35 mm long and 15 mm width. Adult weevil is a large sized ferrugineous brown weevil having an average length of 35 mm and width of 12 mm. Mouth parts are elongated to a slender snout or rostrum bearing a pair of small jaws at the end and pair of antenna at the base. In the males, a tuft of hairs are seen on the dorsal side of the snout (Plate-13). Though the weevils are active both during day and night, generally, the flight and crawling is during day time. The adults are reported to be capable of flying up to three km during night. Adults are believed to feed on plant sap occasionally. Mating takes place at any time during day and many matings take place during the life of adult weevils. Oviposition is confined to soft portions of the palm. The weevil prefers young palms of more succulent nature for oviposition.

Biology

The period during the decline of southwest monsoon (September-November) is favourable for the activity of the pest. Where as, the red palm weevil was least active during June-July when rainfall is high. The female weevil commence oviposition 1 to 7 days after mating. The eggs are laid singly in small hole which are scooped out by the mother weevil in the soft portions of the palm. The preoviposition period lasts for 5 days and the oviposition continues for 25 to 63 days with a post oviposition period of 10 days. A female lays up to 276eggs, the incubation period ranges from 2 to 3 days. The grubs on hatching make tunnels in the soft tissues. Feeding on the growing point by the grubs results in death of the palm, if left unnoticed. Grub period ranges from 36 to 78 days with an average of 55 days. The full grown grub make cocoon winding the fibrous tissues of the palm and pupation takes place inside the cocoons which are elongated and oval. There is a pre pupal stage that lasts for about three days. The pupal stage lasts for about 12 to 33 days. The adults remain in the cocoons for a few days ranging from 4 to 17 days. The females are short lived than the males. The whole life cycle from egg to adult takes about 4 months.

Damage

The damage is caused by the grubs that feed within the stem tissues in varying numbers making tunnels. The symptoms of red palm weevil infestation become very clear in advanced stage by which time the crown of affected palm topples. On close monitoring, it can be seen that the infested palms in early stage show yellowing and latter wilting of leaves of inner and middle whorls. Small circular pencil size holes can be seen on the trunk with a brownish viscous fluid oozing out (Plate-12). The bases of the affected leaves some times split and extrusion of fiber is seen from the cracks/holes. The presence of chewed up fibers/cocoons/weevil, etc. in leaf axil indicates the presence of the pest. Gnawing *and* nibbling sound produced by the grub inside while feeding is audible in many cases. Ultimately the toppling of the crown or falling of palm occurs.

Management

For the management of red palm weevil an integrated approach involving all proven methods of control is quite feasible. This includes:

i. Sanitational and Cultural Methods

The palm crown has to be cleaned periodically to avoid decaying of organic debris in leaf axils. Dead palms which lodge various stages of the weevil should be removed, cut open and burnt so as to destroy all stages of the pest there by preventing spread of the weevil to neighbouring healthy palms. As far as possible avoid making any cuts causing injuries to the palm through agricultural tools and implements as these will attract weevil for egg laying. The cuts or injuries if any, may be treated with coal tar + carbaryl. When fronds are to be removed from the palm, it should be cut leaving a petiole length of 120 cm. This will avoid entry of the pest into the trunk portion. Palms affected by bud rot and leaf rot disease and rhinoceros beetle may be properly treated with respective fungicides and insecticides.

ii. Biocontrol Control

Earwig *Chelisoches moris* F. (Forficulidae) is predator on eggs and early instar grubs of red palm weevil. A highly potent cytoplasmic polyhedrosis virus (CPV) has been detected on various stages of red palm weevil. Presence of microbial pathogens like *Bacillus* sp. and *Serratia* sp. in the nature as probable biocontrol agents of the pest.

iii. Insecticidal Treatment

a. Prophylactic Method

1. Treat wounds with a slurry of mud and insecticide to prevent egg laying by weevil.
2. Fill leaf axils with neem seed kernel powder 100g + fine sand (150g)
3. Treat bud rot infected palms with copper oxy chloride fungicides (Blitox 3 g/litre of water).
4. Fill holes of black beetle with neem seed kernel powder + seed.

b. Curative Treatment

The affected palms in early stages of infestation could be saved by root feeding with monocrotophos as described earlier. Chisel out affected trunk region and burn it and smear the wounded portion with coaltar. After cleaning fill the tunneled portion of trunk with cement and sand mixture to give strength to the palm.

iv. Mechanical

1. Expose different stages of the pest present inside dead trees and burn.
2. Trapping the adults.
3. Use of traps in red palm weevil control is an important tool of Integrated Pest Management. The mass trapping programme of palm weevil using lures helps to capture and destroy a sizeable amount of floating weevils.

v. Log Trapping with Toddy

Fresh coconut logs, 50 cm long, split longitudinally and the cut surfaces smeared with fresh toddy fermented with yeast or acetic acid are effective in attracting the weevils. The traps are set in such a way that the two split halves are placed one above the other with their cut surfaces facing each other. Pieces of fresh coconut petioles smeared with fermented toddy and kept in pots also serve as a weevil trap. Such traps in the garden should be set up in the evening, weevils can be collected and destroyed next day morning. About 10 such traps are to be placed in one hectare area of the garden.

vi. Mud Pot Trapping with Molasses

Mud pot trapping with molasses 2.5 kg/toddy 2.5 lt + acetic acid 5 ml + yeast 5 gm + longitudinally split tender coconut stem/logs of green petioles of leaves, at the rate of 75 numbers in one hectare are effective in trapping adult weevils in large numbers.

vii. Pheromone Trapping

Pheromone lure is hung on palm stems, 1 to 1.5 m above the ground level in the field (1 traps per hectare). Trap bucket is filled with 100 gm pineapple/sugarcane, 2 gm yeast and 2 gm carbaryl in one litre of water as food bait to orient the weevil into the trap.

3. Black Headed Caterpillar, *Opisina arenosella* Walker

Identification

The adult of *Opisina arenosella* is a medium sized moth. Female is 10-15 mm long and 20-25 mm wide (wings expanded), male slightly smaller in size. Head and thorax is light greyish *chreaceous* in color; forewings are elongated with finely scattered blackish scales. Female abdomen is stout and pointed at the tip. Abdomen in male is slender, ending in a short brush or scales and with a conspicuous tuft of hair at the base of the hind wing (Plate-3). The period of activity is more after dusk till dark. The moths are not generally attracted to light. The mating generally takes place during the night, following the day of emergence. The oviposition starts on the day of succeeding pairing and it continues for about two days. Egg is oval, creamy white in color when freshly laid, but gradually turns pinkish before hatching. The Egg measures 0.6 to 0.7 mm long and 0.3 to 0.4 mm wide. Larval body is cylindrical, slightly compressed with a tapering hind end. Head is small, brown or black in colour and curved inwards. Three longitudinal reddish stripes are present dorsally, one median and two laterals are present. Final instar larva measures about 15 mm long. Pre-pupa is inactive, non-feeding, later larva is with silken covering and about 13 mm long. Pupa is light to dark brown in colour, somewhat flattened dorsoventrally with anterior end blunt and posterior end tapering and about 9 mm long.

Biology

Eggs are laid on the lower surface of the leaflet near old larval gallery. Egg period is 5 days, larval period is 42 days in 8 instars including the pre-pupal stage and pupal stage of 12 days. Egg to adult stage period completes in 2 to 2.5 months. Fecundity is 137 eggs per female. Longevity of adult is about 5 days for female and 7 days for male.

Damage

Generally the infestation starts on the outer whorls of leaves and palms of all ages are susceptible to infestation. Cammell *et al.* (1990) found potassium levels were highest in the youngest fronds which are not normally attacked, whereas peak amounts of nitrogen occurred in the fronds which are more susceptible to attack. Due to the attack, the photosynthetic efficiency especially of the lower fronds is impaired therefore the nut development in lower bunches is particularly affected. During sporadic out breaks it feeds on the green surfaces of the petioles, spathes and nuts also. If the pest occurs severely in the nurseries or on newly planted young palms, infestation may lead to the death of seedlings. In the gardens, severe out break of this pest causes button drop resulting in reduced yields. Loss of chlorophyll in the entire foliage ultimately leads to yield losses in subsequent years.

Management

i. Mechanical

Early to mild stages of infestation can be reduced by cutting and burning the badly infested leaves/leaflets.

ii. Biological

Among the parasitoids, the larval parasitoids, *Bracon hebetor, Goniozus nephantidis, the* prepupal parasitoid, *Elasmus nephantidis* Rohw and pupal parasitoid, *Brachymeria nosatoi* are the most promising ones. These parasitoids could be mass multiplied and released in fixed norms in *Opisina arenosella* affected coconut plantations. In addition to above parasitoids other effective parasitoids that are being observed and collected in the infested gardens under east coast conditions are the larval parasitoid, *Apanteles taragamae* and pupal parasitoids, *Xanthopimpla punctata, Trichospilus pupivora, Tetrastichus Israeli* etc., the bacterium *Serratia marcescens* kills the pest during the rainy seasons.

Besides these parasitoids the other natural indeginous mortality factors such as carabid predators *(Parena nigrolineata* and *Calleida splendidula)*, anthocoreid *(Cardiastethus* spp.), chrysopid *(Ankylopteryx octopunctata candida)* and several species of spiders *(Cheiracanthium, Heteropoda laprosa* simon, *Morpisa calcutaenisis* Tikader, *Melanostoma* Thorell, *Olios lamarcki* Latreille, *Rhene* and *Sparassus* spp.) take a heavy toll of different stages of this pest in the filed and exert an appreciable effect on pest population.

iii. Insecticidal

In severely infested gardens spray the palms once with dichlorvos 0.02 per cent. Spray the under surface of the leaves to give a thorough coverage to the galleries of the pest.

Root feeding with Monocrotophos: Expose some roots by excavating the soil at the base of the palm. No need to trace out the tip of the root. Select a mature dark brown root near the base and cut it without damaging the cut end. Insert the cut end of root in a small (15 X 10 cm) polythene cover and place

Monocrotophos 10 ml and 10 ml water. Adjust the cover so that the cut end of the root is completely immersed in the solution. The solution will be absorbed by the palm with in 24 hours (Plate-6). If it is not absorbed by the palm within 24 hours, change the root on the following day of treatment and rearrange, as above with same solution. By this method, the larval stage of the pest black headed caterpillar will be controlled effectively. Before giving the treatment harvest all mature nuts. No harvest should be made (either mature or tender nuts) at least for 45 days after treatment (Ganeswara Rao *et al.,* 1980).

iv. Under Epidemic Populations

1. Removal and burning of one or two badly infested leaves in the outermost whorl.
2. Spray dichlorvos 0.02 per cent initially, if the pest is in its active larval stage.
3. Give root feeding with monocrotophos @ 10 ml + 10 ml water when the pest is in larval stage after harvesting the matured nuts. A safety period of 45 days should be observed.
4. Release of larval parasitoids after safety period and pupal parasitoids after root feeding.
5. Combined with the management techniques, adequate manuring (recommended dose of fertilizers: 1 kg urea, 2 kg single super phosphate and 2 ½ kg muriate of potash/palm) and irrigations are to be resorted for rejuvenating the very severely affected palms.

4. Coconut Eriophyid Mite, *Aceria guerreronis* Keifer

Identification and Biology

The mite is a microscopic, creamy white vermiform organism. It has an anterior cephalothorax and an annulated tapering abdomen with two pairs of legs and needle like mouthparts, which are anteriorly placed. Finely ringed body is covered with bristles arranged in rows. To the naked eye, colonies looks like white powder sprinkled here and there. Adult female mite is 200-250m long and 36-52m wide. The female mite lays 200-250 eggs during its lifetime. The eggs are ovoid, translucent, glassy and measure on an average 35 microns in diameter. The incubation period is 2 – 3 days. The first instar nymph is protonymph, and second instar is deutonymph, which is elongated than protonymph. The mite completes its development cycle from egg to adult within 10 – 12 days. Under favourable conditions these mites multiply enormously. Various stages of the mites are seen in the lobes of the perianth and tender portion of developing nuts.

Damage

When mite infestation is severe drying and shedding of buttons/young nuts may occur. Draining of the sap from young buttons results in poor development of the nut. This leads to reduction in nut size and kernel content. The nuts appear malformed, kernel is not fully developed and at times they are partially developed or even barren. Reduction in nut size leads to almost 25 per cent loss in yield of copra. The husk is also poorly developed. It becomes thickened and hard with loss of fibres, which leads to poor quality fibres. At maturity the husk of infested nut is very tight and shrunken causing difficulty in dehusking.

The mite sucks sap with their needle like mouth parts from the tender regions of nuts covered by perianth. 2 to 4 months nuts are most preferred as they contain soft, young growing tissue underneath the bract (Sujatha *et al.,* 2002). The initial symptom of attack is manifested by appearance of elongated white streaks below the perianth which later appears as pale yellow triangular patch turning gradually to brown colour (Plate-23). As the nut grows this injury leads to warting and longitudinal fissures on the surface. In many cases when the perianth is removed a pinkish band can be seen on its inner side. The discoloration can also be in the shape of stripes. About two months after pollination, the patch develops fully and becomes more conspicuous. During this period, some buttons also show browning and necrosis on the periphery of the perianth. At this stage, the husk develops cracks, cuts, and gummosis.

Management
a. Cultural Methods
Removal of dried and fallen flower parts, nuts etc., and burying in the soil or burning to minimize the pest inoculam.

b. Biological Methods
Natural enemies like predatory mites *(Leptotar sonemus, Bdella indica* and *Amblysieus* sp.) were reported to feed on all stages of mites. The entomopathogenic fungus *Hirsutella thompsonii* is pathogenic to mite. As the pest is under, the perianth, the natural enemies especially the predatory mites fail to enter the colony inside the very tight perianth of younger buttons.

c. Chemical Methods
1. Spraying of Sulphur (80 per cent w.p.) 6g/lt of water on bunches twice at 7 – 10 days interval.
2. Root feeding with monocrotophos @ 10 ml + 10 ml water/palm Fenpyroximate 5 per cent EC 7.5 ml + 7.5 ml urea solution (1 per cent) by observing 45 days safety period.
3. Spraying of profenophos 50 per cent EC @ 5 ml or triazophos 40 per cent EC @ 5 ml or methyl demeton 25 per cent EC @ 4 ml or dicofol 18.5 per cent EC @ 6 ml per litre of water on 2 to 8 months old bunches.
4. Second spraying of sulphur 80 per cent wp @ 6 g/lt of water or Fenpyroximate 5 per cent EC @ 0.75 ml/lt of water on bunches on 10th day of 1st spraying.
5. Spraying of 20 ml neem oil + Extract from 20 g garlic + 5 g soap in 1 litre of water on coconut bunches.

d. Improved Agronomic Practices

1. Organic manuring (application of neem cake @ 10kg/palm/year, FYM, green manure *in situ*, vermy compost etc.,) reduces the mite intensity as well as damage.

2. Intercropping coconut with annuals like banana, yam, colocasia, turmeric, pineapple etc., or perennials like cocoa would decrease the intensity of mite incidence.

3. RDF: urea @ 1 kg/palm, single super phosphate @ 2 kg/palm and muriate of potash @ 2 ½ kg/palm would reduce the incidence of mite.

5. Slug Cterpillar, *Macroplecta nararia* Moore

Damage

The damaging stage of the pest is caterpillar. Early instar caterpillars cause leaf spots due to feeding on the leaf tissue. Grown up caterpillars eat away entire laminar portion of the leaf leaving the mid ribs. Some times, balls of excreta will be seen as a layer on the ground around the coconut palm basin. In severe out break, the pest invades nuts and even leaf stalks. Drying of entire foliage, drooping of leaves and bunches, falling of buttons and nuts are ultimate symptoms of pest attack. The pest was observed causing damage even to intercrops like banana/ cocoa and surrounding hedge plants like agave after drying of coconut crop. In such cases falling of buttons and nuts, drying of total foliage leads to severe yield losses and spathe emergence will be delayed till the palm recovers.

Biology

Adult is a small yellow moth with 6 days longevity. The moth lays tiny scale like eggs singly (60 – 300 nos.) on both sides of the leaf. Eggs will be hatched in 4 – 5 days period. Caterpillar grows up to 8 – 11 mm long. Caterpillar is yellowish green in colour, a series of tubercles are present on the dorsal and lateral sides of the caterpillar. After completion of larval period in 31 days it pupates in the corners of the leaflets or crown region in a small 5.7 mm round brownish shell. Adult emerges in 15 days from the pupa.

Management

1. Installation of light traps with 500 w bulbs for mass trapping and destruction of the adult moths [Sujatha, *et al.*, 2011].

2. Cutting and burning of severely infested and dried leaves will reduce the pest population to a greater extent.

3. Spraying of carbaryl 50 per cent w.p. @ 3 gm/1 lit. of water in the young gardens will check the pest.

4. Root feeding of monocrotophos 36 per cent WSL @ 10 ml + 10 ml water/palm by observing 45 days safety period will control the caterpillar stage of the pest.

5. Immediate application of recommended dose of fertilizers (1 kg Urea, 2 kg Single super phosphate and 2 ½ kg Muriate of potash) and irrigation will boost the palms t o recover early.

6. Checking the pest on non target inter crop also helps to keep the pest under check.

7. Paecilomyces lilacinus as an entomopathogenic fungi on slug caterpillar of coconut [Rao *et al.*, 2012].

6. Coconut Defoliator – *Phalacra* sp. (Drepanidae: Lepidoptera)

Damage

The neonate larvae scrape green material of leaf longitudinally along the veins, giving, a slit like appearance or as a line cut by the blade. Slowly the larvae fed on lamina, leaving the mid rib. A characteristic feature is the caterpillar starts feeding from the tip of the leaf lets on the top portion of leaf and proceeds down wards.

Biology

Adult is a medium sized, ash colored moth. Female moth is larger in size with stout abdomen. After two days of pre-oviposition period, the female laid on an average 92 round and cream colored eggs in clusters in a span of 5 days on all over the leaf surface. Eggs slowly turned to red color before eclosion. Larvae immediately after hatching are in deep red color with hairs over the body and slowly turn to light green color with the advance of age. Larval period ranged from 25 to 28 days. Pupation took place in between leaf fold which is fastened together with strong silken material. Pupal period ranged from 8 to10 days. Life cycle is completed in 40 days [Emmanuel and Rao, 2011].

Management

1. Installation of light traps for mass trapping ofadult moths.

2. Root feeding of monocrotophos @ 10 ml chemical + 10 ml water.

Chapter 27

Pests of Coffee and Tea

1. Coffee White Borer *Xylotrechus quadripes* Ch. (Cerambycidae: Coleoptera)

Damage

The grubs burrow into the stem for 8–9 months and cause wilting of branches and occasionally death of bushes. It is a serious pest of Arabica coffee. Infested plants show external ridges around the stem. Affected plants also show yellowing and willing of leaves.

Biology

The beetle has white cross bands and dark brown elytra. The adults emerge in large numbers at two distinct periods *viz.*, April-May and October-November. A beetle lays about 50–100 eggs in crevices of the bark on stem. The grubs hatch out from the eggs in 8–10 days and pass through the larval stage for 8–9 months. Pupation takes place in the stem itself and pupal stage lasts for 25–30 days.

Management

1. Maintain optimum shade.
2. The wilting branches and bushes should be removed and destroyed.
3. Chlorpyriphos 20 EC may be swabbed over the stem once in April–May and twice at an interval of a month during October – December and for effective control of infestation by the pest or 0.05 per cent monocrotophos or phosalone.

2. Coffee Berry Borer *Hypothenemus hampei* (Ferrari) (Scolytidae: Coleoptera)

Damage

Pin holes at the tip of berries. In severe cases of infestation two or more holes may be seen. Infested berries may fall due to injury or secondary infection. Severe infestation may result in heavy crop loss up to 40–80 per cent.

Biology

Adult females bore a hole in coffee berries and lay their eggs near the two coffee beans found inside the berry. Once the eggs hatch, the larvae feed on the beans rendering them unfit for commerce or lowering their quality. The total life cycle is completed in 25–30 days.

Management

1. Timely and clean harvest. Use mats to prevent gleanings.
2. Remove off-season berries and gleanings.
3. Spot spray 0.07 per cent endosulfan 35 EC when most of the beetles are waiting near the naval region of fruit.
4. Dry coffee to prescribed moisture level (arabica/robusta parchment 10 per cent, arabic cherry 10.5 per cent and robusta cherry 11 per cent).
5. Mass trapping of beetles with coffee fruit extract in 1: 1 combination of ethanol and methanol.

3. Green Scale *Coccus viridis* (Gr.) (Coccidae: Homoptera)

Damage

The scale is 3 mm long, flat, yellowish green ovate and slightly convex and covers the tender leaves and shoots and sucks the sap. Sometimes it becomes serious affecting the vigour of the bushes considerably and sooty mould development is commonly noticed affecting photosynthesis. It causes debilitation of older plants and death of nursery plants.

Management

1. Maintain optimum shade.
2. Spray application of Malathion 0.1 per cent or methyl parathion 0.05 per cent or profenofos 0.05 per cent or phosalone 0.07 per cent.

PESTS OF TEA

1. Tea-mosquito Bug *Helopeltis theivora* Waterhouse (Miridae: Hemiptera)

Damage

The nymphs and adults suck the sap from tender buds, leaves and stem and the toxic saliva of the insect injected at the time of feeding cause brownish patches and curling of leaves and ultimate drying of the shoots.

Biology

It is a slender insect, 6–8 mm long, with a yellowish-brown head and abdomen, a dark–red thorax, and long dark appendages. The prothorax has a prominent and characteristic clubbed horn. The elongate and sausage shaped eggs are also peculiar in possessing two filamentous long processes which remain jutting out from the tender plant-tissue in which the eggs are embedded by the female. The eggs are laid practically in all tender parts of the plant. The incubation period varies within wide limits (5–27 days). The freshly-hatched nymphs are rather spidery in general appearance due to their elongate appendages. They undergo five moults to become adults and the time required for the completion of one generation varies from about two weeks in June to eight weeks or more in the cold weather.

Management

1. Monitoring the infestation level in the field.
2. Removal of stalks containing eggs while plucking.
3. Encouraging the egg parasitoid, *Erythmelus helopeltidis* population to build up.
4. Application of quinalphos 25 EC @ 750 ml/ha or chlorpyrifos 20 EC @ 750 ml/ha or fenthion 80 EC @ 200 ml/ha or quinalphos 25 EC + dichlorvos 76 EC @ 750+250 ml/ha. Spraying may be undertaken during early mornings or evenings when these bugs are active.

2. Looper Caterpillar, *Biston supressaria*

Identification

Larva–dark brown with pale yellowish white lines on the back and sides. Adult–grey wings speckled with light brown or black markings and irregular wavy yellow lines.

Damage

Young caterpillars feed on the tender leaves–making punctures. Mature larvae prefer older leaves.Grown up larvae feed entire leaf. Severe infestation–tea bushes are completely denuded.

Management

Spray NSKE 5 per cent (or) neem based oil formulations 3 per cent. Spray any one of the following insecticides endosulfan (or) phosalone (or) chlorpyriphos (or) fenitrothion (or) malathion @ 2ml/lit.

3. Flush Worm, *Cydia leuocostoma*

Identification

Eggs are pale yellow and lay singly on the under surface of mature leaves. Larva – brown in colour. Adult is very small moth, blackish brown in colour.

Damage

Caterpillar ties up the margin of tender leaves and forms a case enclosing the bud. Feed on the upper epidermis of leaves. Affected leaves–rough, crinkled and leathery. Bud–Shoot growth is arrested.

Management
Spray chlorpyriphos or fenitrothion 2 ml/lit.

4. Tea Tortix, *Homona coffearia*

Identification
Larva is green in colour. Pupae–initially green colour and turns to reddish brown. Adult is brown coloured and bell shaped in outline.

Damage
Caterpillars make leaf nests by webbing the leaves using silken threads. Feed from inside the leaf nest. Young larvae prefer tender leaves and Older larvae are seen in mature leaves.

Management
Spray dimethoate 30 EC (or) chlorpyriphos 20 EC 2 ml/lit.

5. Tea Mite Complex

a. *Red Spider Mite, Oligonychus coffeae*
1. Infest the upper surface of mature leaves.
2. Infestation starts along midrib and veins and spreads to the entire upper surface of leaves.
3. Affected leaves–bronzed, dry and crumpled.
4. Eggs are reddish in colour and spherical in shape.
5. Adult–female is elliptical in shape bright crimson anteriorly and dark puplish brown posteriorly.

b. *Scarlet Mite, Brevipalpus californicus*
1. Mites congregate on the under surface of mature tea leaves.
2. Feeding by scarlet mites leads to brown discolouration of leaves.
3. Eggs–are bright red colour and elliptical in shape.
4. Adult–scarlet red in colour and ovate in shape.

c. *Purple Mite, Calacarus carinatus*
1. Mites feed on the under surface of mature leaves.
2. Assam type of tea–more susceptible to purple mite.
3. Damaged leaves–coppery brown discolouration.
4. Adult–very small, spindle shaped and dark puple in colour
5. Five longitudinal white waxy ridges on the dorsal side.

d. *Pink Mite (or) Orange Mite, Acaphylla theae*
1. Mites are found on the under surface (abaxial) of young leaves.
2. Affected leaves turn pale and upward curling.

3. In severe infestation, leaves become leathery and brown.
4. Damage–restricted to top 10 – 15 cm of tender leaves.
5. Assam type of tea is susceptible.
6. Adult–very minute, orange coloured and carrot shaped.

e. Yellow Mite, *Polyphagotarsonemus latus*

1. mite is a polyphagous pest attacking tomato, pulses, potato, chillies.
2. Mite is seen on young leaves and the bud.
3. Affected leaves become rough and brittle and corky lines.
4. Downward curling.
5. Intermodes get shortened.
6. Shoots–stunted and deformed.
7. Male mites are small and white to pale yellow in colour.
8. Females are yellowish and bigger than the males.
9. Yellow mites are active and fast moving mite.

Management

1. Spray dicofol 18EC 2 ml/lit or ethion 50 EC 2 ml/lit or monocrotophos 1ml/lit or wettable sulphur 80 WP 2g/lit.

6. Thrips, *Scirtothrips bispinosus*

Damage

Adults with brown abdomen. Thrips prefer young leaves and buds. Leaf surface becomes uneven, curled and matty. Feeding marks on the unopened buds–parallel brown lines on the leaves.

Management

Spray dimethoate 30 EC (or) chlorpyriphos 20 EC 2 ml/lit.

7. Aphid, *Toxoptera aurantii*

Damage

Colonies of aphids are seen on tender shoots of tea immidiately after pruning. Leaves curl up and shoot growth is stunted. Honey dew secreted development of sooty moulds.

Management

1. Collect and destroy the infested plant parts, Spray dimethoate 30 EC (or) chlorpyriphos 20 EC 2 ml/lit, (or) phosalone 2 ml/lit.

Chapter 28

Pests of Cocoa, Cashew and Arecanut

A. PESTS OF COCOA

Defoliating Pests

1. Indian Rose Beetle, *Adoretus versutus*

First report from Andhra Pradesh by Emmanuel *et al.,* 2010 b. Adults feed on cocoa plant foliage at night, creating a lace-like or shot with holes appearance on leaves by feeding on plant tissue between leaf veins. In severe cases most leaves are skeletonized. The life cycle (egg–adult) is completed in 6-7 weeks.

2. Black Chaffer Beetle, *Apogonia blanchardi*

First report from Andhra Pradesh by Emmanuel *et al.,* 2010 b. Black leaf chaffer beetle were observed feeding on the cocoa foliage during night times confirming to be nocturnal. Adults feeds on the cocoa leaf from the peripheral region. A adult is deep black species without metallic luster. Length is 9.5 to 11 mm, ovate, glabrous, shining black, the apex of the elytra and the two basal ventral segments, however, opaque; the antennae and palpi pale ferruginous; the under surface and legs sprinkled with pale coloured setae.

3. Bag Worms, *Pteroma plagiophelps* and *Clania* sp.

First report from Andhra Pradesh by Emmanuel *et al.,* 2010 a. The self enclosing bags of *P. plagiophelps* is 1.8 to 2 cm long and 5 to 6mm wide. In the rowing survey, the *P. plagiophelps* larvae with enclosing bags were collected and were reared on cocoa leaves in the laboratory. The caterpillar builds silky bag in narrow cone shape with plant materials live in this mobile case and rest by sticking top opening of the case to the cocoa leaves and hang their bag vertically. Whereas the self enclosing bags of *Clania* sp. is 3.5 to 4.4 cm long and 8 to 9mm wide. This species uses sticks of similar size and there are about 9 – 13 sticks arranged parallel around the silk case except one or two of their sticks used are much longer than the other. The caterpillar lives inside the case and feeds on the cocoa foliage. Both the *P. plagiophelps* and *Clania* sp. feeds the cocoa leaves from the central leaf lamina in a circular to irregular holes.

4. Tussock caterpillar, *Dasychira mendosa*

The larvae of this moth feed on foliage of many plants including cocoa. Therefore, due to this large host range, the breeding continues throughout the year during which there are probably 5 or 6 generations. The moth, which has pale yellow, hind wings and forewings are irregularly patterned with various shades of brown, lays large masses of eggs. The feeding by larvae results in defoliation and the larval period lasts for 21-28 days. The fully-grown hairy larva has a reddish head; the body greyish or yellowish, is striped with red and with long dense dorsal tufts of whitish hairs. They pupate in loose cocoons made of silk and hairs and the pupal period lasts for 11-12 days.

5. Hairy Caterpillar, *Euproctis subnotata*

First report from Andhra Pradesh by Emmanuel *et al.,* 2011 b. Laboratory studies with *Euproctis subnotata*, reared on Cocoa leaves, showed that the average incubation period was 6.7 days, larval period 23.57 days, pupal period 11 days, total life cycle 35.77 days and. Male adult longevity was 6 days, while the female adult longevity was 8 days. Females laid an average of 155 eggs and the larvae underwent 6 instars.

6. Hairy Caterpillar, *Euproctis fraterna*

The mean incubation period was 6.93 days. The larvae passed through 6 instars and completed development in 23.25 days. The pupal stage occupied 11 days. The total life cycle was completed in 41.18 days. Male adult longevity was 6 days, while the female adult longevity was 8 days. Females laid an average of 146 eggs.

7. *Lymantria obfuscata*

First report from Andhra Pradesh by Emmanuel *et al.,* 2011 b. Laboratory studies with *Lymantria obfuscata*, reared on Cocoa leaves, showed that the average incubation period was 8.7 days, larval period 26.57 days, pupal period 11 days, total life cycle 36.77 days and. Male adult longevity was 5 days, while the female adult longevity was 7 days. Females laid an average of 158 eggs and the larvae underwent 6 instars

8. *Euproctis scintillans*

First report from Andhra Pradesh by Emmanuel *et al.,* 2011 b The larvae of this species feed voraciously on leaves. Initially the larvae scrape the green matter resulting in skeletonization. Later the larvae move into the other parts of the plant and defoliate. This hairy caterpillar larval stage lasts for 13-29 days and the pupal period lasts for 9-20 days. The total life cycle lasts for 6-7 weeks.

9. Slug Caterpillar, *Parasa lepida*

First report from Andhra Pradesh by Emmanuel *et al.,* 2011 b Eggs of *Parasa lepida* are laid on leaves which hatch in 3-5 days. Young larvae feed on the leaf epidermis and as they develop, chew up leaves. There are 7 larval instars (often 8 instars for females) which are completed in 35-42 days. Pupation occurs in cocoons often attached to stem or bark and the pupal stage lasts 21-24 days.

10. Looper, *Thalassodes* spp.

First report from Andhra Pradesh by Emmanuel *et al.,* 2011 b. This looper is found feeding on developing leaves. The larval period lasts for 17-18 days. The larva possessing the color of new shoots and assuming a characteristic pose on the twig is often mistaken for a leaf petiole. The pupal period lasts for 7-8 days. The total life cycle was completed in 18-33 days.

11. Management

Amongst the various insecticides tested against the bagworm *[Pteroma plagiophelps],* 100 per cent reduction in the pest population at 1 Day after spraying was caused by Carbaryl 50 per cent WDP, Acephate 75 per cent SP and Quinalphos 25 per cent EC while, 5.88, 60.00 and 61.54 per cent reduction was achieved through the Endosulfan 35 per cent EC, Neem oil Azadirachtin EC 1500 ppm and Profenophos 50 per cent EC. However, to achieve the 100 per cent reduction in the pest population the Neem oil [Azadirachtin EC 1500 ppm] and Endosulfan 35 per cent EC required three and two days after spraying. Similar insecticidal efficacy patterns were observed against the tussock moth caterpillars, *Euproctis fraterna* and the web worm *Acria* sp. [Emmanuel *et al.,* 2011 b].

Fusarium solani (Mart.) was found as naturally occurring pathogen of bagworm Pteroma plagiophelps in cocoa and coconut. [Emmanuel *et al.,* 2011 a].

Sucking Pests

1. Mirid Bugs, *Distantiella theobroma, Sahlbergella singularis* and *Helopeltis* spp.

Feeding by mirids is characterized by dark markings known as lesions, on both pods and shoots, which result from the collapse of plant tissue caused by the toxic saliva. The shape of the lesions is somewhat characteristic of the mirid species. For example, lesions resulting from the feeding of *Helopeltis* spp. are roundish whilst those by *Distantiella* and *Sahlbergella* tend to be elliptical with the long axis parallel with that of the stem. Secondary damage characterized by canker and dieback occurs when the feeding lesions are invaded by parasitic fungi notably *Calonectria rigidiuscula* and *Fusarium decemcellulare*. Mirids can kill only young green shoots and such damage is restricted to periods of flush when this type of tissue is present. Young cocoa is particularly susceptible to mirid attack. Mirids on pods feed largely in the parenchymatous husk tissue. Cherelles may wilt, and pods less than three months old have very little chance of surviving, usually dying from mirid damage or from fungi entering the pods through the lesions. Well-grown pods seldom seem directly affected. A comparison between ripe pods that had been heavily attacked (more than half the surface blackened by feeding) by *S. singularis* in Nigeria and clean pods revealed no significant differences in dimensions, pod weight, number of beans or weight of peeled beans. Mirids do not feed during the heat of the day, but rest at the fork and branch unions on the underside of pod stalks and in other protected situations and are relatively inactive. Feeding does not begin until 5.30 p.m. unless conditions are dull and wet and if the morning is sunless they may feed to 10 a.m.

2. Mealy Bug, *Drosicha mangiferae*

These mealy bugs damage by sucking out plant sap, by excreting honeydew in which sooty mold can grow, and by causing distorted growth with their toxic saliva. They further disfigure plants by secreting cottony wax. Crawlers crawl from an infected to healthy plantMealy bug provide ants with their sugary secretion (honeydew) as food and in return ants help in spreading of mealy bug. *Drosicha mangiferae* Green has 1 generation each year. Eggs hatch in December or January after diapausing in ovisacs in the soil or the duff around the host. First instars move to the leaves and molt 3 times to become adults. Males are indistinguishable from females until the third instar prepupa. The prepupa wanders for a while, forms a waxy test, and molts to the pupal stage. Adults appear in April, mate, and migrate off of the host to the ground where an ovisac is produced and eggs are laid.

3. Thrips, *Selenothrips rubrocinctus*

The female is about 1.20 mm in length and has a dark brown to black body. Females lay up to 50 eggs and live for up to one month. The eggs hatch within four days. The life cycle is completed in about three weeks, and there are several generations a year. Redbanded thrips prefer young foliage and their feeding and causes leaf silvering, leaf distortion, and leaf drop. The thrips destroys the cells on which it feeds, causes some leaf distortion, injury to the fruit, and leaves unsightly dark colored droplets or blotches of excrement on the leaf surface. A more serious injury is leaf drop, which may denude trees. Honeydew excretory products from red-banded thrips and other insect infestations fall to leaves, fruits or objects beneath, giving rise to the objectionable fruit-degrading, black sooty mold.

4. Aphids, *Toxoptera auranti*

This pest congregates on the tender young shoots, flower buds and the undersides of young leaves. They are not known to feed on the older and tougher plant tissues. it causes leaf distortion and malformation of growth of leaves and tips of shoots. It is often more a serious pest in nurseries. Like other soft bodied insects such as leafhoppers, mealybugs and scales, aphids produce honeydew.

This sweet and watery excrement is fed on by bees, wasps, ants and other insects. The honeydew serves as a medium on which a sooty fungus, called sooty mold, grows. Sooty mold blackens the leaf, decreases photosynthesis activity, decreases vigor and causes disfigurement of the host. When the sooty mold occurs on fruit, it often becomes unmarketable or of a lower grade as the fungus is difficult to wash off.

Management

Acephate 75 per cent SP and Cholrpyriphos 20 per cent EC was very effective causing 100 per cent reduction in the pest population at three days after spraying. Whereas, profenophos 50 per cent EC took four days to achieve the same result [Emmanuel *et al.,* 2011 b].

B. PESTS OF CASHEW

1. Cashew Tree Borer, *Plocaederus ferrugineus,* Cerambycidae: Coleoptera

Damage

Adult is a medium sized, dark brown longicorn beetle Eggs are laid under loose bark of the stem and roots. Grub is creamy white, robust and fleshy. Grubs bore into the bark in their early stage and into the wood in their late stages making extensive tunnels within. Both young and old plants are affected. The young plants are killed immediately whereas the older plants gradually become weak and succumb. Infestation results in Yellowing and shedding of leaves and drying of twigs. Chewed up fibre, excreta and gummy secretions seen protruding from the bore holes.

Biology

Larval period is 6-7 months and full grown larva measures up to 7.6 cm. It tunnels downwards and reaches the roots where it pupates in a calcareous pupal chamber. The pupal period is 2 months. Only one life cycle is completed in a year.

Management

1. For trees over seven years age, spraying or swabbing on tree trunks up to 3 feet from ground and exposed roots with 5 per cent neem oil once in 4 months *i.e.* 3 times a year as a prophylactic measure.

2. Removal of dried branches, dead trees and burn them.

3. Removal of grubs, pupae and adults from damaged portion by physically chisteling and destroying them.

4. Chistled trunks and roots treated with carbaryl 4 g/l.

5. Removing the soil around the base of tree up to 1 foot depth and applying 300-500 g of carbaryl dust per tree and covering it with fresh soil.

6. Cleaning the bore holes to insert 1-2 table ts of aluminium phosphide in each hole and plugging the holes.

2. Cashew Shoot and Blossom Webber, *Lamida moncusalis*, Pyralidae: Lepidoptera

Damage

Larva webs inflorescence at the time of flowering and feeds on the floral parts. Apples and nuts are also covered with webs with the caterpillar scraping the upper green layer of tender apples and nuts. It results in cracking of tissue and retardation in nut development. The pest incidence is severe at the time of new flush. The symptoms of damage are Webbed leaves, inflorescence, apples and nuts Full grown larva is reddish brown with yellow lateral longitudinal bands and pinkish dorsal lines and measures up to 26 mm in length.

Biology

The moth lays eggs on the leaves, twigs and inflorescence stalks. Egg period is 4 -7 days. Newly hatched caterpillar is pale white and feeds on the leaves by webbing.

Larval period is 16-22 days. Pupation takes place within the webbed leaves in a silken cocoon. Pupa is dark reddish brown. Pupal period is 9-14 days. Total life history takes 29 -43 days.

Management

1. Three species of *Apanteles* (Braconidae) were observed during January and February parasitizing the larval population.

2. Spraying of carbaryl 3g/l after disturbing the webs.

3.Tea Mosquito Bug, *Helopeltis antonii*, Miridae: Hemiptera

Identification

Adult is a reddish brown bug with black head, red thorax and black and white abdomen; a knobbed process arises mid -dorsally on the thorax.

Biology

Eggs are inserted into epidermis of tender shoots, axis of inflorescence and nuts. The egg is elongated and slightly curved with a pair of filaments on the operculum projecting out. Egg period is 7 days. There are five nymphal instars with duration of 14.9 days. Life cycle is completed in 22.2 days on an average. Peak infestation occurs during summer months and disappears at the onset of monsoon.

Damage

Nymphs and adults feed on petiole, tender shoots and leaf veins causing symptoms like Brownish black necrotic patches on foliage and elongate streaks and patches on shoots Resins exuding from the feeding punctures. Blossom blight and die back symptoms.

Management

1. First spray coinciding with new flush in Oct – Nov with Imidachloprid 0.25ml/l.

2. Second spray during emergence of inflorescence in Dec – Jan with dimethoate 2 ml/l.

4. Cashew Thrips *Rhipiphorothrips cruentatus*–Leaf Thrips, *Rhynchothrips raoensis:* Flower thrips, Thripidae: Thysanoptera

Damage

Leaf thrips: Both nymphs and adults appear in colonies on the lower side of the leaf. They scrape the leaf surface and suck sap. Affected leaves turn pale green, later to pale brown with dark brown spots. Ultimately the affected leaves shrivel and drop off.

Flower thrips: Incidence is severe during December and January. Both nymphs and adults suck sap from flowers, flower stalks, app les and green nuts. This results in flower and fruit drop and development of scab on apple and green nut.

Management

Spraying chlorpyriphos 2.5 ml/l or profenophos 1 ml/l.

5. Cashew Leaf Miner, *Acrocercops syngramma,* Gracillaridae: Lepidoptera

Damage

This pest regularly occurs at the time of new flush generally during June - July and Jan-Feb. The caterpillar mines into the leaves, as a result the thin epidermal layers of the leaf swells up in the mined area and appears as whitish patches on the leaf surface of tender leaves. In older leaves big holes are formed due to the drying and crumbling of the mined areas. Generally young plants are more affected by this pest.

Management

Spraying of carbaryl 2 g/l or profenophos 1ml/l is effective.

C. PESTS OF ARECANUT

1. Spindle Bug, *Carvalhoia arecae*

Identification

Nymphs are light violet brown, greenish yellow with border of the body. Adults are brightly coloured red and black.

Damage

Sap sucking bug – damage the unopened spindle leaf. Inhabit the inner most leaf axils, usually below the spindle. Suck the sap from tender leaflets and spindle. Severe infestation–blackish brown linear lesions on the spindle leaf. Stunted growth and twisted. Leaves become dried and shed.

Management

1. Spray application of dimethoate 0.05 per cent.
2. Filling the inner most leaf axils with phorate 10 per cent G (10g/palm).

2. Root Grub, *Leucopholis burmeisteri*

Identification

Adult beetle is chestnut brown in colour.

Damage

Grubs feed on growing roots. Infested palms show a sickly appearance. Yellowing of leaves. Tapering of stem and reduction in yield.

Management

1. Collection and destruction of adult beetles.
2. Digging and forking of the soil.
3. Addition of organic amendments and anti-feedants (neem, pongamia and oilcake).
4. Application of phorate (Thimet 10G) @ 15g per palm give effective control.
5. Soil application of phorate around the plant twice a year.
6. After the monsoon (Sep-October).
7. In severely infected gardens, the soil should be drenched with eco-friendly insecticides.

3. White Mite *Oligonychus indicus*

Damage

Adults and nymphs feed on the lower surface of leaves and the colony is found under white webs. Mite infested leaves turn Reddish.

Management

1. Removal of heavily infested and dried leaves and burning.
2. Spraying under surfaces of leaves and crown with Dicofol (2 ml/litre) or Rogor (1.5 ml/litre).
3. Repeat spraying at an interval of 15-20 days.

4. Inflorescence Caterpillar, *Tirathaba mundella*

Damage

1.Caterpillars feed on the inflorescences (tender female flowers) and rachillae. Webbing and feeding the inflorescence. Spathe opening is delayed. Yellowing of spadices, Presence of small holes with frass and drying on the spathe.

Management

Infected spadices are forced open and sprayed with malathion 0.05 per cent.

Chapter 29

Pests of Pepper, Cardamom, Cinnamon, Clove, Turmeric and Zinger

1. Pepper Pollu Beetle, *Longitarsus nigripennis* (*Alticidae, Coleoptera*)

Identification and Biology

The adult is a small black beetle measuring about 2.5 mm x 1.5 mm, the head and thorax being yellowish brown and the fore wings black. The eggs hatch in 5-8 days and the grubs bore in to the soft tender berries and feed extensively the internal contents of berry. Fully grown grubs are creamy white in colour measuring 5 mm in length. The grub stage of the insect will be about 20-30 days and fully grown grubs drop down into the soil to pupate. The total life cycle is completed in 23-34 days. The insect completes 4 overlapping generations in a year.

Damage

The grub cause damage by boring into the berries and eating the contents completely in about 10 days. Each grub destroys at least 3-4 berries during its larval period. The attacked berries appear dark in colour, are hallow inside and crumble when pressed. The term 'pollu' denotes the hollow nature of the infested berries in Malayalam. It causes around 13 per cent yield loss.Exit holes are seen in berries. The grubs may also eat into the spike and cause the entire distal region to dry up. The adult feed voraciously on tender leaves and make holes in them. Yield loss due to infestation may range from 30-40 per cent.

Management

1. Regulation of shade in plantation reduces the population of the pest.
2. Pruning of excess vines.
3. Tilling the soil at the base of vines at regular intervals and incorporation of Chloropyriphos 5 per cent D can reduce the population considerably.
4. Spray application of 1.5 litres of dimethoate 30 EC in 500 litres of water OR quinalphos (0.05 per cent each) during June/July and September/October or quinalphos (0.05 per cent each) during July and Neemgold (0.6 per cent) (neem-based insecticide) during August, September.

2. Berry Gall Midge, *Cecidomyia malabarensis (Cecidomyiidae, Diptera)*

Identification

Adult is a small fly and Maggots are whitish in colour.

Damage

Berries harbouring the midge larvae appear larger inb size in the beginning but remain stunted later.Galls are noted on tender stalks and shoots also Full grown maggot falls to the ground and pupates in the soil.

Management

Spray application of Malathion 50 EC @ 2 ml/lit.

3. Cardamom Aphid, *Pentalonia nigronervosa f. caladii* (Aphididae, Hemiptera)

Identification

Adult banana aphids are small to medium sized aphids (1/25 to 1/12 inch), shiny, reddish to dark brown or almost black. They have six-segmented antennae that are as long as the body. Alates have prominent, dark (brown or black) wing veins. Adults start producing young one day after reaching maturity. They can give birth to 4 aphids per day with an average production of 14 offspring per female. Newly emerged nymphs are oval at first and become slightly elongated. They are reddish brown, with four segmented antennae, and measure 1/250 inch in length. The second stage nymphs are similar in appearance and measure approximately 7/250 inch long. The third nymphal stage individuals are light brown, measuring about 9/250 inch in length; the compound eyes are more noticeable beginning with this stage, and the nymphs have five-segmented antennae. The fourth stage nymphs have six-segmented antennae, are light brown in color, and are 1/25 inch long. The first, second, third, and fourth nymphal stages last 2 to 4, 3 to 4, 2 to 4, and 2 to 4 days, respectively.

Biology

P. nigronervosa typically has four nymphal instars and an adult stage. It has a high rate of reproduction, with around 30 generations a year under favourable conditions. In India, a life cycle of 10-15 days was reported, with 26-36 overlapping generations a year. Under favourable laboratory conditions, nymphal development took 9-13 days, adult longevity varying between 9.9 and 12.5 days and fecundities of 3-20 offspring per female occurred. In new colonies, there are around 7-10 generations of apterae (wingless females) before alatae (winged females) start to appear.

Damage

Nymphs and adults suck the sap of leaf sheath and pseudostem. vector of **"Katte disease"** (cardamom mosaic) wehere thin chlorotic flecks on youngest leaves of stem are observed which develop into pale green stripes running from midrib to leaf margin parallel to veins; all leaves emerging subsequently have stripes; symptoms then spread to all tillers.

Management
1. Crop sanitation is to be followed. Infested plants should be removed and burnt.
2. Spray application of Methyl dematon 25EC @ 2 ml/lit., Dimethoate 30 EC @ 1.75 ml/lit., Phosphomidon 85 WSC @ 0.5 ml/lit.

4. Cardamom Thrips, *Sciothrips cardamomi* (Thripidae, Thysanoptera)

Damage

The damage is caused by both nymphs and adults as they colonize and breed in unopened leaves, leaf sheath, panicle, flower buds and capsules. They lacerate the plant tissue and feed on the exuding sap. There will be shedding of flowers and immature capsules. The cardamom pods become shrivelled, undersized, scabby, warty, malformed and lose their characteristic aroma. The seed formation is also affected. The infested capsules are light in weight, inferior in quality and fetch very low market value. It is estimated that 78 per cent by weight and 82 per cent by number of capsules are damaged due to thrips. The adult thrips are tiny, yellowish in colour with fringed wings. They reproduce in large numbers especially during post monsoon seasons. Each female lay around 30 eggs. The incubation and nymphal periods are 8-12 and 10-12 days respectively. Pseudopupal stage takes five days. The complete life cycle from egg to adult takes 25-30 days.

Management
1. Maintain thrips resistant Malabar types of cardamom clumps.
2. Regulation of shade and removal of alternate host plants like *Panicum longipes, Hedychium flavescens* etc. in the vicinity of plantations would help in reducing the build up of this pest.
3. The panicle and 1/3 portion of the base of the clumps are to be treated with insecticides.
4. Spray 5-7 rounds of insecticides like quinalphos 0.025 per cent, phosalone 0.07 per cent, chlorpyrifos 0.05 per cent or profenofos

    0.05 per cent at 35 days interval. Approximately 250-500 ml of spray fluid will be required per clump.

5. Alternatively dust formulations of quinalphos 1.5 D, carbaryl 10 D or phosalone 4D each at 25kg/ha may also be applied.

6. Removal of dried leaf sheaths prior to spraying increases the efficiency of the applied insecticides. It is advisable to do spraying as dusting is found more harmful for honey bees.

6. Cardamom Shoot, Panicle and Capsule Borer, *Conogethes punctiferalis* (Pyraustidae, Lepidoptera)

Biology

    Adult moth lays eggs singly/groups on tender parts of plant. Egg period is 6 to 7 days. Larva is brown in colour and covered with minute hairs arising on warts. Larval period 15-18 days, pupal period 7-10 days. Pupation takes place in loose silken cocoon in larval tunnel. Adult is pale, yellowish with black spots on wings. Life cycle lasts for 3-35 days.

Damage

    The larva bores into the central core of the pseudostems resulting in the death of the central spindle causing characteristic "dead heart" symptom. Larva feeds on the immature capsules and feed on seeds rendering them empty. Oozing out of frass materials at the mouth of the bore hole - very conspicuous on stem/pods.

Management

1. During day time adult moths rest on the lower surface of the cardamom. They may be collected with insect net and killed.

2. The practice of removing the tillers showing 'dead heart' symptoms should be carried with due care.

3. Tillers may be removed if the attack is fresh.

4. The infestation by early stages of larva of this pest in emerging panicle, immature capsule and leaf bud can be controlled effectively with insecticide application.

5. Once the late larvae bore and go deep inside the pseudostem, the chemical spray even in its higher dose becomes ineffective.

6. Spraying fenthion 0.075 per cent is effective in controlling this pest.

6. Cardamom Early Capsule Borers, *Lampides elpis* (Lycaenidae, Lepidoptera)

Identification and Biology

    Adult is a blue butterfly with wings having metallic luster on the upper surface and bordered with a white thin line and black shade. It lays eggs on the buds, flowers and inflorescence. Egg period 10 days. The larva is like slug, flat and pink measuring 2 – 3 cm long, larval period 18 – 20 days. Pupal period 15 days. Total life cycle is 45 days.

Damage
The larva feeds on the buds, flowers and capsule making a circular bore hole on the developing capsules. The capsules become yellowish brown, dried, empty and shed.

Management
Spray application of Malathion 50 EC @ 2 ml/lit.

7. Scale Insects, Pepper Mussel Scale, *Lepidosaphes piperis* and Coconut Scale, *Aspidiotus destructor*

Identification
Pepper mussel scale: Scale is small, dark, boat shaped.

Coconut scale: Circular (about 1 mm in diameter) and yellowish brown.

Damage
This scale insect is found on the lower surface of leaves, leaf sheath, panicles and fruit stalk. As a result of damage, capsules get shrivelled, panicles become dry and the leaves become yellow. The pest is mostly seen during summer months.

Management
1. Systematic stripping of the affected plant parts reduce the pest incidence.
2. Spray dimethoate (0.1 per cent) repeat spraying after 21 days to control the infestation completely.

8. Cinnamon Butterfly, *Chilasa clytia*

Identification
Adults large butterflies with black-brown wings and white markings; larvae pale yellow caterpillars with black markings.

Damage
New growth damaged; entire tree defoliated with only leaf veins remaining; adult insect is a large swallowtail butterfly with black-brown wings and white markings; young larvae are dark green or black, velvety caterpillars which mature to pale yellow with dark stripes.

Management
Spraying quinalphos 0.05 per cent on tender and partly mature leaves.

9. Turmeric Rhizome Fly, *Mimegralla coerruleifrons (Micro-pisidae, Diptera)*

Identification
The flies are fairly large with slender body and long legs. The body is black

in colour and wings are transparent with ashy spots. The wing expansion of flies varies from 13 to 15mm. The eggs are small, white, cigar shaped, tapering at either side. The full grown larva is creamy white in colour, apodous and measures 9.5mm in length and1.95mm in breadth.

Biology

Fliers are noticed in the fields in the months of August and September. Female fliers lay eggs singly or in clusters of 6-10 near the base of the plants under small lumps of soil, in cracks and on the surface or soil. The incubation period lasts for 2-5 days. The larval period lasts for 13.18 days. The full-grown maggots pupate into rotten rhizomes. The pupal period lasts for 110-15 days. The pupal period lasts for 10-15 days. The total period of life cycle is about 4 weeks.

Damage

The maggots feed on the rhizome as a result of which yellowing of plants and rotting of rhizomes takes place.

Management

1. Deep ploughing should be carried out.
2. More drainage facilities should be provided.
3. 10 kg Carbaryl 4G/ha in soil or Chlorophyriphos 5 per cent D @ 20 kg/ha.

Chapter 30

Pests of Flower Crops

1. Bud Borer, *Helicoverpa armigera (Noctuidae, Lepidoptera)*

Damage

This pest damages the flowers. Larvae bore into buds and flowers. Feed on the internal contents.

Management

1. Collection and destruction of damaged buds reduces the damage.
2. Setting up of light traps to attract and kill the adults.
3. Spraying of methyl parathion 0.05 per cent at appearance of eggs on buds and tender foliage stage.
4. Neem oil 0.5 per cent to repell various stages of pest.

2. Rose Aphid, *Macrosiphum rosaeformis (Aphididae, Hemiptera)*

Identification and Biology

Rose aphids are a pinkish brown colour. Wingless females are often seen in large number on stems and buds of roses. Aphids are soft-bodied with sucking mouthparts. They have a pair of tubes on the abdomen which secrete a waxy fluid. When they start to get overcrowded or run out of food, females produce winged young which fly away to find new food sources. Adults reproduce parthenogenetically and in a short duration.

Damage

Nymphs and adults are found in clusters on the tender shoots, flowers and buds and suck the sap.Withering of tender shoots.Buds fall off prematurely and the flowers show fading.Each aphid produces several punctures, producing wounds, which leave their mark as the flower opens. Sooty mould also develops on the honey dew excreted by these insects giving an ugly appearance to the plant. Cluster of aphids are seen in tender shoots, buds and flowers. Yellowing and drying of tender shoots is observed.

Management

1. Spray methyl dematon 25 EC@ 1ml/lit or phosphomidon 85 WSC @ 2.5 ml/lit. or Dimethoate 30 EC @ 2 ml/lit or Monocrotophos 36 WSC @ 2 ml/lit.

3. Rose Thrips, *Rhipiphorothrips cruentatus (Thripidae, Thysanoptera)*

Identification

Nymphs are found on the under surface of leaves and are reddish. adults are 2 mm long with fringed wings. Body dark reddish brown, antennae and legs largely yellow, fore wing pale with yellow veins. Head with complex irregular sculpture, with a transverse ridge near posterior and basal reticulate collar; cheeks sharply incut behind eyes and constricted to basal neck. Antennae 8-segmented.

Damage

Thrips feed by sucking the contents from individual plant cells. R. cruentatus feeds almost exclusively on the lower surface of leaves, and the larvae often occur in groups. The feeding damage includes silvering that gradually turns brown, and the leaves become coated with the brown spots of the thrips excreta. The leaves shrivel under heavy attack, and fruit develops a rough surface.

Management

Spray application of acephate 1.5 g (or) Fipronil @ 2ml or Spinosad 0.3 ml/lit.

4. Rose Red Scale, *Lindingaspis rossi (Diaspididae, Hemiptera)*

Identification

The female scale insect has a circular, brownish-red cover about 1.8 millimetres in diameter. It is firmly attached to the surface when the female is moulting or reproducing. The insect itself is visible through the cover and has an oval body which becomes kidney-shaped at the last instarstage. The female moults twice, exuding the material from which the cover is formed and developing a concentric ring in the center each time.There is a characteristic whitish coating on the underside of the body which separates it from the host plant.The female is viviparous with the eggs hatching internally.She produces 100 to 150 young altogether and live nymphs or crawlers emerge from under their mother's cover at the rate of two to three per day. When they first hatch the nymphs are a yellowish colour and search for a suitable place to settle in depressions on twigs, leaves or fruits. They then start feeding by inserting their mouthparts deep into the plant tissue and sucking sap from the parenchymacells. The male scale insect develops similarly until after the second moult when it becomes oval and darker than the female, measuring about one millimetre in diameter with an excentric cover. The adult male is a small, yellowish two-winged insect that emerges from under its elongated cover after four moults. It lives for about 6 hours and its sole purpose is to mate.It locates unmated females by detecting the pheromones they release.

Damage

Adults and nymphs and suck the sap from green shoots. They are present at the basal portion of the stem causing chlorosis resulting in drying of the entire plant.

Management

1. Rub off the scales with cotton soaked in kerosine or diesel.

2. Spray application of 500 ml of dimethoate 30 EC in 500 lit. of water per ha.

5. Leaf Cutter Bee, *Magachile anthracina* (Megachilidae, Hymenoptera)

Damage

Leafcutter bees are important as pollinators. They are not aggressive and have a mild sting that is used only when they are handled. Adult bee cuts the leaves in a circular or semicircular fashion to construct nests. The newly emerged females begin constructing nests after they emerge in Spring. In each cell they will lay a single egg, and supply it with pollen upon which the larva can feed once it hatches. The larvae pupate and develop inside these cells. They will over-winter in their cells as mature larvae, and emerge as adults the following spring or early summer.

Management

Spray application of Malathion 50 EC @ 2 ml/lit.

6. Chrysanthemum Leaf Miner, *Liriomyza trifolii* (Agromyzidae, Diptera)

Biology

Liriomyza species are leaf mining flies. The adults have a characteristic yellow spot on the back (the scutellum) It is an introduced pest from California through chrysanthemum seedlings. Eggs are laid in upper or lower surface of the leaf. Eggs hatch in 2-4 days. Larva feeds between the epidermal layers. Larval stage is completed in 6-7 days. Larva on maturity, makes a longitudinal slit in the leaf, comes out and pupates in the leaves or in the soil.

Damage

They cause damage by puncturing the leaf surface to feed on the leaf tissue and also to lay eggs. When the eggs hatch, the larvae tunnel within the leaf tissue forming damaging and disfiguring mines. Leaf mines and punctures reduce the quality of high value horticultural crops in addition to reducing the photosynthetic ability of the plant.

Management

Spray application of neem based formulations or Chlorpyriphos @ 2 ml/lit.

7. Jasmine Bud Worm, *Hendecasis duplifascialis (Pyralidae, Lepidoptera)*

Identification

Eggs are round, white in colour and fastened to the flower buds. The larva is green with pale body hairs, black head and pro thoracic shield. Head and prothoracic shield solid black or brown,Long and pointed spinneret.Adult is a small creamish white moth with black wavy lines on the hind wings.

Damage

The infested flower turns violet and eventually dries out. Caterpillar makes hole on the flower bud and feeds on the inner content.Larva attacks 2 -3 budsPetals are eaten by the larvae.

Management

1. Collect and destroy the damaged buds with larvae Proper pruning and hygienic maintenance of bushes. 2. Spray NSKE 5 per cent or malathion 50EC @ 2ml/lit or 3. dimethoate 30 EC or 200 ml of cypermethrin 25 EC in 500 litres of water per ha.

8. Jasmine Leaf Web Worm, *Nausinoe geometralis* (Pyraustidae, Lepioptera)

Biology and Identification

Average larval period is 10days and pupal period was 8 days. Adult period was 4 days for male and 6 days for female. The total life cycle was completed in 26 days for male and 28 days for female. The female laid average 55 eggs during its life span. Young caterpillars are light yellow in colour but as they grow they become darker. Adult is a medium sized moth with light brownish wings having hyaline patches. Their abdomen is purplish brown.

Damage

Caterpillars attacks leaves of the plant mostly in the lower bushy and shaded portions. The leaves are webbed in an open and loose manner. The silk threads are seen as a cobweb on the surface of the leaves. Larvae skeletonize the leaves by eating away the parenchyma.

Management
Spray 500 ml of Dimethoate 30 EC in 500 lit of water per ha.

9. Jasmine Eriophyid Mite, *Aceria jasmini* (Eriophyidae, Acari)

Identification
Female is cylindrical and vermiform with two pair of legs and measures about 150-160 µ long and 44 µ thick.

Damage
Feeding causes felt-like hairy out growth (Erineum) on the surface of leaves, tender stem and flower buds. Makes web which look like felt and appear to be a white hairy growth on the leaf surface, tender stems and flower buds.

Management
Spray application of Dicofol @ 2ml/lit.

10. Lily Leaf Caterpillar *Polytela gloriosae* (Noctuidae, Lepidoptera)

Identification
Adult is brownish with red, yellow and black markings on fore wings. The hind wings are black. Caterpillar has a brown head and has red, black and white mosaic patterns on the body.

Damage
Caterpillar feed on leaves resulting in absolute defoliation of the plant.

Management
Spray application of Malathion 50 EC 1 per cent.

11. Gladiolus Thrips, *Taeniothrips simplex* (Thripidae, Thysanoptera)

Identification
Adults emerge milky-white, but soon turn brown and begin feeding. The female is approximately 1.65 mm long and slightly larger than the male. The antennae are dark brown except for the 3rd segment which is light brown. The wings have a light tranverse band near the base. The first vein of the forewing has seven setae on the distal half. The egg is about 0.3 mm long, opaque white, smooth, and bean-shaped. Eggs are deposited in the leave tissue and corms. The two larval stages are light yellow and are usually found beneath the leaves or bracts. The fully developed second instar larva is about the size of the adult. The first pupal stage is distinguished from the second pupal stage by having forward projecting antennae and short wing pads. The 2nd pupal stage, which is a quiescent period, has the antennae folded over the back and much longer wing pads.

Damage

The thrips feed and reproduce primarily on gladiolus flower spikes and corms. The gladiolus thrips causes deformities and discoloration of gladiolus flowers, and corms (bulbs) become soft and are prone to decay.

Management

1. Corms should be stored in an area where there is little air circulation. Sprinkle napthalene flakes among the corms in storage. Cover the corms with a light canvas or wrapping paper. 2. Spray Monocrotophos 36WSC @1 ml/lit.

12. Carnation Totrix Moth, *Tortrix pronubana* (Torticidae, Lepidoptera)

Identification

The adult moth is 7-10 mm long, the wingspan is 15-17 mm. The color of the forewings is from warm ochre to greyish brown, lightly striped. Hindwings are ochre.The larva is polyphagous.Larvae is yellowish green and Adult is a small moth with greyish brown forewings.

Damage

The young larva pulls together some leaves or petals in its web or sometimes bores a short tunnel into the plant tissue. Cuttings may fall onto the ground and flowerbuds stop developping and dry out. Larva bores into the flower buds also.

Management

Spray application of Phosphamidon 85 WSC @ 1 ml/lit.

13. Tuberose Bulb Mite, *Rhizoglyphus echinopus (Tenuipalpidae, Acarina)*

Identification

The adult bulb mites are yellowish-white, often tinged with pink, with reddish-brown mouth parts and legs. The female is minute, approximately 1/25-inch long. The male is usually smaller; a 10X hand lens is necessary to detect it. The mature forms move around freely. The pest is most active when the humidity is high and the temperature is between 60° and 80°F. At temperatures below 50°F and above 90°, the mite becomes inactive. The female may live for about a month, while the male dies soon after mating. The eggs are white, minute and laid singly on the bulbs. They hatch in 2 to 7 days. A female may lay 50 to 100 eggs at the rate of six to eight per day. Under favorable conditions, an entire life cycle may be completed in 2 to 4 weeks. In addition to the larval stage and the two nymphal stages, the bulb mite has an additional nymphal stage called the hypopus stage due to overcrowding. The mite in this stage is minute, eight-legged, oval and does not feed. It is covered with a thick brown cuticle, and is well adapted to withstand unfavorable environmental conditions. Newly hatched larva are translucent white and only have six legs. The adult mite is is creamy white with 4 pairs of brown legs. Shines like bead.

Damage

Bulb mites (*Rhyzoglyphus* species) infest bulb crops (amaryllis, crocus, freesia, gladiolus, hyacinth, lily, narcissus and tulip) and especially those with loose fleshy scales. The two most common species of bulb mites that cause damage are *Rhyzoglyphus echinopus* and *R. robini*. Infested bulbs show reddish-brown discoloration on the fleshy scales and frequently rot prior to or after planting. The leaves will be stunted, distorted and soon turn yellow. The flower stalk will be destroyed or will fail to develop. In storage at temperatures above 50°F, the mite will continue to multiply and will migrate to other bulbs, thus infesting the entire stored lot. Lightly infested bulbs held in cold storage below 50°F may still become seriously infested during shipment, on store shelves or after planting, whenever temperature conditions become favorable.

Management

Application of profenophos @ 2/lit.

14. Gerbera Mite, *Polyphagotarsonemus latus* (Acarina: Tarso-nemidae)

Identification

Female mites are about 0.2 mm long and oval in outline. Their bodies are swollen in profile and a light yellow to amber or green in color with an indistinct, light, median stripe that forks near the back end of the body. Males are similar in color but lack the stripe. The two hind legs of the adult females are reduced to whip-like appendages. The male is smaller (0.11mm) and faster moving than the female. The male's enlarged hind legs are used to pick up the female nymph and place her at right angles to the male's body for later mating.The eggs are colorless, transclucent and elliptical in shape. They is about 0.08 mm long and are covered with 29 to 37 scattered white tufts on the upper surface. Young broad mites [Larvae] have only three pairs of legs. They are slow slow moving and appear whitish due to minute ridges on the skin. As they grow they range in size from 0.1 to 0.2 mm long (Anonymous a). The quiescent stage appears as an immobile, engorged larva (Baker, 1997). After one day, the larva becomes a quiescent nymph that is clear and pointed at both ends. The nymphal stage lasts about a day. Nymphs are usually found in depressions on the fruit, although female nymphs are often carried about by males.

Damage

Nymphs and adults lacerates the leaves and feeds on the exuding sap. The toxic saliva of mite causes terminal leaves and flower buds to become malformed. Terminal portions of the plants show twisted, hardened and distorted growth. Mites are usually seen on the newest leaves. Leaves turn downward and turn coppery or purplish. Internodes shorten and the lateral buds break more than normal.

Management

Spraying of acaricides like Dicofol (Kelthane) @ 2 ml/lit.

15. Dahlia Aphid, *Brachycaudus helichrysi* (Aphididae: Homoptera)

Identification

Brachycaudus helichrysi apterous adults are 1.4-2 mm long; egg-shaped; usually green to yellowish- or brownish-green; tip of tibiae and tarsi are black; antennae short, as long as half the body; siphunculi are short and flanged. The tail (cauda) is pale, short and blunts. The winged form is 1.1-2.2 mm long, with dark antennae, legs and siphunculi. Head and thorax are dark brown; abdomen is with a large, irregular, pigmented spot.

Biology

The overwintering eggs are laid on host plants. Unusually these eggs hatch early in November/December and the subsequent nymphs then feed on dormant buds. In spring successive generations feed on young foliage, until in May winged forms migrate to numerous summer hosts. They have been recorded on some 120 plant species, with a notable preference for Compositae such as asters, chrysanthemums, yarrow and groundsel. The migration from winter host to summer host is usually complete by early July. The return migration to *Prunus* spp. begins in the latter half of August and continues to mid October.

Damage

The aphid cause leaves to roll up tightly perpendicular to the mid-rib, thus severely damaging leaves at a time of rapid growth, and so early as to be before natural enemies are active. This species is able to transmit a number of viruses including Plum pox virus, Cucumber mosaic virus, Dahlia mosaic virus, and a mosaic virus disease of cineraria. It is also a notable pest of glasshouse crops and house plants. Although a relatively poor vector of Potato virus Y, a non-persistent virus, it can in some years fly in such large numbers as to become an important vector even on potato crops, which it does not truly colonise. A similar role for this species has been insinuated in the sugar beet crop with respect to its infection by non-persistent Beet mosaic virus.

Management

Spray application of 0.03 per cent dimethoate or oxydemeton methyl.

16. Orchid Weevil, *Diorymerellus laevimargo* (Curculionidae, Coleoptera)

Identification and Damage

The weevil is 1/8" long, shiny-black weevil feeds on roots, tender leaves, flower bud sheaths and bulbs. It will also feed on flower petals prior to opening, allowing secondary pathogens to enter and destroy the blooms. The white, legless grub is even smaller and hollows out new roots. This insect is only a problem where large numbers of orchids are grown.

Management

Spray application of Methoxychlor @ 0.02 per cent.

17. Mealy Bug in Crotons, *Planococcus lilacinus*

Identification

Body of the scale is covered with white wax like mealy substance.Small, soft bodied insects.Female is wingless, oval body clothed with white threads.

Damage

Infestation is found under side of leaves, base of petioles and nodes of tender shoots. Infested plants give withering appearance.

Management

1. Maintain adequate shade.

2. Destroy nets of red ants and cocktailed ant.

3. Dust Quinolphos 1.5 per cent or methyl parathion 2 per cent or malathion 5 per cent.

4. Spray the affected patches with Quinolphos 25 EC 2ml or fenthion 100 EC 1ml per lit of water.

5. Conserve natural enemies like predatory lady bird beetle *Cryptolaemus montrouzieri and* parasitoid *Leptomastix dactylopii.*

Chapter 31

Rodent and Bird Pests of Horticultural Crops

31.1 Rodent Pests

Rodents, as one of the major important vertebrate pests are directly related to the production, storage and processing of the agricultural crops and their eventual utilization by man and its livestock for food, fiber and protection. In India, where malnutrition and starvation are best known to exist due to disparity between human population and available food, the rodents eat about 10 percent of agricultural production. Moreover, as India is situated in tropical and subtropical regions of the world with green vegetation available throughout the year, the turn over rate of rodents is much faster than other biomes of the world. Characteristics of rodents that makes them difficult to manage are:

1. Active by night.
2. With keen sense of hearing, taste, smell, and touch.
3. With poor vision although very sensitive to motion.
4. Besides crops, feeds on weeds, insects, and small animals *(e.g.* frogs, snails, etc.).
5. Engages in cannibalism when food is scarce.
6. Continually chews to sharpen its teeth; chews through electric wires, lead sheathing, aluminum siding, glass and half-cured concrete.
7. Sexually active; a few males can mate with almost all the females in the area.

8. A very social animal, with one or a few highly dominant individuals.

9. Exhibits a temporary "fear" (neophobia) when there is a change in an otherwise familiar condition, protecting if from consequences of curiosity.

31.1.1 Commensal Rodents

They contaminate 20 times the material actually they eat. A rat eats 15-25 gm/day and 25-150 pelletdroppings per day/rat. They regularly shed hairs @ 100-200 hair fragments per day/rat. They bite some times human beings. They spread disease. They are social animal. They share same food source and common run way. They live closely to one another. They are most active at dusk and during calm period. Rats become conditioned to eating a particular food and are suspicious in nature. Taste the food cautiously and develop bait shyness. House mouse is not suspicious of new food. Eagerly tastes all. In single night mice tastes and feed on many different foods, hence difficult to get them to take a lethal dose of poisoned bait. Mice readily accept water baits.

House Mouse *(Mus musculus)*

It is quick, tends to nibble and run rather than stay longer at food source. They can pass through a hole slightly less than 1.25 cm. They live mostly in houses. They produce 6-10 litter per year with 6-10 young ones per litter. They can climb easily and also can swim when necessary. They are distributed all over India and are omnivorous. Total length including tail is 8-22 cm with pointed snout. They are brownish grey above and whitish to light grey on belly.

House Rat *(Rattus rattus)*

Lives in close association with human beings. Excellent climber and good swimmer. 4-6 litters/year and tail length 31-43 cm with pointed snout. Dark brownish above (dorsal) and dirty white on belly.

Field Mouse *(Mus musculus booduga)*

The body of Indian field mouse is about 5 to 8 cm long with 5 cm long tail. It is brown in color with a white belly. It burrows in field bunds causing extensive damage to bunds and wastage of water. It produces 3 to 9 young ones per litter.

Norway Rat *(Rattus norvegicus)*

Closely associated with the activity of man. Good climber and swimmer. Prefers Wet or damp locations. Do not close the burrow openings. Length from nose to tail 35-41 cm with blunt snout. Brownish above, white on belly.

31.1.2 Management of Commensal Rodents

Traps

Trapping the rats using cage traps. Killing rats by sticks. Use of Snap trap kills the rat instantly.

Chemicals

Chemicals are of two types.

1. Acute Poison [used in single dose]

☆ **Zinc phosphide:** To be used only in fields not in houses. Commonly used acute rodenticide in India Recommended at 2.5 per cent technical grade in bait material Broken cereal could be used as bait material with vegetable oil as binding medium.

☆ **Pre-baiting** is compulsory for effective results 95 per cent flour + 1 to 2 per cent Zinc phosphide + 2 per cent groundnut oil + 1 per cent sugar. Prebaiting 2-3 days without Zinc phosphide and then bait is mixed with zinc phosphide.

2. Anticoagulants [multiple doses]

☆ **1st generation anticoagulants: Warfarin, Fumarin, Toumarin, Recumin.** These poisons prevent blood clotting and break cell wall of blood capillaries leading to **haemorrhage**. Rats normally die in aerated areas. House rat and house mouse die after 2-5 days of continuous feeding. Antidote is Vitamin K-1.

☆ **2nd generation anticoagulants: Bromadiolone** is only registered Recommended @ 0.005 per cent ai used in pulsed baiting technique.

Fumigants

Aluminium phosphide (CP) solid., Ethelen dibromide (EDB)., Ethelene dichloride carbontetrachloride (EDCT).

Natural Enemies

Cats, dogs, owl, hawks and snakes.

31.1.3 Field Rodents

Rodents attack rice at all stages of growth from planting to harvest and if there is opportunity, even they will continue to attack the grain in store. Freshly sown seed may be dug up and the seed eaten. On young rice plants, rodents attack the heart of the stem discarding the leaves. The rodents make the rice stems fall by gnawing 5-15 cm above ground level. Some rodent species may store grain in their burrows. Large rodents, besides feeding on the crop may cause serious damage to the bunds.

1. Indian Mole Rat or Lesser Bandicoot Rat *(Bandicota bengalensis)*

It is an excellent swimmer, often living in flooded rice fields and bunds. Also occurs in the wheat crop fields and godowns. It is nocturnal and fossorial. They hoard large amounts of food in its burrows. Breeds commonly twice a year with 8-10 young ones in each litter. Adult weight is 325 gm. Length from nose to tip of tail is 36-48 cm. Tail is18-20 cm; less than or some times equal to length of head and body together, 160-170 rings clearly seen on scaly tail. Ear 2.5 to 2.6 cm in length, thick and opaque. Snout – short, stumpy, pig like. Fur and colour – thick, short and harsh, spines present, dark brown, pale brown or reddish above.

2. Soft Turred Field Rat or Grass Rat *(Millardia meltada)*

It occurs in irrigated fields but observed in pastures also. It is nocturnal and lives in simple burrows. It breeds through out the year with litter size of 2-10 young ones. It is small and slender. Adult weight is 100 gm. Total length including tail is 19-29 cm, tail length is 9-14 cm either equal or little shorter than head and body, moderately to poorly haired. The tail is dark above and pale below.

31.1.4 Management of Rodents

Rodent management should be taken up on community basis well before sowing of the crop. The rodent burrows should be marked and the burrow opening is closed with moist soil. The burrows opened out on the next day are active burrows. Then pre baiting has to be done on the 1st and 3rd day. On 5th day 2 per cent zinc phosphide is added and baits distributed in the field. 70-80 per cent kill of rodent population can be secured by the operation. The remaining population can be controlled by fumigating the burrows. On 6th day in those reopened burrows, aluminium phosphide @ 1.5 gm/should be placed in the active burrow and this will take care of residue rodent population.

31.2 Bird Pests

Phylum	:	Chordate
Subphylum	:	Vertebrata or Craniate
Division	:	Gnathostomata
Super class	:	Tetrapoda
Class	:	Aves

General Characters of Birds

1. Birds are warm blooded vertebrates covered by non conducting feathers lack sweat glands.
2. Bones light hollow and pneumatic.
3. Sense of sight and hearing highly developed but sense of small practice absent.
4. Fore limbs modified as wings and hind limbs covered by scales.
5. Jaws modified into a beak.
6. Birds are agriculturally important as they feed on ripening grains in fields and stored grains in godowns.

Crop loss in horticulture due to birds is an on-going and increasing cost to growers. Estimates of damage vary but are generally reported in research literature as 30 per cent to 35 per cent of small berry production, 7 per cent for wine and table grapes, 13 per cent for apples and pears, 16 per cent for stone fruits, and 22 per cent in the nut crops.[1] This includes whole fruit being consumed, fruit knocked off bushes or canes, and unsalable fruit (pecks, holes, slashes). One of biggest mistakes growers make is waiting until damage becomes obvious before taking action. Discouraging bird feeding becomes difficult, if not impossible, once a feeding pattern has been established and birds recognize the crop as a food source. The time when fruit matures appears to influence the amount of damage. Bird damage tends to be worse on early-maturing cultivars. Fruit that ripens early may be damaged more often because it matures when other food is not available or is less desirable.

Growers also tend to limit their control to a single method, like a banger or propane gun, and leave it operating without checking its effectiveness. Birds exposed to a frequently repeated stimulus will habituate quickly and the tool will lose its effectiveness at scaring them away. Some bird species are sedentary and live within a small area; others actively move around within a region or seasonally migrate into a region. Seasonally migratory species are not strongly attached to a territory in later summer or fall as nesting is finished. This coincides with the time when fruit is vulnerable to attack. Hence, mobile and non-sedentary species *(e.g.* robins) should be easier to scare away than sedentary species *(e.g.* sparrows) which are strongly attached to their territory and will often have nowhere else to go if all neighbouring territories are occupied. Not all bird species present will damage fruit and no control techniques will be effective against all species. Similarly, different species may cause damage in different years depending on environmental factors such availability of alternate food sources because of drought, frost, etc. The abundance of insects and weather influence the number of birds and feeding behaviour and subsequently bird damage levels. Scaring will also be more successful when alternative appealing feeding sites are available.Growers are often unaware of the specific bird species responsible for damage and the need for different management approaches for different species. This is in contrast to our attitude to weed and insect pests, where we distinguish between species and adjust pest control actions accordingly.

31.2.1 Major Bird Species Affecting different Crops

1. Crow *(Corvus Splendens* Vieillot)

Crows cause considerable damage to ripe fruits in orchards and also ripening grains of maize and fruits. The crows are particularly attracted to the grains when they are exposed on a cob. They may prove a menace to the successful growth of field crops as well as harvest of fruits. They are often seen in flocks in maize and other fields.

2. Sparrow *(Passer domesticus)*

The flocks of sparrows is a great menace to various field crops like Jowar, bajra, wheat, maize, etc. mainly in the seed setting stage. They also threaten mulberry and many other small sized juicy fruits and fruit buds. They visit the ripening fruit fields, particularly those of wheat in the spring season, and cause much damage both by feeding and causing the grains to shed. House sparrows consume grains in fields and in storage. They do not move long distances into grain fields, preferring to stay close to the shelter of hedge rows. Localized damage can be considerable since sparrows often feed in large numbers over a small area. Sparrows damage crops by pecking seeds, seedlings, buds, flowers, vegetables, and maturing fruits.

3. Parrot *(Psittacula* spp.)

About eight species of parrots have been recorded in India. Out of these species, Large Indian parakeet *(P. eupatria)* is very common in Maharashtra. This species causes heavy damage to orchards by eating fruits and also spoiling the fruits by cutting it with beak. The parakeets are among the most wasteful a destructive birds. They gnaw at and cut into bits all sorts of near-ripe fruits such as guava, ber, mango, plums, peaches, etc. In sunflower when the seeds are soft the parrots cause extensive damage by feeding on the seed thus reducing the yield.

4. Blue Rock Pigeon (*Columba livia*)

Bill is Short and slightly curved with a white crop at the base. Size id13- 14 inches long with 25-inch wingspan, stocky body and pointed. Typical pigeons are a blue gray overall with an iridescent neck that reflects blue, green and purple. Birds may have thick black wing bars and most pigeons are light underneath the wings. Eyes and legs are orange or reddish. Additional color variations include white, brown, tan or mottled birds. Feeds on Grass, seeds, grains, berries, scraps, trash.

31.2.3 Management of Birds

1. Start Management early at 10 to 30 days prior to the fruit colouring.
2. Trapping the birds in nets or catching them with the help of sticky substance 'Lassa'.
3. A piece of Chapatti dipped in 0.04 per cent parathion is a good bait for crows.
4. Parrots and sparrows are repelled by spraying 0.6 per cent thiurum' on crops.
5. Scaring devices using mechanical, acoustic and visual means are normally employed *e.g.* Beating of drums. Birds quickly get accustomed to noise that is stationary, shoots at regular intervals or fires very rapidly. Scaring devices should only be used during the day and at times when birds are expected to feed.
6. Fire crackers placed at regular intervals along a cotton rope. The rope burns from one end and ignites the crackers at regular interval which produce sounds and scare away the birds.
7. Loud sounds due to the burning of acetylene gas produced at intervals are utilized to scare away birds and small animals.
8. Birds may be scared by display of scare crows, Reflective tapes, dead birds and visually attractive flags etc.
9. The reflective ribbon is popularly called as bird scaring tape method.

Objective Questions for Competitive Examinations [JRF/SRF/NET/ARS/SAU Exams]

Integrated Pest Management

1. **In neem, the insecticidal property is chiefly due to the bioactive principle called**   **[b]**
    a. Citrinellol
    b. Azadirachtin
    c. Garaniol
    d. Eugenol

2. **Inundative release is an activity associated with**   **[d]**
    a. Chemical Control
    b. Mechanical Control
    c. Cultural Control
    d. Biological Control

3. **DDT was discovered in**   **[b]**
    a. 1947
    b. 1939
    c. 1941
    d. 1942

4. **Malarial parasite was first discovered by Sir Ronald Ross in**   **[d]**
    a. Paris
    b. Madras
    c. Calcutta
    d. Hyderabad

5.   Commercial formulation of *Bacillus thuringiensis* is   [b]
  a. DOOM   b. DIPEL

  c. GIPLURE   d. VAPONA

6.   Chlorpyriphos is a   [c]
  a. Synthetic pyrethroid   b. Carbamate

  c. Organophosphate   d. Chlorinated hydrocarbon

7.   A common antidote for insecticide poisoning is   [d]
  a. Nicotine   b. Methoprene

  c. Moban   d. Atropine

8   Genetic Engineering Appraisal Committee [GEAC] works under   [a]
  a. Ministry of Environment & Forests   b. Ministry of Agriculture

  c. Ministry of Commerce   d. Ministry of Science

9.   In India, *Bt* cotton is released for commercial cultivation by GEAC during the year   [c]
  a. 1998   b. 2000

  c. 2002   d. 2005

10. Directorate of Plant Protection, Quarantine and Storage is located at   [a]
  a. Faridabad   b. Dehradun

  c. Jodhpur   d. Hyderabad

11. Biodiversity Act was passed in the year   [c]
  a. 1998   b. 2000

  c. 2002   d. 2005

12. In which year GOI passed Insecticide Act ?   [d]
  a. 1959   b. 1969

  c. 1958   d. 1968

13. Destructive Insect and Pests Act [DIPA] was passed in the year [b]
  a. 1903   b. 1914

  c. 1929   d. 1968

14. National Bureau of Agriculturally Important Insects [NBAII] is located at   [c]
  a. New Delhi   b. Chennai

  c. Bangalore   d. Mau

15. Antixenosis term was proposed by which scientist ? **[b]**
a. R.H. Painter
b. Kogan and Ortman
c. M. S. Swaminathan
d. E. F. Knipling

16. On contact with moisture, celphos tablets releases which gas ? **[c]**
a. Phosgene gas
b. CO_2
c. Phosphine
d. Sulphur-dioxide

17. Which of the following insecticide belong to phenyl pyrazole group? **[a]**
a. Fipronil
b. Indoxacarb
c. Etofenprox
d. Imidachloprid

18. Active principle of BHC is **[b]**
a. Hexa chloro
b. Hexa chloro hexane
c. Hexa hexane
d. Hexa chloro pentane

19. Microsomal enzyme responsible for metabolism of Organo-Phosphate insecticides **[d]**
a. Esterases
b. Cytochrome P 450
c. Glutathion S Transferase
d. All the above

20. White grubs can be controlled by **[b]**
a. Ha–NPV
b. *Metarhizium anisopliae*
c. *Trichogramm chilonis*
d. *Apanteles obliqua*

21. Which of the following insects is controlled by NPV **[d]**
a. *Spodoptera litura*
b. *Helicoverpa armigera*
c. Red hairy caterpillar
d. All of the above

22. First time quarantine started in India at **[b]**
a. New Delhi
b. Mumbai
c. Kolkotta
d. Chennai

23. Biological control agent of *Eichornia crassipes* [water hyacinth] is **[d]**
a. *Trichogramma japonicum*
b. *Bacillus thuringiensis*
c. *Zygogramma bicolorata*
d. *Neochetina bruchi*

24. Which of the following is not correct match **[a]**
a. Phorate–Dursban
b. Aldicarb–Temik
c. Indoxacarb–Avaunt
d. Triazophos–Hostathion

25. An insecticide which is used as seed treatment against sucking insect pests   **[a]**

  a. Imidachloprid   b. Endosulphan

  c. Deltamethrin   d. Monocrotophos

26. National Institute of Plant Health Management is located at   **[c]**

  a. New Delhi   b. Mumbai

  c. Hyderabad   d. Chennai

27. Tobacco is a trap crop in castor cultivation against   **[a]**

  a. *Spodoptera litura*   b. *Helicoverpa armigera*

  c. *Agrotis ipsilon*   d. *Aceria cajani*

28. The fungus, Verticillium lecanni is well documented entomopathogen of   **[d]**

  a. Leaf footed bugs   b. Leaf and tree hoppers

  c. Stem borers   d. Thrips, White fly and Scale insects

29. The nymphal–adult parasitoid, *Epiricarinia melanoluca* parasitize   **[d]**

  a. *Nilaparvata lugens*   b. *Alerolobus barodensis*

  c. *Amrasca bigutulla*   d. *Pyrilla perpusilla*

30. Trichocard is used for the control of   **[c]**

  a. Sucking insect pests   b. Coleopteran borers

  c. Lepidopetran borers   d. Forest insects

31. The common inhibitory neurotransmitter substance in the insect nervous system is   **[a]**

  a. Acetyl choline   b. Serotonin

  c. Gama-aminobutric acid   d. Dopamine

32. A mechanical device pointed iron hook is used for collecting   **[c]**

  a. Coconut black headed caterpillar   b. Red palm weevil

  c. Rhiniceros beetle   d. Lady bird beetles

33. The most commonly used chitin synthesis inhibitors for hemipterans   **[a]**

  a. Buprofenzin   b. Diflubenzuron

  c. Flufenozuron   d. Teflubnzuron

34. Marigold is used as a trap crop against   **[b]**

  a. *Spodoptera litura*   b. *Helicoverpa armigera*

  c. Red hairy caterpillar   d. All of the above

35. Token stimulus theory was given by **[d]**
   a. R.H. Painter   b. Kogan and Ortman
   c. Kennedy and Booth   d. Fraenkel

36. Resting potential of a nerve cell is **[c]**
   a. + 70 mv   b. – 60 mv
   c. –70 mv   d. + 50 mv

37. Attractant is type of **[d]**
   a. Kairamone   b. Pheromone
   c. Allomone   d. Both (a) and (b)

38. The weed prickly pear, *Opuntia dilleni* is controlled by **[b]**
   a. *Teleonomus scrupulosa*   b. *Dactylopius opuntiae*
   c. *Zygogramma bicolorata*   d. *Neochtina bruchi*

39. Biological control agent for *Parthenium hysterophorus* is **[c]**
   a. *Ophiomyia lantanae*   b. *Dactylopius opuntiae*
   c. *Zygogramma bicolorata*   d. *Neochtina bruchi*

40. Timely planting of sorghum escapes the damage by **[a]**
   a. Shoot fly   b. Shoot bug
   c. Stem borer   d. Sorghum midge

41. Neem seed kernel extract [NSKE] contains **[b]**
   a. Nicotie   b. Triterpinoids
   c. Hexanotriterpinoids   d. Neonicotinoids

42. The intracellular crystal proteins of *Bacillus thurienginsis* are soluble in **[d]**
   a. Neutral medim   b. Acidic medium
   c. Saline medium   d. Alkaline medium

43. The head quarters of Locust warning organization is located at **[c]**
   a. Bikaner   b. Jaisalmer
   c. Jodhpur   d. Jaipur

44. Methyl eugenol is used as attractant against **[a]**
   a. Oriental fruit fly   b. House fly
   c. Melon fruit fly   d. Carrot fly

45. Mode of action of insecticide Fipronil is **[a]**
   a. Blockage of Chloride ion conduction
   b. Stimulation of Chloride ion conduction
   c. Disruption of ATP Production
   d. Inhibition of sodium ion entry

46. Antagonistic crop against nematode pest is   **[a]**
a. Marigold
b. Castor
c. Coriander
d. Maize

47. Host plant resistance where many minor genes are involved and termed quantitative is   **[d]**
a. Immunity
b. Vertical resistance
c. Induced Resistance
d. Horizontal resistance

48. Compounds that antagonizes the insect Juvenile hormone activity is   **[b]**
a. Plumbagin
b. Ecdysone
c. Juvabione
d. Precocene

49. *Bacillus thuringiensis* is a   **[d]**
a. Gram negative, aerobic bacteria
b. Gram negative, anaerobic bacteria
c. Gram positive, aerobic bacteria
d. Gram positive, anaerobic bacteria

50. Second generation neo-nicotinoid insecticide, acetamiprid is   **[c]**
a. Ach esterase agonist
b. Dopamine inhibitor
c. Ach receptor agonist
d. Octapamine inhibitor

51. Anti –juvenile hormone, isolated from plant *Ageratum houstonianum* is a   **[d]**
a. Protocene
b. Trichocecne
c. Dicosene
d. Precocene

52. Chemical released by an organism which is advantageous to the recipeint   **[b]**
a. Allomone
b. Kairomone
c. Apneumones
d. Synamones

53. Need based application of insecticides is based on   **[d]**
a. Population density of the insect
b. Economic injury level
c. Extent of damage
d. Economic threshold level

54. Protection and maintenance of existing natural enemies under biological control is termed as   **[a]**
a. Augmentation
b. Conservation
c. Natural Control
d. Both a and b

55. Following is reported to be resistant to practically to all group of conventional insecticides **[b]**
a. *Spodoptera litura*
b. *Helicoverpa armigera*
c. *Bemisia tabaci*
d. *Plutella xylostella*

56. Mode of action of Avermectins **[d]**
a. Blocking the Na + gates
b. Inhibition of Ach esterase
c. Stimulation of octopomine receptors
d. Agonist at GABA binding sites

57. Indoxacarb belongs to the group of **[d]**
a. Organiphosphates
b. Synthetic pyrethoids
c. Neonicotinoids
d. Oxadiazine

58. Egg larval -Parasitoids **[c]**
a. *Cotesia plutella*
b. *Diadegma semiclasum*
c. *Chelonus blackburni*
d. *Ooencyrtus* sp.

59. What is the mechanism of resistance to synthetic pyrethroids **[c]**
a. Reduced penetration
b. Decreased nerve sensitivity
c. Enhanced metabolism
d. Storage and excretion

60. Immature of the Neuroptera would be classified as **[a]**
a. Predators
b. Parasites
c. Herbivores
d. Scavengers

61. Insecticide responsible for death of over 100 persons in kerala in 1958 **[b]**
a. Carbaryl
b. Endosulfan
c. Synthetic Pyrethroids
d. Methyl Parathion

62. The Bt crystal toxin are classified based on the **[a]**
a. Lipid sequence
b. Sterol requence
c. Amino acid sequence
d. N–Sequence

63. Chitin synthesis inhibitor of plant origin **[b]**
a. Dimilin
b. Plumbagin
c. Teflubenzuron
d. Buprofezin

64. Single dose anticoagulant rodenticide is **[b]**
a. Warfarin
b. Zinc phosphide
c. Bromodiolone
d. All the above

65. Zero ETL is used for the management of **[b]**
a. Stem borers
b. Vectors
c. Pests
d. All the above

66. The international centre for insect physiology and ecology is located at **[d]**
a. USA
b. Germany
c. India
d. Kenya

67. Endogram is the trade name of **[b]**
a. *Bt* formulation
b. *Trichogramma chilonis*
c. Ha–NPV
d. Sl – NPV

68. Number of Aluminum Phosphide tablets required for 100 cubic meter space is **[c]**
a. 120
b. 160
c. 140
d. 180

69. Which of the following insecticide shows negative temperature correlation **[c]**
a. Malathion
b. Parathion
c. DDT
d. All the above

70. Which of the following is used as biocontrol agent against management of Parthenium **[a]**
a. *Zygogramma bicolorata*
b. *Apantles partheni*
c. *Trichogramma chilonis*
d. *All the above*

71. Which of the following is used as an antidote for organochlorines **[a]**
a. Phenobarbitol
b. Atropine
c. Phytostigmine
d. Charcoal

72. The chemical present in mosquito coils is **[a]**
a. Allethrin
b. Endosulfan
c. Cypermethrin
d. Deltamethrin

73. Insecticide Act came into force during **[a]**
a. 1971
b. 1981
c. 1968
d. 1969

74. Which of the following is synthetic analogue of neries toxin **[d]**
a. Hydropene
b. Methoprene
c. Naled
d. Cartap

75. Resurgence of whitefly is largely associated with **[b]**
   a. E xcessive use of DDT
   b. Indiscriminate use of synthetic pyrethroids
   c. Excessive use of Endosulfan
   d. Excessive use of Monocrtophos

76. Which of the following is an emulsifier in insecticidal formulation **[b]**
   a. Teepol   b. Triton X–100
   c. Polyethethylene Glycol   d. PBO

77. The dose of HaNPV required for the control of *Helicoverpa armigera* **[a]**
   a. 250 LE   b. 500 LE
   c. 750 LE   d. 1000 LE

78. The relation between the following two parameters is a straight line **[d]**
   a. Corrected mortality and dose
   b. Probit per cent mortality and square rootof dose
   c. Dose and Probit per cent mortality
   d. Log dose and Probit per cent mortality

79. Inclusion bodies are characteristic feature of **[c]**
   a. *Bacillus thuringiensis*   b. *Beauveria bassiana*
   c. *Baculovirus*   d. *Metarhizium anisopliae*

80. Antibiosis is **[c]**
   a. Bad Host   b. Expelling guests
   c. Adverse effect of host plant on insect d. All the above

81. Particle size of aerosols in microns is **[a]**
   a. 1- 50   b. 0.1–1
   c. 51–100   d. 0.01 – 0.1

82. Which of the following insecticide is ecdysone agonist that effect moulting **[c]**
   a. Plumbagin   b. Juvenoid
   c. Methoxyfenozide   d. Spinosad

83. Avermectin B₁ a is effective against **[d]**
   a. Insects   b. Mites
   c. Nematodes   d. All the above

84. First synthesis inhibitor developed as insecticide is [a]
a. Diflubenzuron
b. Triflumuron
c. Flufenoxuron
d. Hexaflumuron

85. Which of the following insecticide is banned for use in vegetable crops [d]
a. Cypermethrin
b. Methyl demeton
c. Profenophos
d. Monocrotophos

86. Acceptable daily intake of pesticide is expressed in [a]
a. mg/kg body weight/day
b. mg/person/day
c. ppm
d. None of these

87. Which transgenic crop occupies largest area under cultivation in the world [b]
a. Cotton
b. Soybean
c. Corn
d. Canola

88. The transgenic crop with protease inhibitor gene inhibit [a]
a. Trypsin
b. Chymotrypsin
c. Pepsin
d. Cathepsin

89. The parasporal crystalline toxin in the *Bacillus thruringiensis* is [a]
a. Protien
b. Carbohydrate
c. Protease inhibitor
d. Amylase inhibitor

90 Bacillus thuringiensis was first isolated by Berliner from [a]
a. Mediterranean flour moth
b. American bollworm
c. European corn worm
d. Bonbyx mori

91. Dose killing 50 per cent of treated organism is termed as [d]
a. DT_{50}
b. LC_{52}
c. IP_{50}
d. KD_{50}

92. The book "Silent Spring" was written by Rachel Carson in the year [c]
a. 1952
b. 1972
c. 1962
d. 1982

93. Indian Standards for Commercialization of HaNPV and SLNPV is [a]
a. 1×10^9 POB/ML
b. 4×10^9 POB/ML
c. 1×10^4 POB/ML
d. 5×10^9 POB/ML

94. Indian Standards for Commercialization of Entomopathogenic bacteria **[b]**

a. Minimum of 1 X 10 ^{8} CFU/ml or gm

b. Minimum of 2 X 10^8 CFU/ml or gm

c. Minimum of 2 X 10 ^{9} CFU/ml or gm

d. None of the above

95. Indian Standards for Commercialization of Entomopathogenic fungi **[b]**

a. Minimum of 1 X 10 ^{5} CFU/ml or gm

b. Minimum of 1 X 10 ^{8} CFU/ml or gm

c. Minimum of 1 X 10 ^{7}CFU/ml or gm

d. *None of the above*

96. Spinosad, a mixture of spinosyn A and spinosyn D, secondary metabolites of **[a]**

a. *Saccharopolyspora spinosa*　　　b. *Bacillus thuringiensis*

c. *Azadiracta indica*　　　d. *Beaveria bassiana*

97. Milbemectin [mixture of milbemycin A3 and milbemycin A4, natural products isolated from the fermentations of *Streptomyces hygroscopicus* subsp. *aureolacrimosus*] is **[b]**

a. Insecticide　　　b. Acaricide

c. Fungicide　　　d. Herbicide

98. Bagging/wrapping of pomegranate and mango fruits in paper bags avoids the infestation of pomegranate butterfly *Virachola isocrates* and mango fruit fly *Bactrocera dorsalis is a* **[c]**

a. Cultural method　　　b. Physical Method

c. Mechanical Method　　　d. None of the above

99. Use of an alkathene band around the tree trunks of mango to check the migration of **[a]**

a. Mealybugs and red ants　　　b. Only Mealy bugs

c. Only Red ants　　　d. None of the above

100. Light trap arrangement for mass trapping and destruction of insect pest is a **[b]**

a. Cultural method　　　b. Physical Method

c. Mechanical Method　　　d. None of the above

Pests of Horticultural Crops

1. **Withered terminal shoots, shedding of flower buds, bore holes on brinjal fruits plugged with excreta is characteristic feature of**   **[a]**
   a. *Leucinodes arbonalis*
   b. *Euzophera perticella*
   c. *Urentius histricellus*
   d. *Amrasca devastans*

2. **Pusa purple long is a resistant variety of brinjal against**   **[d]**
   a. *Amrasca devastans*
   b. *Bemisia tabacii*
   c. *Earias vitella*
   d. *Leucinodes arbonalis*

3. **A lady bird beetle with 28 black spots on elytra**   **[c]**
   a. *Batocera rufomaculata*
   b. *Cylas formicarius*
   c. *Henospilachina vigintioctopunctata*
   d. *Coccinella septumpunctata*

4. **Ladder like window and skeletonization of brinjal leaf is due to infestation of**   **[b]**
   a. *Earias vitella*
   b. *Henospilachina vigintioctopunctata*
   c. *Earias insulana*
   d. *Leucinodes arbonalis*

5. **Yellow sticky traps are installed in brinjal field against**   **[c]**
   a. *Euzophera perticella*
   b. *Leucinodes arbonalis*
   c. *Amrasca devastans*
   d. *Henospilachina vigintioctopunctata*

6. **Vector of " little leaf of brinjal" is**   **[b]**
   a. *Amrasca devastans*
   b. *Hishimonas phycitis*
   c. *Thrips tabacii*
   d. *Bemisia tabacii*

7. **A serious pest of Bhendi where moth is buff coloured forewing with a green wedge**   **[a]**
   a. *Earias vitella*
   b. *Helicoverpa armigera*
   c. *Earias insulana*
   d. *Spodoptera litura*

8. **Adult moth with completely green fore wing is pest of bhendi**   **[c]**
   a. *Earias vitella*
   b. *Diacrisia obliqua*
   c. *Earias insulana*
   d. *Spodoptera litura*

9. Red spider mite infesting okra is **[b]**
a. *Aceria cajani*

b. *Polyphagotarsonemus latus*

c. *Aceria guerreronnis*

d. *Tetranichus cinnabarinus*

10. Bhendi Yellow vien mosaic virus is transmitted by **[d]**
a. *Amrasca devastans* b. *Hishimonas phycitis*

c. *Thrips tabacii* d. *Bemisia tabacii*

11. Scientific name of diamond back moth is **[c]**
a. *Agrotis ipsilon* b. *Helicoverpa armigera*

c. *Plutella xylostella* d. *Pieris brassicae*

12. Greyish brown moth with diamond – like white patches dorsally when wings are folded over back at rest **[c]**
a. *Spodoptera litura* b. *Trichoplusia ni*

c. *Plutella xylostella* d. *Pieris brassicae*

13. Scientific name of cabbage butterfly is **[b]**
a. *Helicoverpa armigera* b. *Pieris brassicae*

c. *Plutella xylostella* d. *Spodoptera litura*

14. Adult butterflies with snow white forewings with black distal margins **[b]**
a. *Bactrocera dorsalis* b. *Pieris brassicae*

c. *Myzus persicae* d. *Agrotis ipsilon*

15. Mustard is sown as trap crop in cabbage for the control of **[b]**
a. *Spodoptera litura* b. *Trichoplusia ni*

c. *Plutella xylostella* d. *Pieris brassicae*

16. Common name of *Hellula undalis* is **[c]**
a. Cabbage butterfly b. Cabbage semilooper

c. Cabbage head borer d. Cabbage aphid

17. Deformed heads of cabbage are due to damage by **[c]**
a. Cabbage butterfly b. Cabbage semilooper

c. Cabbage head borer d. Cabbage aphid

18. *Athalia lugens proxima* is pest on **[d]**
a. Rice b. Brinjal

c. Cotton d. Cabbage

19. *Athalia lugens proxima* **has _____number of prolegs** **[c]**
   a. 4   b. 6

   c. 8   d. 2

20. Adults are reddish brown flies with yellow markings on thorax and brown bands on the wings **[b]**
   a. *Musca domestica*   b. *Bactrocera cucurbitae*

   c. *Liriomyza trifolii*   d. *Bemisia tabacii*

21. Methyl eugenol is used as attractant against **[a]**
   a. *Bactrocera cucurbitae*   b. *Pieris brassicae*

   c. *Plutella xylostella*   d. *Bemisia tabacii*

22. Frequent raking up of the soil under the cucurbit vines controls **[b]**
   a. Bag worms   b. Fruit flies

   c. White flies   d. Leaf miners

23. Early sowing of pumpkin will control **[c]**
   a. Bag worms   b. Fruit flies

   c. Pumpkin beetles   d. Leaf miners

24. Scientific name of pumpkin leaf caterpillar **[b]**
   a. *Plusia peponis*   b. *Diaphania indica*

   c. *Bactrocera cucurbitae*   d. *Plutella xylostella*

25. The maggots bore inside the tender stem of *coccinia indica* forming galls in distal shoots **[c]**
   a. Diaphania indica   b. *Alacophora cincta*

   c. *Neolasioptera cephalandrae*   d. *Plusia peponis*

26. *Phytomyza horticola* **is** **[c]**
   a. Pea pod borer   b. Pea stem fly

   c. Pea leaf miner   d. Pea blue butterfly

27. Infested field looks like a cattle grazed field due to attack of **[b]**
   a. *Etiella zinkenella*   b. *Agrotis segetum*

   c. *Phytomyza horticola*   d. *Plusia peponis*

28. *Lampides boeticus* **is a** **[d]**
   a. Pea pod borer   b. Pea stem fly

   c. Pea leaf miner   d. Pea blue butterfly

29. The larva mines into the developing Potato tubers and makes galleries **[c]**
   a. *Spilosoma obliqua*   b. *Pthorimaea operculella*

   c. *Cylas formicarius*   d. *Agrotis segetum*

30. Adult weevil is small, ant like bluish black in colour with reddish brown prothorax and long snout **[b]**
    a. *Oryctes rhinoceros*   b. *Cylas formicarius*
    c. *Rhyncophorous ferrugenius*   d. *Myllocerus discolor*

31. Scientific name of Amaranthus web worm **[a]**
    a. *Hymenia recurvalis*   b. *Trichoplusia ni*
    c. *Phytomyza horticola*   d. *Agrotis segetum*

32. *Eupterote mollifera* is **[c]**
    a. Moringa bud midge
    b. Moringa pod fly
    c. Moringa hairy caterpillar
    d. Moringa leaf eating caterpillar

33. Shedding of moringa flower buds is due to infestations of **[a]**
    a. Moringa bud midge
    b. Moringa pod fly
    c. Moringa hairy caterpillar
    d. Moringa leaf eating caterpillar

34. Drying, splitting of moringa fruits from tips and gummy exudates oozing from fruits is due to infestations of **[c]**
    a. *Noorda blitealis*   b. *Contarina moringae*
    c. *Gitona distigma*   d. *Euptrote mollifera*

35. Presence of ribbons of wood chips, frass and silken threads over moringa bark surface is due to **[b]**
    a. *Gitona distigma*   b. *Indarbella tetraonnis*
    c. *Euptrote mollifera*   d. *Noorda blitealis*

36. *Hypolixus truncatulus* is **[c]**
    a. Amaranthus caterpillar   b. Amaranthus thrips
    c. Amaranthus stem weevil   d. Amaranthus aphid

37. Scientific name of Rose aphid **[d]**
    a. *Aphis gosypii*   b. *Myzus persicae*
    c. *Aphis craccivora*   d. *Macrosiphum rosaeformis*

38. The rose bud turns brown in colour and do not open due to attack of **[a]**
    a. Aphids   b. Thrips
    c. Whiteflies   d. Red scale

39. The larva bores into the rose flowers and feeds on contents [b]
   a. *Spodoptera litura*
   b. *Helicoverpa armigra*
   c. *Plutella xylostella*
   d. *Etiella zinkenella*

40. Adult bees cut the rose leaves in circular/seri circular fashion to construct nests is [d]
   a. *Apis dorsata*
   b. *Apis cerana indica*
   c. *Apis mellifera*
   d. *Megachile anthracina*

41. *Hendecasis duplifascialis* is a _____ bud worm [b]
   a. Rose
   b. Jasmine
   c. Lilly
   d. Crossandra

42. *Elasmopalpus jasminophagus* is jasmine [c]
   a. Budworm
   b. Galleryworm
   c. Bagworm
   d. Cutworm

43. Scientific name of Lily leaf caterpillar [b]
   a. *Spilosoma obliqua*
   b. *Polytela gloriosa*
   c. *Cacoecomorpha pronubana*
   d. *Hendecasis duplifascialis*

44. Scientific name of Tuberose bulb mite [a]
   a. *Rhizoglyphus echinopus*
   b. *Polyphagotarsonemus latus*
   c. *Aceria cajani*
   d. *Taeniothrips simplex*

45. Common name of *Brachycaudus helichrysi* [c]
   a. *Rose aphid*
   b. *Jasmine aphid*
   c. *Dahlia aphid*
   d. *Orchid aphid*

46. Scientific name of orchid weevil [a]
   a. *Diorymerellus laevimargo*
   b. *Cylas formicarius*
   c. *Rhyncophorous ferrugenius*
   d. *Myllocerus discolor*

47. The attacked pepper berries appear darker, hollow inside and crumble when pressed due to infestation of [b]
   a. *Diorymerellus laevimargo*
   b. *Longitarsus nigripennis*
   c. *Rhyncophorous ferrugenius*
   d. *Myllocerus discolor*

48. Gall formation in tender shoots of pepper is due to infestation of [d]
   a. Pollu beetle
   b. Flea beetle
   c. Chaffer beetle
   d. Top shoot borer

49. Katte disease of cardamom is transmitted by

[c]

a. *Amrasca devastans* b. *Hishimonas phycitis*

c. *Pentalonia nigronervosa* d. *Bemisia tabacii*

50. Shoot, panicle, capsule borer of cardamom is [a]

a. *Conogethes punctiferalis* b. *Helicoverpa armigera*

c. *Pentalonia nigronervosa* d. *Spodoptera litura*

51. Upward curling of chilli leaves is due to [b]

a. *Polyphagotarsonemus latus* b. *Scirtothrips dorsalis*

c. *Thrips tabcii* d. *Frankleniella schulzii*

52. Faded chili pericarp with seeds intact is due to infestation of [c]

a. *Helicoverpa armigera* b. *Spodoptera litura*

c. *Utethesia pulchella* d. All the above

53. Murda disease of chilli is transmitted by [a]

a. *Polyphagotarsonemus latus* b. *Hishimonas phycitis*

c. *Pentalonia nigronervosa* d. *Bemisia tabacii*

53. Common name of *Delia antiqua* is [a]

a. Onion fly b. Melon fruit fly

c. Cucurbit fruit fly d. Mango fruit fly

54. Silvery appearance of onion leaf is due to infestation of [c]

a. Onion fly b. Onion ear wig

c. Onion thrips d. Onion aphid

55. Early instar larva resembling birds droppings is [a]

a. Citrus butterfly b. Citrus fruit sucking moth

c. Citrus bark eating caterpillar d. *Citrus psylla*

56. Scientific name of mango red banded caterpillar [b]

a. *Batocera rufomaculata* b. *Deanolis albizonnalis*

c. *Amritodes atkinsoni* d. *Ideoscopus clypealis*

57. Which one of the following is monophagous pest of Mango [a]

a. Mango nut weevil b. Mango mealy bug

c. Mango stem borer d. All the above

58. Citrus butterfly lays eggs on/in [b]

a. Soil b. Singly on leaves

c. Root d. Stem

59. Polythene banding of Mango tree trunk is effective against [c]
  a. Bark eating caterpillar   b. Mango leafhopper
  c. Mango mealy bug   d. Mango stem borer

60. *Citrus tristeza* virus is transmitted by [b]
  a. Aphid   b. Psillid
  c. Leafminer   d. Butterfly

61. Mango stem borer belongs to the family [c]
  a. Scotylidae   b. Scarabaeidae
  c. Cerambicidae   d. Curculionidae

62. Mango mealy bug lays its eggs in/on [d]
  a. Leaves   b. Stem
  c. Leaf axils   d. Soil

63. Grubs of mango nut weevil feed on and damage [b]
  a. Pericarp of the fruit   b. Cotyledons of mango seed
  c. Flowers   d. Shoot

64. In mango withering and shedding of inflorescence is a symptom of damage due to [a]
  a. Leaf hoppers   b. Stem borers
  c. Gall midges   d. Fruit fly

65. The pomegranate fruit borer Duedorix Isocrates belongs to the family [c]
  a. Pyralidae   b. Noctuidae
  c. Lycaenidae   d. Gelichidae

66. Mites are serious problem in the plantations of [a]
  a. Tea   b. Cardamom
  c. Betel vine   d. Pepper

67. Geometrical cutting on coconut fronds is due to [a]
  a. Rhinoceros beetle   b. Redpalm weevil
  c. Coconut mite   d. Black headed caterpillar

68. Gummosis and crown topping in coconut is characteristic symptom of [b]
  a. Rhinoceros beetle   b. Redpalm weevil
  c. Coconut mite   d. Black headed caterpillar

69. The immature and adult stages of exotic predator, *Platymeris laevicollis* consume eggs and early instar larvae of [a]
  a. Rhinoceros beetle   b. Redpalm weevil
  c. Coconut mite   d. Black headed caterpillar

70. Which of the following is serious pest on Nuts/Fruits of coconut **[a]**

a. *Aceria guerreronis*

b. *Oryctes rhinoceros*

c. *Rhyncophoros ferruginus*

d. *Opisina arenosella*

71. Appearance of elongated white streaks below the perianth which later appears as pale yellow triangular patch turning gradually to brown colour and As the nut grows this injury leads to warting and longitudinal fissures on the surface due to feeding by **[b]**

a. *Oryctes rhinoceros*

b. *Aceria guerreronis*

c. *Rhyncophoros ferruginus*

d. *Opisina arenosella*

72 The entomopathogenic fungus *Hirsutella thompsonii* is pathogenic to **[b]**

a. *Oryctes rhinoceros*

b. *Aceria guerreronis*

c. *Rhyncophoros ferruginus*

d. *Opisina arenosella*

73. *Macroplecta nararia* is **[d]**

a. Rhinoceros beetle

b. Redpalm weevil

c. Coconut mite

d. Coconut slug caterpillar

74. In coconut root feeding technique is used for the management of **[d]**

a. Coconut slug caterpillar

b. Red palm weevil

c. Black headed caterpillar

d. All the above

75. *Bracon hebetor* is larval parasiotoid of **[c]**

a. Rhinoceros beetle

b. Redpalm weevil

c. Black headed caterpillar

d. Coconut slug caterpillar

76. Periodical spraying of *Metarrhizium anisopliae* on manure heaps controls **[d]**

a. Coconut slug caterpillar

b. Red palm weevil

c. Black headed caterpillar

d. Rhinoceros beetle

77. Scientific name of Mango Leaf Gall midge **[b]**

a. *Oryctes rhinoceros*

b. *Procystiphora mangiferae*

c. *Rhyncophoros ferruginus*

d. *Batocera rufomaculata*

78. Bunch of dried fruits with a black entry hole of larvae at beak region of the mango fruit is the typical symptom of **[d]**

a. Mango nut weevil

b. Mango mealy bug

c. Mango stem borer

d. Mango fruit borer

79. The full grown Mango fruit borer larva pupates   **[a]**
a. dead wood on the tree or cracks   b. Fruit
    and crevices of the bark

c. Soil   d. Leaves

80. The larva is cream in colour with brown head having red bands on the dorsal side   **[b]**
a. Mango nut weevil   b. Mango fruit borer

c. Mango stem borer   d. Mango mealy bug

81. Corky scabs formation on Guava fruits is due to   **[a]**
a. *Helopeltis antonii*   b. *Bactrocera correcta*

c. *Conogethes punctiferalis*   d. *Nephopteryx eugraphella*

82. The caterpillar webs and feeds on the sapota leaves   **[d]**
a. *Spodoptera litura*   b. *Metanastria hyrtaca*

c. *Conogethes punctiferalis*   d. *Nephopteryx eugraphella*

83. The grubs and adults bore into the banana rhizome and cause stunting of rhizome development   **[c]**
a. *Odoiporus longicollis*   b. *Pentalonia nigronervosa*

c. *Cosmopolites sordidus*   d. *Oryctes rhinoceros*

84. Scientific name of Papaya mealy bug   **[b]**
a. *Drosicha mangifera*   b. *Paracoccus marginatus*

c. *Ferrisia virgata*   d. *Maconellicoccus hirsutus*

85. Hymenopteran parasitoid of Papaya mealy bug   **[a]**
a. *Acerophagous papaya*   b. *Trichogramma chilonis*

c. *Apanteles obliqua*   d. *All of the above*

86. Sthenias grisator is   **[c]**
a. Grape vine thrips   b. Grape vine chaffer beetle

c. Grape vine stem girdler   d. Grape vine Mealybug

87. Scientific name of ber fruit fly   **[b]**
a. *Dacus dorsalis*   b. *Carpomyia vesuviana*

c. *Bactrocera cucurbitae*   d. *Helopeltis antonii*

88. The beetle is pest of coffee with white cross bands and dark brown elytra   **[c]**
a. *Hypothenemus hampei*   b. *Oryctes rhinoceros*

c. *Xylotrechus quadripes*   d. *Rhyncophoros ferruginus*

89. Pin holes at the tip of coffee berries is a symptoms of **[a]**
a. *Hypothenemus hampei*
b. *Oryctes rhinoceros*
c. *Xylotrechus quadripes*
d. *Rhyncophoros ferruginus*

90. Brown patches and curling of leaves and ultimate drying of the shoots of Tea is due to **[d]**
a. *Dacus dorsalis*
b. *Carpomyia vesuviana*
c. *Bactrocera cucurbitae*
d. *Helopeltis antonii*

91. The Indian rose beetle, *Adoretus versutus* is a serious defoliator pest of **[a]**
a. Cocoa
b. Coconut
c. Oilpalm
d. All the above

92. The Bag worm, *Pteroma plagiophelps* is defoliator pest of **[d]**
a. *Cocoa*
b. Coconut
c. *Oilpalm*
d. All the above

93. Yellowing and shedding of leaves and drying of Cashew twigs, Chewed up fibre, excreta and gummy secretions seen protruding from the bore holes are the a symptoms of **[b]**
a. *Hypothenemus hampei*
b. *Plocaederus ferrugineus*
c. *Xylotrechus quadripes*
d. *Rhyncophoros ferruginus*

94. Blossom blight and dieback symptoms in cashew are due to **[a]**
a. *Helopeltis antonii*
b. *Plocaederus ferrugineus*
c. *Xylotrechus quadripes*
d. *Rhyncophoros ferruginus*

95. Pest of quarantine importance in Mango is **[a]**
a. Mango nut weevil
b. Mango fruit borer
c. Mango stem borer
d. Mango mealy bug

96. Black fly is serious pest of **[b]**
a. Grape
b. Mango
c. Citrus
d. Coconut

97. A Serious pest of citrus nursery **[a]**
a. Leaf miner
b. Leaf roller
c. Lemon butterfly
d. Citrus psylla

98. An introduced pest from Italy **[a]**
a. Potato tuber moth
b. Serpentine leaf miner
c. Coconut mite
d. Diamond back moth

99. Which of the following pest that migrates from field to storage **[a]**

 a. *Cylas formicarius*   b. *Anthonomis grandis*

 c. *Corcyra cephalonica*   d. All the above

100. Oozing out of brown liquid from the holes made by **[b]**

 a. Rhinoceros beetle   b. Redpalm weevil

 c. Coconut mite   d. Coconut slug caterpillar

References

Abe, K. and Arai, S. 1985. Purification of a cysteine proteinase inhibitor from rice, *Oryza sativa* L. japonica. *Agriculture Biology and Chemistry*, vol. 49, p. 3349-3350.

Abe, K., Emori, Y., Kondo, H.,Suzuki, K. and Arai, S. 1987. Molecular cloning of a cysteine proteinase inhibitor of rice (oryzacystatin). Homology with animal cystatins and transient expression in the ripening process of rice seeds. *Journal of Biological Chemistry*. vol. 262, no. 35, p. 16793-16797

Adams, L. A., Liu, C. L., McIntosh, S. C. and Stranes, R. L. 1996. Diversity and biological activity of *Bacillus thuringiensis*. In: L. G. Copping (ed). *Crop protection agents from nature: Natural Products and analogues.* The Royal Society of Chemistry, Cambridge, UK, p. 360-380.

Agnihotri, N. P. 2000. Pesticide consumption in agriculture in India – An update. *Pestic. Res. J.* 12 (1) : 150-155.

Akhtar,M. S., Javed,I.,Hayat, C. S and Shah, B. H. 1985. Efficacy and safety of *Caesalpinia crista*, Linn, seeds its extracts in water and methanol against natural *Neoascaris vitulorum* infection in buffalo-calves. *Pakistan Veterinary Journal*, **5**: (4): 192-196.

Akhurst, R. J. 1993. Bacterial symbionts of entomopathogenic nematodes— The power behind the throne. *In* "Nematodes and the Biological Control of Insect Pests" (R. Bedding, R. Akhurst, and H. Kaya, Eds.), pp. 127–135. CSIRO Publications, East Melbourne.

Ananthakrishnan, T. N. 1992. *Dimensions in Insect plant interactions.* Oxford and IBH Publishing Co. Pvt. Ltd., New Delhi, India.

Anwar,T., Jabbar,A., Khalique,F., Tahir,S. and Shakeel, M. A. 1992. Plants with insecticidal activities against four major insect pests in Pakistan. *Tropical Pest Management*, **38**: 4, 431-437.

Applebaum, S. W. Biochemistry of digestion. In: Kerkot, G. A. and Gilbert, L. I., eds. *Comprehensive insect physiology, Biochemistry and Pharmacology*. New York, Pergamon Press, 1985, vol. 4, p. 279-311.

Arai, S., Watanabe, h., Kondo, h., Emori, y. and Abe, K. 1991. Papain inhibitory activity of oryza cystatin, a rice seed cysteine proteinase inhibitor, depends on the central Gln-Val-Val-Ala-Gly region conserved among cystatin superfamily members. *Journal of Biochemistry* Vol. 109, p. 294-298.

Armes, N. J., Jadav, D. R and DeSouza, K. R. 1996. A Survey of insecticide resistance in *Helicoverpa armigera* in the Indian sub continent. *Bull. Ent. Res.* 86: 499-514.

Aronson, A. I. 1994. *Bacillus thuringiensis* and its use as a biological insecticide. *Pl. Breed. Rev.* 12: 19-45.

Attri, B. S. 1975. Utility of neem oil extractives as feeding deterrents to locust. *Indian Journal of Entomology*, **37**: 417-419.

Bajpai, N. K. and Sehgal,V. K. 2003. Effect of botanicals on oviposition behavior of *Helicoverpa armigera* moths at Pantnagar, India. *Indian Journal of Entomology*, **65**: 4, 427-433.

Baker, j. e., Woo, S. M. and Mullen, M. A. 1984. Distribution of proteinases and carbohydrates in the midgut of the larvae of the sweet potato weevil *Cyclas formicarius* and response of proteinase to inhibitors from sweet potato. *Entomologia Experimentalis et Applicata.* vol. 36, p. 97-105.

Barrett, A. J. The classes of proteolytic enzymes. In: DALLING, M. J ed. *Plant proteolytic enzymes,*Florida, CRC Press Inc., 1986, vol. 1, p. 1-16.

Battu,G. S., Arora, R. and Dhaliwal, G. S. 2001. Prospectus of baculovirus in integrated pest management. In: O. Koul and G. S. Dhaliwal (eds). Microbial Biopestide, Harwood Academic Publishers, Amsterdam, pp. 215 – 238.

Battu, G. S., Ramakrishnana, N., Dhaliwal, G. S. 1993. Microbial pesticides in developing countries: Current status and future trends. In : G. S. Dhaliwal and B. Singh (eds). Pesticides : Their Ecological Impact in Developing Countries. Commonwealth Publishers, New Delhi, India, pp. 270-334.

Bedding, R. A. 1984. Large-scale production, storage and transport of the insect-parasitic nematodes *Neoaplectana* spp. and h*eterorhabditis. Ann. Appl. Biol.* **101,** 117–120.

Begley, J. W. 1990. Efficacy against insects in habitats other than soil. *In* "Entomopathogenic Nematodes in Biological Control" (R. Gaugler and H. K. Kaya, Eds.), pp. 215–231. CRC Press, Boca Raton, FL.

Behera, U. K. and Satapathy, C. R. 1996. Screening of indigenous plants for their insecticidal properties against *Spodoptera litura* Fab. *Insect Environment*, **2** : 43 -44.

Berry, R. E., Liu, J., and Groth, E. 1997. Efficacy and pesistence of *Heterorhabditis marelatus* (Rhabditida: Heterorhabditidae) against root weevils (Coleoptera: Curculionidae) in strawberry. *Environ. Entomol.* **26,** 465–470.

Blancolabra, A., Martínez-gallardo, N. A., Sandoval-cardoso, l. and Délano-frier. 1996. J Purification and characterization of a digestive cathepsinD proteinase isolated from *Tribolium castaneum* larvae (Herbst). *Insect Biochemistry and Molecular Biology.* vol. 26, p. 95-100.

Bohnenstengel, F. I., Wray, V., Witte, L., Srivastava, R. P. and Proksch, P. 1999. Insecticidal meliacarpins (C-secolimonoids) from *Melia azedarach.* *Phytochemistry*, **50**(6): 977-982.

Bonning, B. C. and Hammock, B. D. 1996. Development of recombinant baculoviruses for insect control. *A. Rev. Ent.* 41: 191-210.

Bown, D. P., Wilkinson, H. S. and Gatehouse, J. A. 1997. Differentially regulated inhibitor sensitive and insensitive protease genes from the phytophagous insect pest, *Helicoverpa armigera*, are members of complex multigene families. *Insect Biochemistry and Molecular Biology.* vol. 27, no. 7, p. 625-638

Broadway, R. M. and Duffey, S. S. 1986a. The effect of dietary protein on the growth and digestive physiology of larval *Heliothis zea* and *Spodoptera exigua. Journal of Insect Physiology.* vol. 32, p. 673-680.

Broadway, R. M. and Duffey, S. S. Plant proteinase inhibitors: mechanism of action and effect on the growth and digestive physiology of larval *Heliothis zea* and *Spodoptera exigua. Journal of Insect Physiology,* 1986b, vol. 32, p. 827-833.

Brovosky, D. Proteolytic enzymes and blood digestion in the mosquito *Culex nigripalpus. Archives of Insect Biochemistry and Physiology,* 1986, vol. 3, p. 147-160.

Brück, E., Elbert, A., Fischer, R., Krueger, S., Kühnhold, J., Klueken A. M., Nauen, R., Niebes, J. F., Reckman, U., Schnorbach, J. J., Steffens, R. and van Waetermeulen, X. (2009). Movento®, an innovative ambimobile insecticide for sucking insect pest control in agriculture: biological profile and field performance. *Crop Protection*, 28, 10, 838-844,

Cabanillas, H. E., Poinar, G. O., Jr., and Raulston, J. R. 1994. *Steinernema riobravis* sp. nov. (Rhabditida: Steinernematidae) from Texas. *Fund. Appl. Nematol.* **17,** 123–131.

Campos, F. A. P., Xavier-Filho, J., Silva, C. P. and Ary, M. B. 1989. Resolution and partial characterization of proteinases and alpha-amylases from midgets of larvae of the bruchid beetle *Callosobruchus maculatus* (F.). *Comparative Biochemistry and Physiology-B.* vol. 92, p. 51-57.

Casida, J. E. and Quistad, G. B. (1998). Golden age of insetiicde research: past, present, or future? *Annual Review of Entomology*, 43, 1-16.

Chari, M. S. and Muralidharan,C. M. 1985. Neem (*Azadirachta indica*) as feeding deterrent of castor semilooper (*Achaea janata*). *Journal of Entomological Research*, **9**: 243-245.

Chelliah, S. 1987. Insecticide-induced resurgence of rice brown plathopper, Nilaparvata lugens (Stal). In : S. Jayaraj (ed). Resurgence of Sucking Pests. Tamil Nadu Agricultural University, Coimbatore, India, pp. 1-10.

Chelliah, S. and Bharathi, M. 1993. Insecticide-induced resurgence of insect pests of crop plants. In : G. S. Dhaliwal and B. Singh (eds). Pesticides: Their Ecological Impact in Developing Countries. Commonwealth Publishers, New Delhi, India, pp. 51-80.

Chiu, S. F. and Zhang, Y. G. 1984. Effects of some plant materials of Meliaceae on fifth instar larvae of *Spodoptera litura* as feeding inhibitors. *Neem Newsletter*, **1**(3): 23-24.

Chiu,S. F., Schmutterer, H, and Ascher, K. R. S. 1987. Experiments on the practical application of chinaberry, *Melia azedarach*, and other naturally occurring insecticides in China. Natural pesticides from the neem tree (*Azadirachta indica* A. Juss) and other tropical plants. *Proceedings of the 3ʳᵈ International Neem Conference*, Nairobi, Kenya, 10-15 July 1986., 661-668.

Chueca, P., Garcera, C., Molto, E., Jacas, J. A., Urbaneja, A. and Pina, T. (2010). Spray deposition and efficacy of four petroleum-derived oils used against *Tetranychus urticae. Journal of Economic Entomology*, 103, 2, 386-393,

Copping, L. G. and Duke, S. O. (2007). Natural products that have been used commercially as crop protection agents – a review. *Pest Management Science*, 63, 6, 524-554,

Couey HM. 1989. Heat treatment for control of postharvest diseases and insect pests of fruits. *Hort Science,* Vol. 24(2): 198-202.

DeBach, P. and Rosen, D. 1991. Biological Control by Natural Enemies. Cambridge University Press, Cambridge, UK.

Dekeyser, M. A. (2005). Acaricide mode of action. *Pest Management Science*, 61, 2, 103-110,

Denholm, J. and Rowland, M. W. 1992. Tactics for managing pesticide resistance in arthropods. *A. Rev. Ent.* 37: 91-112.

Dent, D. 1991. Insect pest Management. CAB International, Wallingford, UK.

Dhaliwal, G. S and R. Arora. 2004. Integrated pest Management – Concepts and Approaches. Kalyani Publishers, New Delhi, India.

Dhingra, S., Hegde, R. S., Vohra, S. and Parmar, B. S. 2002. Bioefficacy of neem oil microemulsions against gram pod borer, *Helicoverpa armigera* (Hub). *Pesticide Research Journal*, **14** (2): 224-248.

Dhingra, S. 1996. Effect of different vegetable oils on the toxicity of cypermethrin in mixed formulations to the adults of *Mylabris pustulata* Thunb. *Journal of Entomological Research*, **20** (1): 19-22.

Dilawari,V. K., Singh,K and Dhaliwal,G. S. 1994. Sensitivity of diamondback moth, *Plutella xylostella* to *Melia azedarach*. *Pesticide Research Journal,* **6**(1): 71-74.

Douresammy, S., Chandra Mohan,M., Sivaprakasam, N., Subramanian, A., and Surendrababu, P. C. 1997. Management of Spiralling Whitefly. Indian Silk, October : 15-16

Dunaevskii, Y. E., Gladysheva, I. P., Pavlukova, E. B., Beliakova, G. A., Gladyschev, D. P., Papisova A. I., Larionova N. I. and Uelozersky U. 1997. The anionic protease inhibitor BBWI - 1 from buckwheat seeds. Kinetic properties and possible biological role. *Physiologia Plantarum.* vol. 100, p. 483-488.

Dunn, B. M. *Proteolytic enzymes: A practical approach.* Oxford, IRL Press, 1989. 57-81 p. ISBN 0-19-963058-5.

Eguchi, M., Iwamoto, A. and Yamaguhi, K. Interaction of proteases from the midgut lumen, epithelial and peritrophic membrane of the silkworm *Bombyx mori* L. *Comparative Biochemistry and Physiology-A,* 1982, vol. 72, p. 359-363.

Emmanuel Nathala and Swaran Dhingra. 2005 b. Antifeedant activity of seeds and leaves of *Caesalpinnia crista* and *Melia azadirach* against *Helicoverpa armigera*. *Annals of Plant Protection Sciences.* 14(1): 218-219.

Emmanuel Nathala and Swaran Dhingra. 2006 a. Biological effects of *Caesalpinia crista* seed extracts on *Helicoverpa armigera* (Lepidoptera: Noctuidae) and its predator, *Coccinella septumpunctata* (Coleoptera : Coccinellidae). *Journal of Asia Pacific Entomology.* 9(2): 1-6.

Emmanuel Nathala and Swaran Dhingra. 2005 a. Chronic effects of *Melia azedarach* seed extracts on survival, feeding, growth and development of *Helicoverpa armigera* (Lepidoptera: Noctuidae). *Annals of Plant Protection Sciences.*

Emmanuel Nathala and Swaran Dhingra. 2006 b. Antifeedant activity of aqueous extracts from the seeds and leaves of *Caesalpinia crista* (Lin.) and *Melia azedarach* against *Helicoverpa armigera* (Hubner). *Annals of Plant Protection Sciences.* 14 (1) : 235-236

Emmanuel Nathala, Swaran Dhingra and. Suresh Walia, Shankar Ganesh K. 2008. Antifeedant and toxic activity of *Melia azedarach* seeds extracts against *Helicoverpa armigera* (Lepi: Noctuidae). *Pesticide Research Journal.* Volume : 20, pp 16- 20.

Emmanuel, N., and S. Suresh. 2003. Gallic acid- A major biochemical factor imparting resistance in rice genotypes against white backed plant hopper and effect of insecticides. In*: Proceedings of the National Symposium on Frontier Areas of Entomological Research* 5-7 Nov, 2003. p: 389-390.

Emmanuel, N., S. Suresh and P. Ashok. 2002. Biochemical basis of resistance in rice hybrids and conventional varieties against white backed plant hopper *Sogatella furcifera* (Horvath.). *Ann. Pl. Protec. Sci.* Vol. 10 (2): 212-215.

Emmanuel. N, A. Sujatha and B. Gautam. 2010 a. Occurrence of Bag Worms *Pteroma plagiophelps* Hamps and *Clania* sp. on Cocoa Crop. *Insect Environment.* Vol. 16 (2) pp: 60-61

Emmanuel. N, and N. B. V. Chalapathi Rao. 2011. *Phelera sp* (Drepanidae : Lepidoptera) – an emerging defoliator of coconut in Andhra Pradesh. *Insect Environment,* Vol. 17(1), pp. 19-20.

Emmanuel. N, N. B. V. Chalapathi Rao and A. Snehalatha Rani. 2011 a. Fusarium solani (Mart.) as a naturally occurring pathogen of bagworm Pteroma plagiophelps in coconut. *Insect Environment,* Vol. 17(1), pp. 11-12.

Emmanuel. N, N. B. V. Chalapathi Rao and B. Gautham. 2011 b. New record of defoliator pests of Cocoa in Godavari Districts of Andhra Pradesh and their management. *The Cashew and Cocoa Journal.* Vol. III,. pp. 21-28

Emmanuel. N, Sujatha. A and B. Gautham. 2010 b. Record of Leaf Chafer Beetles *Adoretus versutus* Harold and *Apogonia blanchardi* Ritsema on Cocoa (*Theobroma cacao* L.) in Andhra Pradesh. *Insect Environment,* Vol 16 (1), April-June 2010.

FAO 1967. Report of the First Session of the FAO Panel of Experts of Integrated Pest Control. Food and Agriculture Organization, Rome.

FAO. 1967. Report of the first session of the FAO Panel of Experts on Integrated Pest Control, Rome (Italy), Sept. 18-22, 1967, 19 pp.

FAO. 1967. Report of the first session of the FAO Panel of Experts on Integrated Pest Control, Rome (Italy), Sept. 18-22, 1967, 19 pp.

FAO. 1980. Research Summary. Integrated pest management. EPA-600/8-80-044. 28 pp.

Farina,S. R. 1994. Problems on WSCA and strategies for its control at Takakar sugar plantations. Report to the manager of Takakar sugar plantations [Unpublished report] p. 9.

Fernandes, K. V. S., Sabelli, P. A., Barratt, D. H. P., Richardson, M., Xavier-Filho, J. and Shewry, P. R. 1993. The resistance of cowpea seeds to bruchid beetles is not related to level of cysteine proteinase inhibitors. *Plant Molecular Biology.* vol. 23, no. 1, p. 215-219.

Food Quality Protection Act. 1998. Food Quality Protection Act of 1996, P. L. 104- 170, Title II, Section 303, Enacted August 3, 1996. Codified in: Title 7, U. S. Code, Section 136r-1. Integrated Pest Management.

Forgash A. J. 1994. History, evolution, and consequences of insecticide resistance. *Pesticide Biochemistry and Physiology,* 22 178-186.

Frazao, C., Bento, I., Costa, J.,Soares, C. M., Verissimo, P., Faro, C., Pires. E., Cooper. J and Carrondo, M. A. 1999. Crystal structure of cardosin A, a glycosylated and Arg-Gly-Asp- containing aspartic proteinase from the flowers of *Cyanara cardunculus* L. *Journal of Biological Chemistry,* September 1999, vol. 274, no. 39, p. 27694-27701.

Garciaolmedo, S. F., Sanchez. M. G., Gomez, R. L., Royo, J and Carbonero, P. 1987. Plant proteinaceous inhibitors of proteinases and α-amylases. *Oxford Survey Plant Molecular and Cell Biology.* vol. 4, p. 275-334.

Gatehouse, A. M. R. and Boulter, D. 1983. Assessment of the antimetabolic effects of trypsin inhibitors from cowpea (*Vigna unguiculata*) and other legumes on development of the bruchid beetle *Callosobruchus maculatus. Journal of the Science of Food and Agriculture.* vol. 34, p. 345-350.

Gatehouse, A. M. R., Davidson, G. M., Newell, C. A., Merryweather, A., Hamilton, W. D. O., Burgess, E. P. J., Gilbert, R. J. C. and Gatehouse, J. A. 1997. Trangenic potato plants with enhanced resistance to the tomato moth, *Lacanobia oleracea*: growth room trials. *Molecular Breeding,*1997, vol. 3, p. 49-63.

Gatehouse, J. A., Hilder, V. A. and Gatehouse, A. M. R. 1991. Genetic engineering of plants for insect resistance. In : D. Grierson (ed.). *Plant Biotechnology Series.* Vol. 1. Plant Genetic Engineering. Blackie Publications, London, UK, pp. 105-135.

Gaugler, R. 1997. Alternative paradigms for commercializing biopesticides. *Phytoparasitica* **25,** 179–182.

Geier, P. W. and Clark, L. R. 1961. An ecological approach to pest control. In: Proc. Eighth Technical Meeting. International Union for Conservation of Nature and Natural Resources, 1960, Warsaw, Poland, pp. 10-18.

Geier,P. W and and Clark,L. R. 1961. An ecological approach to pest control. In: Proc. Eighth Technical meeting. International Union of Conservation of Nature and Natural Recourses. 1960, Warsaw, Poland,pp. 10-18.

Georghiou, G. P. 1983. Pesticide resistance in time space. In: G. P. Georghiou and T. Saito (eds). Pest Resistance to Pesticides. Plenum Press, New York, USA, pp. 769-792.

Georgis, R. 1997. Commercial prospects of microbial insecticides in agriculture. *In* "Microbial Insecticides: Novelty or Necessity?" (H. F. Evans, chair). *Proc. Br. Crop Prot. Council Symp.* **68,** 243–252.

Georgis, R., and Manweiler, S. A. 1994. Entomopathogenic nematodes: A developing biological control technology. *In "Agricultural Zoology Reviews"* (K. Evans, Ed.), Vol. 6, pp. 63–94.

Gill, J. S and Lewis, C. T. 1971. Systemic action of an insect feeding deterrent. *Nature,***232**: 402-403.

Gourinath, S., Alam, N., Srinivasan, A., Betzel, C. and Singh, T. P. 2000. Structure of the bifunctional inhibitor of trypsin and alpha-amylase from ragi seeds at 2. 2 A resolution. *Acta Crystallography D Biology Crystallography.* vol. 56, no. 3, p. 287-293.

Greenblatt, H. M., Ryan, C. A. and James, M. N. G. 1989. Strucuture of the complex *Streptomyces griseus*proteinase B and polypeptide chymotrypsin inhibitor-I at 2. 1 A resolution. *Journal of Molecular Biology.* vol. 205, p. 201-228.

Gujar, G. T. and Mehrothra, K. N. 1988. Biological activity of neem against the red pumpkin beetle, *Aulacophora foveicollis. Phytoparasitica,***16** : 293 – 302.

Havkioja, E. and Neuvonen, L. Induced long-term resistance to birch foliage against defoliators : defense or incidental. *Ecology,* 1985, vol. 66, p. 1303-1308.

Hawtin, R. E. 1993. The chitinase of Autographa calofornica nuclear polyhedrosis virus. Ph. D. Thesis, Oxford Brookes University, Oxford, UK.

Heinrichs, E. A. 1998. IPM in the 21st century : Challenges and opportunities. In : G. S. Dhaliwal and E. A. Heinrichs (eds). Critical Issues in Insect Pest Management. Commonwealth Publishers, New Delhi, India, pp. 267-276.

Hilder, V. A., Gatehouse, A. M. R., Sheerman, S. E., Barker, R. F. and Boulter, D. 1987. A novel mechanism of insect resistance engineered into tobacco. *Nature.* vol. 300, p. 160-163.

Hilder,V. A., Gatehouse, A. M. R., Sheerman, S. E., Barker, R. F. and Boulter, D. 1987. A novel mechanism of insect resistance engineered into tobacco. *Nature,* 1987, vol. 300, p. 160-163.

Hollander-Czytko, H., Andersen, J. L. and Ryan, C. A. 1985. Vacuolar localization of wound-induced carboxy peptidase inhibitor in potato leaves. *Plant Physiology.* vol. 78, p. 76-79.

Horowitz, A. R., Ellsworth, P. C. and Ishaaya, I. (2009). Biorational Pest Control – An Overview, In: *Biorational Control of Arthropod Pests*, Ishaaya, I. and Horowitz, A. R. (ed.), 1-20, Springer, ISBN 978-90-481-2315-5, Dordrecht, the Netherlands.

Houseman, J. G. and Downe, A. E. R. 1983. Cathepsin D-like activity in the posterior midgut of Hemipteran insects. *Comparative Biochemistry and Physiology-B.* vol. 75, p. 509-512.

Houseman, J. G., Downe, A. E. R. and Philogene, B. J. R. 1989. Partial characterization of proteinase activityin the larval midgut of the European corn borer *Ostrinia nubilalis* Hubner (Lepidoptera: Pyralidae). *Canadian Journal of Zoology.* vol. 67, p. 864-868.

Hoy, M. A. 1994. Parasitoids and predators in management of arthropod pests. In: R. L. Metcalf and W. H. Luckmann (eds). Introduction to Insect Pest Management. John Wiley and Sons, Inc., New York, pp. 129-198.

Huber, R. and Carrell, R. W. 1989. Implications of the three-dimensional structure of alpha I -antitrypsin for structure and function of serpins. *Biochemistry.* vol. 28, p. 8951-8966.

Jeyakumar,P. and Gupta,G. P. 1999. Effect of neem seed kernel extract (NSKE) on *Helicoverpa armigera. Pesticide Research Journal,* **11**(1): 32-36.

Joshi, B., Sainani, M., Bastawade, K., Gupta, V. S. and Ranjekar, P. K. 1998. Cysteine protease inhibitor from pearl millet: a new class of antifungal protein. *Biochemical and Biophysical Research Communications*. vol. 246, p. 382-387.

Joshi, B. G., RamPrasad, C. and Rao, S. N. 1984. Neem seed kernel suspension as an antifeedant for *Spodoptera litura* in a planted flue cured virginia tobacco crop. *Phytoparasitica,* **12**: 3-12.

Juan, A., Sans, A. and Riba, M. 2000. Antifeedant activity of fruit and seed extracts of *Melia azedarach* and *Azadirachta indica* on larvae of *Sesamia nonagrioides*. *Phytoparasitica,* **28** (4): 311-319.

Kalauni,S., Awale, S.,Yasuhira, T., Banskota, A. H., Zawhinn, T. and Kadota, S. 2004. Methyl migrated cassane- type Furano diterpenes of *Caesalpinia crista* from Mayanmar. Institute of Natural Medicine, Toyama Medical and Pharmaceutical University, Sugitami, Toyama, Japan.

Kareem, A. A., Saxena, R. C. and Palanginan, E. L. 1998. Effect of neem (*Azadirachta indica*) seed bitters (NSB) and neem seed kernel extract (NSKE) on pests of mung bean following rice. *International Rice Research Newsletter,***13** (6) : 41-42.

Karel Sláma. 1969. Plants as a source of materials with insect hormone activity. Entomologia experimentalis et applicata,volume 12, issue 5, pages 721–728.

Karlson, P and Butenandt, A. 1959. Pheromones (ectohormones) in insects. *Annual Review of Entomology*. vol. 4: 39-58

Kaur, V and Singh, G. 2003. Antifeedant activity of *Melia azedarach* Linn. from three locations against *Plutella xylostella* Linn. *Pesticide Research Journal,* **15**(1): 17-18.

Kaya, H. K., and Stock, S. P. 1997. Techniques in insect nematology. *In* "Manual of Techniques in Insect Pathology" (L. A. Lacey, Ed.). pp. 281–324. Academic Press, London.

Keifer, H. H. 1965. *Eriophyid Studies*. Department of Agriculture, Bureau of Entomology, B -14 California. p. 20.

Keilova, H. and Tomasek,V. 1979. Isolation and properties of cathespin D inhibitor from potatoes. *Collection of Czechoslovak Chemical Communication*. vol. 41, p. 489-497.

Kervinen, J., Tobin, G. J., Costa, J., Waugh, D. S., Wlodawer, A. and Zdanov, A. 1999. Crystal structure of plant aspartic proteinase prophytepsin: inactivation and vacuolar targeting. *Journal of European Molecular Biology Organization*. vol. 18, no. 14, p. 3947-3955.

Khan, K. H. 1996. "Integrated pest management and sustainable agriculture". *Farmers and Parliament,* 30 (2): 15-17.

Kim, S. S. and Yoo, S. (2002). Comparative toxicity of some acaricides to the predatory mite, *Phytoseiulus persimilis* and the twospotted spider mite, *Tetranychus urticae. BioControl,* 47, 5, 563-573,

Kimura, M., Ikeda, T., Fukumoto, D., Yamasaki, N. and Yonekura, M. 1995. Primary structure of a cysteine proteinase inhibitor from the fruit of avocado (*Persea americana* Mill). *Bioscience Biotechnology and Biochemistry.* vol. 59, no. 1, p. 2328-2329.

King L. A., Possec, R. D., Hughes, D. S., Atkinson, A. E., Palmer, C. P., Marlow, S. A., Pickesing I. M., Yoyce, K. A., Lawrie, A. M., Miller, D. P. and Beadle, D. J. 1994. Advances in insect virology. *Adv. Insect Physiol.* 25: 1-73.

Kitch, L. W. and Murdock, L. L. 1986. Partial characterization of a major gut thiol proteinase from larvae of*Callosobruchus maculatus* (F.). *Archives of Insect Biochemistry and Physiology.* vol. 3, p. 561-575.

Klein, M. G. 1990. Efficacy against soil inhabiting pests. *In* "Entomopathogenic Nematodes in Biological Control" (R. Gaugler and H. K. Kaya, Eds.), pp. 195–231. CRC Press, Boca Raton, FL.

Kogan, M. and Ortman, E. F. 1978. Antixenosis – A new term proposed to define Painter's 'Non-preference' modality of resistance. *Bull. Ent. Soc. Am.* 24: 175-176.

Kogan, M. 1994. Plant resistance in pest management. In: R. L. Metcalf and W. H. Luckmann (eds). Introduction to Insect Pest Management. John Wiley and Sons, New York, USA, pp. 73-127.

Kogan. M. 1998. Integrated Pest Management: Historical Perspectives and Contemporary Developments. *Annu. Rev. Entomol.* 43: 243 – 270.

Koiwa, H., Bressan, R. A. and Hasegawa, P. M. 1997. Regulation of protease inhibitors and plant defense. *Trends in Plant Science.* vol. 2, p. 379-384.

Kontsedalov, S., Gottlieb, Y., Ishaaya, I., Nauen, R., Horowitz, R. and Ghanim, M. (2008). Toxicity of spiromesifen to the developmental stages of *Bemisia tabaci* biotype B. *Pest Management Science*, 65, 1, 5-13,

Koppenho¨fer, A. M., Grewal, P. S., and Kaya, H. K. 2000. Synergism of imidacloprid and entomopathogenic nematodes against white grubs: The mechanism. *Entomol. Exp. Appl.* **94,** 283–293.

Krämer, W. and Schirmer, U. (2007). *Modern Crop Protection Compounds, Vol. 3,* WILEY – VCH Verlag GmbH and Co., ISBN 978-3-527-31496-6, Weinheim, Germany.

Kunitz, M. Crystallization of a trypsin inhibitor from soybean. *Science*, 1945, vol. 101, p. 668-669.

Ladd, T. L and Jacobson, M. 1980. Neem seed extracts as feeding deterrents to Japaneese beetles. 41 (En) SEA, USDA, Japaneese Beetle Research Laboratory. OARDC, Ohio.

Laskowski, M. JR. and Kato, I. Protein inhibitors of proteinases. *Annual Review of Biochemistry*, 1980, vol. 49, p. 685-693.

Lee, H. - K., Cheong, H. and Gill, S. S. 1998. Microbial control of insects: Use of bacterial insecticides. In: G. S. Dhaliwal and E. A. Heinrichs (eds). Critical

Issues in Insect Pest Management. Commonwealth Publishers, New Delhi, pp. 87-117.

Lee, S. I., Lee, S. H., Koo, J. C., Chun, H. J., Lim, C. O., Mun, J. H., Song, Y. H. and Cho, M. J. 1999. Soybean Kunitz trypsin inhibitor (SKTI) confers resistance to the brown planthopper (*Nilaparvata lugens*Stal) in transgenic rice. *Molecular Breeding.* vol. 5, p. 1-9.

Lewis, E. E., Gaugler, R., and Harrison, R. 1993. Response of cruiser and ambusher antomopathogenic nematodes (Steinernematidae) to host volatile cues. *Can. J. Zool.* **71,** 765–769.

Li, M., Liu, C. L., Li, L., Yang, H., Li, Z. N., Zhang, H. and Li, Z. M. (2010). Design, synthesis and biological activities of new strobilurin derrivatives containing substituted pyrazoles. *Pest Management Science*, 66, 1, 107-112.

Lipke, H., Fraenkel, G. S. and Liener, *I. E.* 1954. Effects of soybean inhibitors on growth of *Tribolium confusum. Journal of the Science of Food and Agriculture.* vol. 2, p. 410-415.

Macphalen, C. N. and James, M. N. G. 1987. Crystal and molecular structure of the serine proteinase inhibitor CI-2 from barley seeds. *Biochemistry.* vol. 26, p. 261-269.

Mane, S. D. 1968. Neem seed spray as a repellant against some of the foliage feeding insects. M. Sc. Thesis, I. A. R. I, New Delhi.

Marèiæ, D., Periæ, P., Prijoviæ, M., Ogurliæ, I. and Andriæ, G. (2007). Effectiveness of spirodiclofen in the control of European red mite (*Panonychus ulmi*) on apple and pear psylla (*Cacopsylla pyri*). *Pesticides and Phytomedicine*, 22, 3, 301-309.

Martinez-Villar, E., Saenz-de-Cabezon, F. J., Moreno-Grijalba, F., Marco, V. and Perez-Moreno, I. (2005). Effects of azadirachtin on the two-spotted spider mite, *Tetranychus urticae* (Acari: Tetranychidae). *Experimental and Applied Acarology*, 35, 3, 215-222

Mehrotra, K. N. and Phokela, A. 2000. Insecticide resistance in insect pests: Current status and future strategies. In: G. S. Dhaliwal and B. Singh (eds). Pesticides and Environment. Commonwealth Publishers, New Delhi, India, pp. 39-85.

Meisner,J., Melamed, M. J., Tam,S. and Ascher, K. R. S. 1986. The effect of azadirachtin on the larvae of European corn borer, *Ostrinia nubilalis. Zeitschrift fur Pflanzenkrankheiten und Pflanzenschutz,* **93** (6) : 585 - 589.

Melander A. L. 1914. Can insects become resistant to sprays ? *Journal of Economic Entomology,* 7,164-166.

Metcalf, R. L. 1994. Insecticides in pest management. In: R. L. Metcalf and W. H. Luckmann (eds). Introduction to Insect Pest Management. John Wiley and Sons, New York, USA, pp. 245-314.

Michelbacher,A. E and Bacon, O. G. 1952. Walnut insect and Spider mite control in Northern California. *J. Econ. Ent.* 45 : 1020 – 1027.

Michelbacher, A. E. and Bacon, O. G. 1952. Walnut insect and spider mite control in Northern California. *J. Econ. Ent.* 45: 1020-1027.

Mickel, C. E. and Standish, J. 1947. Susceptibility of processed soy flour and soy grits in storage to attack by *Tribolium castaneum*. *University of Minnesota Agricultural Experimental Station Technical Bulletin*. vol. 178, p. 1-20.

Mikolajizak, K. L., Ziloroski, B. W. and Bartell, R. J. 1989. Effect of Meliaceous seed extracts on growth and survival of *Spodoptera frugiperda. Journal of Chemical Ecology*, **15** (1): 121-128.

Miller, L. K. 1998. Genetically improved baculoviruses for insect pest management. In: V. L. Chopra, R. B. Singh and A. Verma (eds). Crop Productivity and Sustainability-Shaping the Future. Oxford and IBH Publ. Co. Pvt. Ltd., New Delhi, pp. 269-276.

Mukhopadhyay, D. 2000. The molecular evolutionary history of an winged bean alpha-chymotrypsin inhibitor and modeling of its mutations through structural analysis. *Journal of Molecular Evolution*, vol. 50, p. 214-223.

Murdock, L. L., Brookhart, G., Dunn, P. E., Foard, D. E. and Kelley, S. 1987. Cysteine digestive proteinases in Coleoptera. *Comparative Biochemistry and Physiology-B*. vol. 87, p. 783-787.

Murdock, L. L., Shade, R. E. and Pomeroy, M. A. 1988. Effects of E. 64 a cysteine proteinase inhibitor on cowpea weevil growth development and fecundity. *Environmental Entomology*. vol. 17, p. 467-469.

N. B. V. C. Rao, A. Snehalatha Rani, and N. Emmanuel. 2012. New Record of Paecilomyces lilacinus as an entomopathogenic fungi on slug caterpillar of coconut. *Insect Environment*, Vol. 17(4), pp151-153.

Nagata, K., Kudo, N., Abe, K., Arai, S. and Tanokura, M. 2000. Three dimensional solution structure of oryzacystatin-I, a cysteine proteinase inhibitor of the rice, *Oryza sativa* L. japonica. *Biochemistry*. vol. 39, p. 14753-14760.

National Academy of Science. 1969. Insect-pest management and control. pp. 448- 449. Vol. 3. *Principles of plant and animal pest control*. Natl. Acad. Sci. Pub. 1695. 508 pp.

Nauen, R. and Smagghe, G. (2006). Mode of action of etoxazole. *Pest Management Science*, 62, 5,379-382

Novillo, C., Castañera, P. and Ortego, F. 1997. Inhibition of digestive trypsin like proteases from larvae of several lepidopteran species by the diagnostic cysteine protease inhibitor E64. *Insect Biochemistry and Molecular Biology*. vol. 27, p. 247-254.

Ochiai, N., Mizuno, M., Mimori, N., Miyake, T., Dekeyser, M., Canlas, L. J. and Takeda, M. (2007). Toxicity of bifenazate and its principal active metabolite, diazene, to *Tetranychus urticae* and *Panonychus citri* and their relative toxicity to the predaceous mites, *Phytoseiulus persimilis* and *Neoseiulus californicus. Experimental and Applied Acarology*, 43, 3, 181-197.

Ogura, N. 1993. Control of scarabaeid grubs with an entomogenous nematode, *Steinernema kushidai. Jpn. Agric. Res. Q.* **27,** 49–54.

Painter, R. H. 1951. Insect Resistance in Crop Plants, The MacMillan Co., New York, USA.

Panda, N. and Khush, G. S. 1995. Host Plant Resistance to Insects. CAB International, Wallingford, UK.

Park, H., Yamanaka, N., Mikkonen, A., Kusakabe, I. and Kobayashi, H. 2000. Purification and characterization of aspartic proteinase from sunflower seeds. *Bioscience Biotechnology and Biochemistry.* vol. 64, p. 931-939.

Patel, J. R., Patel, J. I., Mehta, D. M. and Shah, B. R. 1993. Integrated management of *Amsacta moorie* with botanical insecticides. *Botanical Pesticides in Integrated Pest Management*, pp. 343-350.

Pedigo, L. P. 1991. Entomology and Pest Management. MacMillan Publishing Co., New York, USA.

Pedigo,L. P. 1991. Entomology and Pest Management. Mac Millan Publishing Co.,New York, USA.

Prabhakar, N., Cowdriet, D. L., Kishaba, A. N. and Meyerdirk, D. E. 1986. Laboratory evaluation of neem seed extract against larvae of the cabbage looper and beet army worm (Lepidoptera: Noctuidae). *Journal of Economic Entomology*, **79** (1) : 39-41.

Pradhan, S. 1983. Agricultural Entomology and Pest Control. Indian Council of Agricultural Research, New Delhi, India.

Pradhan, S., Jothwani, M. G. and Rai, B. K. 1962. The neem seed deterrent to locusts. *Indian farming*, **12**: 7-11.

Pradhan, S., Jotwani, M. G. and Rai, B. K. 1962. The neem seed deterrent to locusts. *Indian Fmg.* 12: 7-11.

Pradhan, S. 1983. Agricultural Entomology and Pest Control. Indian Council of Agricultural Research, New Delhi, India.

Prakasan, C. B., Sreedharan, K., Reddy, A. G. S and Gokuldas, M. 2001. Mass trapping A new component in the IPM of coffee berry borer. In: IPM in horticultural crops : Emerging trends in the new mellinium (Eds. Verghese,A and Parvata Reddy p.) IIHR, Bangalore pp. 117-118.

R. F. Smith and R. van den Bosch. 1967. Integrated Control. pp. 295 - 340. In: Pest control: biological, physical and selected chemical methods, Wendell W. Kilglore and Richard L. Doutt (eds.), Academic Press, New York. 477 pp.

Rancour, J. M. and Ryan, C. A. 1986. Isolation of a carboxypeptidase B inhibitor from potatoes. *Archives of Biochemistry and Biophysics.* vol. 125, p. 380-382.

Rao, G. R and Dhingra, S. 1997. Synergistic activity of some vegetable oils in mixed formulation with cypermethrin against different instars of *Spodoptera litura*. *Journal of Entomological Research,* **21**(2): 153-60.

Ravichandaran, S., Sen, U., Chakrabarti, C. and Dattagupta, J. K. 1999. Cryocrystallography of a Kunitz type serine protease inhibitor: 90 K structure of winged bean chymotrypsin WCI) at 2. 13 A resolution. *Acta Crystallography D Biology Crystallography*. vol. 55, p. 1814-1821.

Reeck, G. R., Kramer, K. J., Baker, J. E., Kanost, M. R., Fabrick, J. A. and Behnke, C. A. 1997. Proteinase inhibitors and resistance of transgenic plants to insects. In: Carozzi, N. and Koziel, M., eds. *Advances in insect control: the role of transgenic plants*. London, Taylor and Francis, 1997, p. 157-183.

Rogers, B. L., Pollock, J., Klapper, D. G. and Griffith, I. J. 1993. Sequence of the proteinase inhibitor cystatin homolog from the pollen of *Ambrosia artemisiifolia* (short ragweed). *Gene*. vol. 133, p. 219-221.

Ryan, C. A. Insect-induced chemical signals regulating natural plant protection responses. In: Denno, R. F. and Mc Clure M. S., eds. *Variable plants and herbivores in natural and managed systems*. New York, Academic Press, 1989.

Ryan, Clarence A. Protease inhibitors in plants: genes for improving defenses against insects and pathogens. *Annual Review of Phytopathology*, 1990, vol. 28, p. 425-449.

Sandhu,G. S. and Singh,D. 1975. Studies on antifeedant and insecticidal properties of neem *Azadirachta indica*. A Juss and Dharek *Melia azedarach* Linn. *Indian Journal of Plant Protection*, **3**: 177-180.

Sankaram, T. 1974. Natural enemies introduced in recent years for biological control of agricultural pests in India. Indian J. agric. Sci. 44(7) : 425-433.

Sardana, R. K., Ganz, P. R., Dudani, A. K., Tackaberry, E. S., Cheng, X. and Altosaar, I. 1998. Synthesis of recombinant human cytokine GMCSF in the seeds of transgenic tobacco plants. In: Cunningham, C. and Porter A. J. R., eds. *Recombinant proteins from plants. Production and isolation of clinically useful compounds*. Totowa NJ, Humana Press, p. 77-87.

Sarode, S. V. and Gobhane, A. T. 1994. Performance of neem seed kernel extract with reduced insecticidal dosage on the infestation of okra fruit borer, *Earias vitella*. *Journal of Entomological Research*, **18** (4): 327-330.

Sathe, T. V. 1999. White fly, Aleurodicus disperses – A new pest of guava, Psidium guava in Kohlapur, Maharastra. *Indian Journal of Entomology* 16 (2) : 195 -196.

Sathiamma,B., Radhakrishnan Nair, C. P.,and Koshy, P. K. 1998. Out break of a nut infesting *Eriophyid* mite in coconut plantations in India. *Indian Coconut Journal*. 29: 1-3

Saxena, R. C. and Khan, Z. R. 1984. Neem oil disrupts *Nephotettix virescens* feeding. *Neem Neewsletter*, **1**: 28-29.

Saxena, R. C. 1982. Note on the use of neem kernel for the protection of dew gram against *Amsacta moorie* Butler. *Indian Journal of Agricultural Sciences*, **52**: 51-52.

Schmutterer, H. and Ascher, K. R. S. (eds). 1987. Natural Pesticides from Neem Tree (Azadirachta indica A. Juss) and Other Tropical Plants. *Proceedings 3rd International Neem Conference*, July 10-15, 1986, Nairobi, Kenya, Dt. Ges. Techn. Zusammenarbeit (GTZ), GmBH, Eschborn, FRG.

Sharma, D and D. V. Rao. 2012. A Field Study of Pest of Cauliflower, Cabbage and Okra in some areas of Jaipur. *International Journal of Life Sciences Biotechnology and Pharma Research*, 1(2): 2250-3137.

Shulke, R. H. and Murdock, L. L. 1983. Lipoxygenase trypsin inhibitor and lectin from soybeans: effects on larval growth of *Manduca sexta* (Lepidoptera: Sphingidae). *Environmental Entomology*. vol. 12, p. 787-791.

Simmonds, M. S. J., Jarvis,A. P., Johnson,S., Jones,G. R. and Morgan, E. D. 2004. Comparison of anti-feedant and insecticidal activity of nimbin and salannin photo-oxidation products with neem (*Azadirachta indica*) limonoids. *Pest Management Science*, **60** (5): 459-464.

Singh, J., Sukhiya, H. S. and Singh, P. 1990. Evaluation of neem oil against rice leaf folder and stem borer. *Proceedings Symposium Botanical Pesticides in IPM*, Rajahmundry,288-290.

Singh, K. and Sharma, U. L. 1986. Studies on the antifeedant and repellant qualities of neem (*Azadirachta indica*) against aphid (*Brevicoryne brassicae*) on cauliflower and cabbage. *Res. Dev. Reptr*, **3**: 33-35.

Singh, R. P.,Devakumar, C. and Dhingra, S. 1988. Activity of neem (*Azadirachta indica* A. Juss) seed kernel extracts against the mustard aphid, *Lipaphis erysimi*. *Phytoparasitica*, **16** (3) : 225-230.

Smith R. F., and H. T. Reynolds. 1966. Principles, definitions and scope of integrated pest control. *Proc. FAO Symposium on Integrated Pest Control* 1: 11-17.

Smith. R. F., and R. van den Bosch. 1967. Integrated Control. pp. 295 - 340. In: Pest control: biological, physical and selected chemical methods, Wendell W. Kilglore and Richard L. Doutt (eds.), Academic Press, New York. 477 pp.

Song, I., Taylor, M. and Baker, K. 1995. Inhibition of cysteine proteinases by *Carica papaya* cystatin produced in *Escherichia coli*. *Gene*. vol. 162, no. 2, p. 221-224.

Srinivasan, K., Viraktmath, C. A., Gupta, M., and Tewari, G. C. 1995. Geographical distribution, Host range and parasitoids of serpentine leaf miner in south India. *Pest Management in Horticultural Ecosystems* 1(2): 93-100

Stienhaus, E. A. 1949. *Principles of Insect Pathology*. McGrawHill,New York.

Stern,V. M., Smith,R. F., van den Bosch, R. and Hagen,K. S. 1959. The integrated control concept. *Hilgardia*,29, 81-101.

Stern, V. M., R. F. Smith, R. van den Bosch, and K. S. Hagen. 1959. The integrated control concept. *Hilgardia*, 29: 81-101.

Sujatha A. and Zaheruddeen, 2006. Survey on the occurrence of mango fruit borer, Deanolis albizonalis in different mango tracts of Andhra Pradesh and its preference to mango varieties. *Indian J. Appl. Ent.* 20 (1) : 33 – 37.

Sujatha, A., Chalam, M. S. V., Vijayalakshmi, P, and Gautam, B. 2008. Coconut Slug Caterpillar, Macroplectra nararia Moore – Out break in Godavari districts of Andhra Pradesh. *Indian Coconut Journal.* March, 2008 : 2-4

Sujatha. A, N. Emmanuel and S. Arul Raj. 2011. Light trap-induced suppression of coconut slug caterpillar,Macroplecta nararia Moore menace in east coast of India. *Journal of plantation Crops.* Vol. No. 3. pp390-395

Swaminathan, M. S. 1999. "The challenges ahead". Survey of Indian Agriculture. The Hindu Group of Publications, Chennai.

Tanada, Y., and Kaya, H. K. 1993. "Insect Pathology. " Academic Press, New York.

Taylor, M. A. J., Baker, K. C., Briggs, G. S., Connerton, I. F., Cummings, N. J., Pratt, K. A., Revell, D. F., Freedman, R. B. and Goodenough, P. W. 1995. Recombinant proregions from papain and papaya proteinase IV are selective high affinity inhibitors of the mature papaya enzymes. *Protein Engineering.* vol. 8, no. 1, pp. 59-62.

Terra, W. R., Ferreira, C. and Jordao, B. P. Digestive enzymes. In: Lehane M. J., ed. Billin London, Chapman and Hall, 1996, pp. 153-194.

Thakar, A. V., Ameta, O. P.,Tripathi, N. N., Vyas, A.,Tauro, P. (ed) and Narwal, S. S. (ed) 1992. Relative efficacy of indegenous plant extracts against gram pod borer, *Helicoverpa armigera* (Hb). **In** : *Proceedings of first National Symposium on Allelopathy in Agroecosystem. (Agric. and For.),* Feb. 12-14, 1992, held at CCS Haryana Agricultural University, India: 162.

The Children's Health Act of 2000. 2000. The Children's Health Act of 2000, PL 106-310, Title V, Section 511, enacted October 17, 2000. Codified in: Title 42, U. S. Code, Section 300w-3(a)(1)(E).

Turk, V. and Bode, W. The cystatins: protein inhibitors of cysteine proteinases. *FEBS Letters*, 1991, vol. 285, pp. 213-219.

University of California State-wide Integrated Pest Management Project. 1997. Annual Report.

Van Leeuwen, T., Witters, J., Nauen, R., Duso, C. and Tirry, L. (2010). The control of eriophyoid mites: state of the art and future challenges. *Experimental and Applied Acarology*, 51, 1-3, 205-224,

Venzon, M., Rosado, M. C., Molina-Rugama, A. J., Duarte, V. S., Dias, R. and Pallini, A. (2008). Acaricidal efficacy of neem against *Polyphagotarsonemus latus* (Banks) (Acari: Tarsonemidae). *Crop Protection*, 27, 3-5, 869-872, ISSN 0261-2194.

Vimala Devi, P. S. and Prasad, Y. G. 1997. The Entomofungal Pathogen Nomuraea rileyi. *Information Bulletin*, Directorate of Oilseeds Research, Hyderabad.

Viraktamath, C. A., Tewari, G. C., Srinivasan, K., and Gupta, M. 1993. American serpentine leaf miner is a new threat to crops. *Indian Farming*. 43 (2) : 10-12.

Waldron, C., Wegrich, L. M., Owens P. A. and Walsh, T. A. 1993. Characterization of a genomic sequence coding for potato multicystatin, of eight-domain cysteine proteinase inhibitor. *Plant Molecular Biology*. vol. 23, p. 801-812.

Walker, A. J., Ford, L., Majerus, M. E. N., Geoghegan, *I. E.,* Birch, A. N. E., Gatehouse, J. A. and Gatehouse, A. M. R. Characterisation of the midgut digestive proteinase activity of the two-spot ladybird (*Adalia bipunctata* L.) and its sensitivity to proteinase inhibitors. *Insect Biochemistry and Molecular Biology*, 1998, vol. 28, p. 173-180.

Wieman, K. F. and Nielsen, S. S. 1988. Isolation and partial characterization of a major gut proteinase from larval *Acanthoscelides obtectus* Say (Coleoptera: bruchidae). *Comparative Biochemistry and Physiology-B*. vol. 89, p. 419-426.

Williams, C. M., and Law, J. H. (1965). *J. Insect Physiol.,* **11**, 569

Williamson,V. M and Hussey, R. S. Nematode pathogenesis and resistance in plants. *Plant Cell,*October 1996, vol. 8, p. 1735-1745.

Wolfson, J. L. and Murdock, L. L. 1987. Suppression of larval Colorado beetle growth and development by digestive proteinase inhibitors. *Entomologia Experimentalis et Applicata*, 1987, vol. 44, p. 235-240

Wolfson, J. L. and Murdock, L. L. 1990. Diversity in digestive proteinase activity among insects. *Journal of Chemical Ecology*. vol. 16, p. 1089-1102.

Zimmermann, G. 1993. The entomopathogenic fungus *Metarhizium anisopliae* and its potential as a biocontrol agent. *Pestic. Sci.* **37,**375-379.

Kraiss, H. Cullen, E. M. 2008. Insect growth regulator effects of azadirachtin and neem oil on survivorship, development and fecundity of *Aphis glycines* (Homoptera: Aphididae) and its predator, *Harmonia axyridis* (Coleoptera: Coccinellidae). *Pest Management Science*, 64(6): 660-668.

Kulkarni, N. S. Sawant, I. S. Sawant, S. D. Adsule, P. G. 2008. Bio-efficacy of neem formulations (Azadirachtin 1 per cent and 5 per cent) on important insect pests of grapes and their effect on shelf life. *Acta Horticulturae*, (785): 305-311.

Misra, H. P. 2009. Evaluation of neem pesticides against major head damaging insect pests of cabbage. *Indian Journal of Entomology*. 71(3): 240-243.

Rafiq, M. Dahot, M. U. Naqvi, S. H. Mahadev Mali Nadir Ali. 2012. Efficacy of neem (*Azadirachta indica* A. Juss) callus and cells suspension extracts against three lepidopteron insects of cotton. *Journal of Medicinal Plants Research,* 2012. 6(40): 5344-5349.

Resources Retrieved from World Wide Web

http://cyberjournalist.org.in/pollution.html

http://pl.wikipedia.org/wiki/Chrysoperla_carnea

http://ucanr.org/repository/cao/landingpage.cfm?article= ca.v065n01p21 &fulltext=yes

http://www.ipm.ucdavis.edu/PMG/NE/trichogramma_spp.html

http://ponent.atspace.org/fauna/ins/fam/chalcididae/chalcididae.htm

http://aggie-horticulture.tamu.edu/galveston/beneficials_intros/beneficials-

http://web.utk.edu/~jurat/Btresearchtable.html

http://www.microbiologybytes.com/virology/kalmakoff/baculo/baculo.html

http://susveg-asia.nri.org/susvegasiabrinjalipm4.html

http://www7.inra.fr/hyppz/IMAGES/7031870.jpg

http://agbsc.blogspot.in/2011/06/bendi-insect-pests-pictures.html

http://www.vegedge.umn.edu/MNFruit&VegNews/vol5/vol5n7.htm

http://agropedia.iitk.ac.in/content

http://ipmworld.umn.edu/chapters/straub.htm

http://www.whatsthatbug.com

http://www.cals.ncsu.edu/course/ent425/library/tutorials/applied_ entomology

http://www.pestnet.org

http://news.nationalgeographic.com/news/2008/08/080826-jamaica-coffee

http://www.infonet-biovision.org/default/ct/124/crops

http://www.ipcnet.org/n/bpd/shootBorer/shootBorer_pathogen.htm

http://www.celkau.in/Crops/Spices/Pepper/pests.aspx

http://www.ozanimals.com

http://www.itsnature.org

http://invasives.wsu.edu/defoliators/host_plants/apple.html

http://www.entomology.umn.edu/cues/inter/inmine/Aphidsgp.html

Index

O

Odoiporus longicollis 302
ODV 57
Oecophylla smaragdina 290
Oligonychus coffeae 340
Oligonychus indicus 350
Omphisa anastomosalis 269
Onion 281
Ophiomyia phaseoli 257
Opisina arenosella 32, 330
Opius exigua 35
Organic insecticides 103, 106
Organochlorine insecticides 158
Organochlorine 103
Organophosphates 119
Orius spp 44
Orthaga exvinacea 290
Oryctes rhinoceros L. 24, 325
Othreis fullonica 294
Ovicides 104
Oxadiazines 135
Oxidation 214
Oxidative phosphorylation disruption 161

P

Paecilomyces species 61
Papaya 305
Papilio demoleus 293
Paracoccus marginatus 305
Parasa lepida 345
Parasite 31
Parasitism 32
Parasitoid 31
Partitioning 211
Passer domesticus 371
Pathogenicity 60, 63

Pathogensis 65
PCMBS 92
PCR techniques 98
Peach 317
Pear 317
Peas and beans 255
Pelleted insecticides 151
Penetration 63
Pentalonia nigronervosa 303, 352
Pepper 351
Permethrin 128
Pest damage 16
Pest management 10, 14, 23, 87
Pest population 17
Pest residue 16
Pest resurgence 192
Pesticide compatibility 57
Pesticide consumption 101
Pesticide definition 102
Pesticide poisoning 6
Pesticide regulations 203
Pesticide residue 6, 200, 217, 219
Pesticide residue analysis 209
Pesticide selection 161
Pesticide standards 210
Petroleum oils 181
Phaenopsitylenchidae 66
Phalacra sp. 335
Phenolics compounds 79
Phenyl pyrazoles 132
Phorate 123
Phosolone 121
Phosphine gas 176
Phosphomidon 122
Phthorimaea operculella 263
Phycita orthoclina 315